Naval Construction Forces Manual

How the SEABEES Do It

Prepared by
United States Navy

Fredonia Books
Amsterdam, The Netherlands

Naval Construction Forces Manual:
How the SEABEES Do it

Prepared by
United States Navy
Civil Engineer Corps

ISBN: 1-4101-0056-1

Reprinted from the 1968 edition

Fredonia Books
Amsterdam, The Netherlands
http://www.fredoniabooks.com

TABLE OF CONTENTS

CHAPTER II - COMMAND AND STAFF (Continued)

CHAPTER V - <u>BATTALION OPERATIONS</u> (Continued)

CHAPTER VI - BATTALION LOGISTICS (Continued)

CHAPTER VII - COMPANY COMMANDERS

CHAPTER X - STINGER SUPPORT FUNCTIONS

TABLE OF ILLUSTRATIONS

TABLE OF ILLUSTRATIONS

PAGE

Chapter I
INTRODUCTION TO THE NAVAL CONSTRUCTION FORCES

SECTION A - HISTORY

1. <u>World War I</u>. Military construction forces of the Navy were formed in World War I when the 12th Naval Construction Regiment was established to supplement the Public Works Department in the construction of recruit training facilities at Great Lakes. Only a small detachment went overseas to France, where they built some communication facilities. After the war, this regiment was decommissioned, but it was the forerunner of the massive naval construction effort of World War II and the creation of the world famous SEABEES.

The planners at the Bureau of Yards and Docks knew that the Navy would need military construction forces to support the fleet if a war of movement developed and the experience of World War I served as a basis for war planning between wars.

2. <u>Formation of World War II NCF Units</u>. As a result of the tense international situation both in Europe and Asia during the late 1930s the Congress authorized an expansion of the naval shore activities as recommended by the Hepburn Board. This work followed the peacetime pattern of construction with projects awarded to private contractors who utilized civilian labor to perform the work. BUDOCKS administered these contracts through Officers in Charge of Construction. Work was started in late 1939 in the Caribbean and the Central Pacific. As the German offensive swept through Europe and Japan expanded her island empire, the pace of construction quickened and additional facilities at new bases were planned. When Japan attacked Pearl Harbor and our outlying bases, the United States got a clear indication of the undesirable effects of using civilians in war zones. Under military law, the contractor's forces could not offer resistance to the invading forces, since a civilian bearing arms would have been considered a guerrilla, and as such, would have been liable to summary execution if captured. In addition, they lacked the training, discipline and mobility of a military unit. The Japanese did execute some of the construction workers and imprisoned the others. The families of those captured became destitute because the contractor could not continue to pay their salaries nor did the Government have any obligation to them.

It was in this atmosphere that Admiral Moreell, the Chief of the Bureau of Yards and Docks, found favorable reception for his plan

for military construction forces for the Navy. In January 1942, three Construction Battalions were authorized. When the Secretary of the Navy signed an order giving the Civil Engineer Corps command of these units, since they "spoke the language of the construction types" the problem of shifting command from a CEC officer during the work day to a line officer for the remainder of the time (as was done in World War I) was overcome. The other problem was one of recruiting skilled construction men from high-paying draft-exempt jobs. Allowance was made for construction experience, and therefore, most men were enlisted directly to Petty Officer status. Some physical defects were overlooked as long as they would not interfere with a man's ability to get a job done. However, patriotism was the main motivating force and many volunteers were too old for the draft. The initials of these new Construction Battalions led to the designation of these forces as SEABEES and the date that this term was authorized is considered the official SEABEE Birthday, 5 March 1942.

3. <u>World War II Accomplishments</u>. The accomplishments of these men during the greatest construction war of history are legend. The NCF swelled to a peak strength of 12,000 officers and almost one-quarter of a million men. They built more than 400 advanced bases along the five major highways to victory; the Northern Atlantic from Newfoundland to England and Normandy; the Southern Atlantic from the Caribbean to Africa, Sicily, and up the Italian peninsula; the Northern Pacific from Alaska out the Aleutian chain; the Central Pacific from Hawaii to Midway to Wake and Guam; and the Southern Pacific from the Solomons through the Philippines. Some of these bases were large enough to house 50,000 men. Not only did these SEABEES build, but they participated in almost every amphibious assault, assembled pontoons and causeways, handled underwater demolition (now a UDT function) and ship off-loading (now a CHB function) constructed floating drydocks, and performed combat engineer functions for the Marine Corps. The record of their exploits may be found in Volume II of the history of BUDOCKS and the CEC titled "Building the Navy's Bases in World War II." A visit to the SEABEE Museum at Port Hueneme, California, would add to an understanding of why the SEABEES became so famous in such a short time.

4. <u>Korea to Present</u>. After World War II, the Naval Construction Forces were reduced and almost went out of existence. Many men who served in World War II were recalled for Korea, where the SEABEES were again writing history. They put the landing force ashore at Inchon and built three major airfields for the First Marine Air Wing. Some of the battalions were recommissioned as Mobile Construction Battalions. These units were smaller than the World War II Construction Battalions and were designed for rapid movement in support of highly mobile combat forces. After Korea, the NCF was stabilized at 10 Mobile Construction Battalions (MCBs) and 2 Amphibious Construction Battalions (ACBs). These fleet units actively engaged in training and preparation for their combat support role by engaging in Fleet and FMF training exercises and by conducting peacetime deployments throughout the Atlantic and Pacific area. In May 1965 the Pacific Alert Battalion on Okinawa, MCB-10, deployed to Chu Lai, RVN to support the landing of the Marine Amphibious Force. From May 1965 to May 1968, the strength of the NCF grew from 10 under strength to 21 augmented battalions. The MCBs, under the leadership of the Third Naval Construction Brigade, supported Marine, Army, Air Force and Navy operations from the Mekong Delta to Khe Sanh. In addition to the MCBs, 13-man SEABEE Teams, the Navy's "Peace Corps," have continuously deployed throughout South America, Africa and Southeast Asia since 1961.

SECTION B - ORGANIZATION

1. <u>Introduction</u>. The Naval Construction Forces are integral commissioned units of the Naval Operating Forces, and as such, are under the control of the Chief of Naval Operations. (See Figure 1). The CNO commissions naval construction units, assigns them to the fleet and approves the deployment of individual units. CNO also defines general mission, approves allowance lists and approves the establishment of detachments.

The Commanders in Chief of the Atlantic and Pacific Fleets are charged with ensuring routing deployment schedules and assigned projects are in consonance with CNO policies. They exercise both operational control and administrative control of the assigned units of the NCF.

The type commander for the Amphibious Construction Battalions is the Amphibious Force Commander. He exercises administrative control through the Commander Naval Beach Group.

Since the Mobile Construction Battalions are part of the logistical support provided to other fleet units, the type commander for the MCBs is the Service Force Commander. The Service Force Commander has delegated many of the type commander functions to the Commander Naval Construction Battalions (COMCBLANT/COM-

CBPAC) because of the differences which exist between the MCBs and other auxiliaries under the administrative control of the Service Force Commander.

The Naval Construction Forces have two basic units; the Amphibious Construction Battalion (ACB) and the Mobile Construction Battalion (MCB), both commanded by officers of the Civil Engineer Corps. While operational regiments and brigades may be formed from MCBs to meet a variety of command requirements, the ACB is designed specifically for support of amphibious landing operations by the Fleet Marine Force. Standard units in support of the MCB are the Naval Construction Regiment (NCR), and the Construction Battalion Maintenance Unit (CBMU).

2. <u>Amphibious Construction Battalion</u>. The Amphibious Construction Battalion (See Figure 2) is a commissioned naval unit organized administratively into a headquarters company, equipment company, two pontoon companies and a construction company. The wartime complement of 31 officers and 638 men includes 14 CEC officers and 220 Group VIII ratings. The size and composition of the ACB is based on providing support for a reinforced infantry division landing over four battalion size beaches. The Battalion is not intended for prolonged use in the field; its mission is accomplished when the work of the Naval Beach Group is finished.

Tactically the ACB operates as small task elements of an amphibious task force under the Transport Group Commander. These task elements are tailored teams and detachments each manned and equipped to handle one phase of the assault over a numbered, battalion size, beach. In the construction support of amphibious assault operations, the ACB provides methods for troop and equipment movement from ship to shore. In addition, it also provides related back-up measures such as tanker-to-shore POL pipelines and methods and techniques for unhampered beach crossing. In some instances, there may be UDT personnel working with these units for removal of underwater obstacles which may jeopardize ship-to-shore transport.

3. <u>The Naval Beach Group</u>. The Naval Beach Group is an administrative organization of the amphibious force, organized and equipped to provide support to a reinforced division during the assault phase of an amphibious operation and for such additional periods as requirements of the operation may dictate. It is composed of a commander, his staff consisting of administrative, intelligence, operations, logistics and communications sections and three commissioned units, the ACB, a beachmaster unit and an assault boat unit. The Commander Naval Beach Group is responsible for the proper training and readiness of all assigned units. He therefore conducts inspections and monitors training programs as necessary to assure the units under his command are capable of performing their mission. Upon receipt of the

FIGURE 1
ADMINISTRATIVE CHAIN OF COMMAND FOR MCB & ACB UNITS

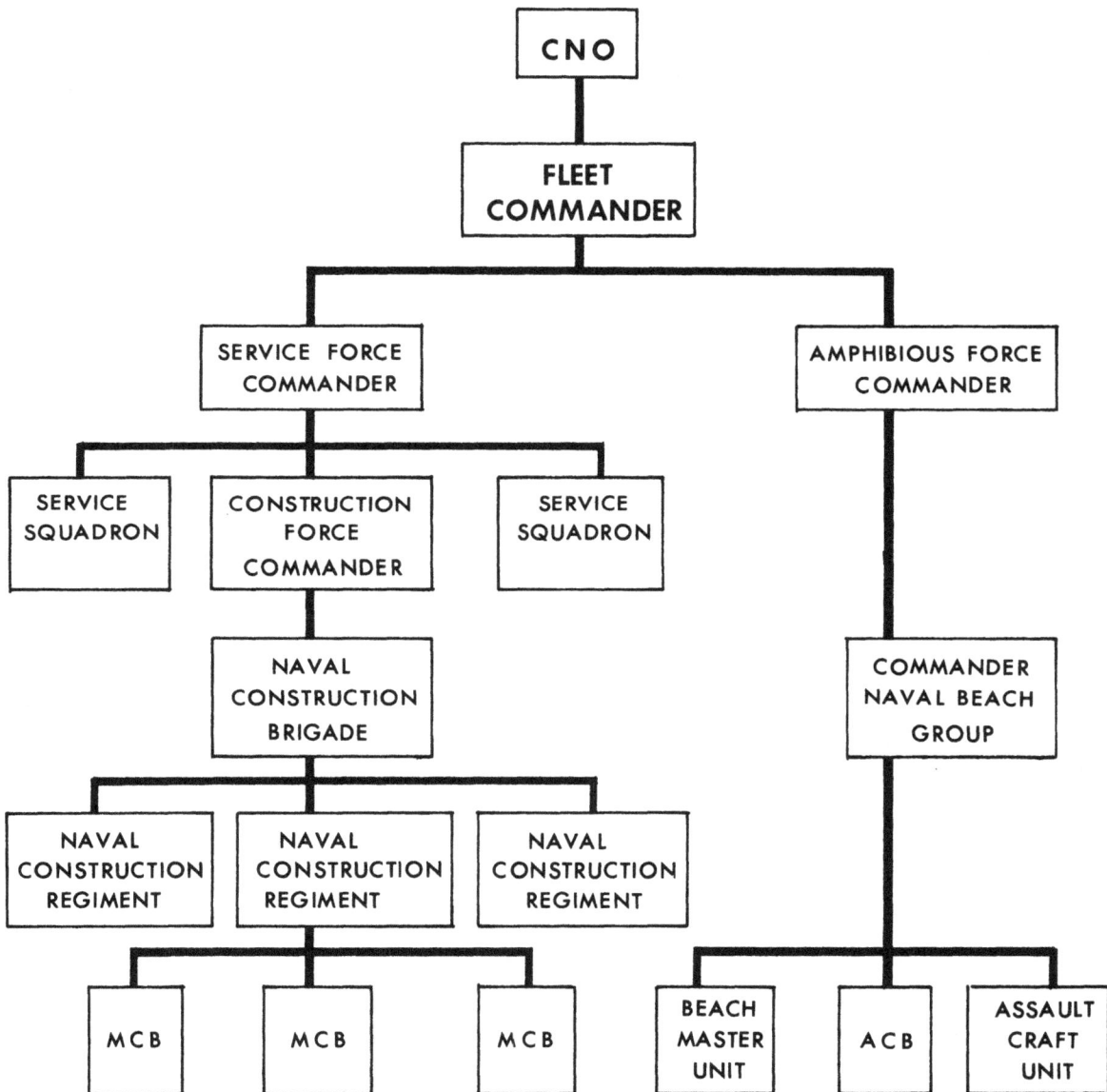

FIGURE 2
BASIC ACB ORGANIZATION

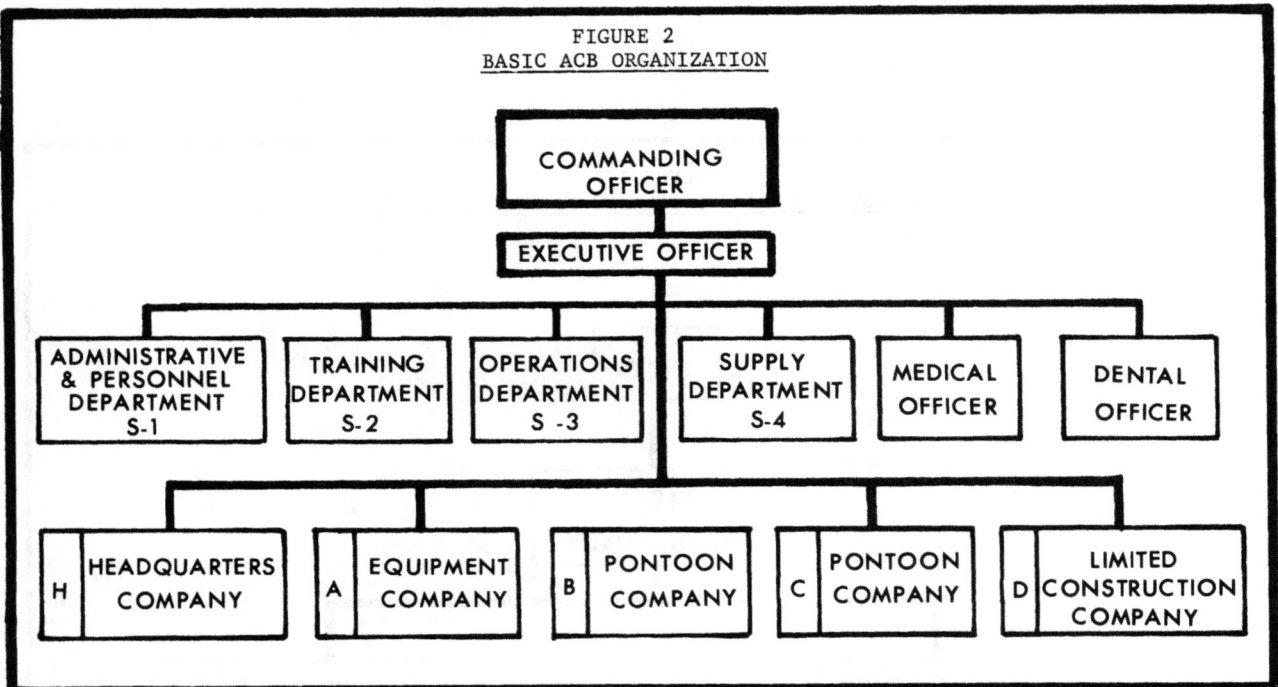

```
                    COMMANDING
                     OFFICER
                        |
                    EXECUTIVE
                     OFFICER
```

| ADMINISTRATIVE & PERSONNEL DEPARTMENT S-1 | TRAINING DEPARTMENT S-2 | OPERATIONS DEPARTMENT S-3 | SUPPLY DEPARTMENT S-4 | MEDICAL OFFICER | DENTAL OFFICER |

| H HEADQUARTERS COMPANY | A EQUIPMENT COMPANY | B PONTOON COMPANY | C PONTOON COMPANY | D LIMITED CONSTRUCTION COMPANY |

amphibious task force commander's directive he activates required tactical elements and assigns them to the amphibious task force commander for operational control. The Commander Naval Beach Beacn Group reports to the amphibious task force commander for planning and may be assigned additional duty on the landing force commander's staff.

During the consolidation phase he normally regroups his tactical elements when the operational control of each is no longer required by the command to which the element was assigned, and assumes operational control of the elements of the naval beach group. A further discussion of the Naval Beach Group can be found in NWIP 22-5.

4. Mobile Construction Battalion. The Mobile Construction Battalion is designed for dual construction/military support operations to build advanced base facilities in support of United States and allied military activities, and to provide engineering support for Fleet Marine Units (See Figure 3). The MCB function also includes corresponding missions of repair and operation of facilities and line of communications during emergencies or contingency operations. The outfitted MCB is a large, self-sufficient unit, requiring only that all classes of consumables be provided to it. The MCB is capable of performing earthmoving, weight-lifting, quarrying, equipment repair, reinforced concrete work, asphaltic concrete work, structural steel fabrication and erection, pipeline installation, well drilling, water purification sewage disposal, electri-

cal power and lighting installation, carpentry, diving (limited capability), hauling, and surveying operations. The MCB also has the capability for disaster recovery operations during both natural disasters and those caused by NBC attack. As a self-sustaining unit, the MCB is capable of self defense for a limited time, performing internal communications, messing and billeting and providing the necessary administrative, personnel, medical, dental, supply and chaplain functions.

Every battalion sub-division has a construction/military support assignment and every officer and enlisted man fills a construction/military support billet. Command channels are the same for both construction and military support, permitting rapid transition from one situation to another. However, the use of MCB units in support of infantry-type operations detracts from the battalion's construction mission and should be avoided. The wartime complement of the MCB is 21 officers and 563 enlisted of which 16 are CEC officers and 475 are Group VIII ratings. The battalion is organized into one headquarters and four construction/rifle companies. Two of the construction/rifle companies each have a weapons platoon containing M60 machine guns and 3.5" rocket launchers. All platoons are organized into work squads which correspond to the weapons or rifle squad organization. Work crews and work squads of construction platoons are also trained as disaster control teams. Each battalion may organize the squads of each platoon to meet its particular needs. The construction/military companies retain

4

FIGURE 3
BASIC MCB ORGANIZATION

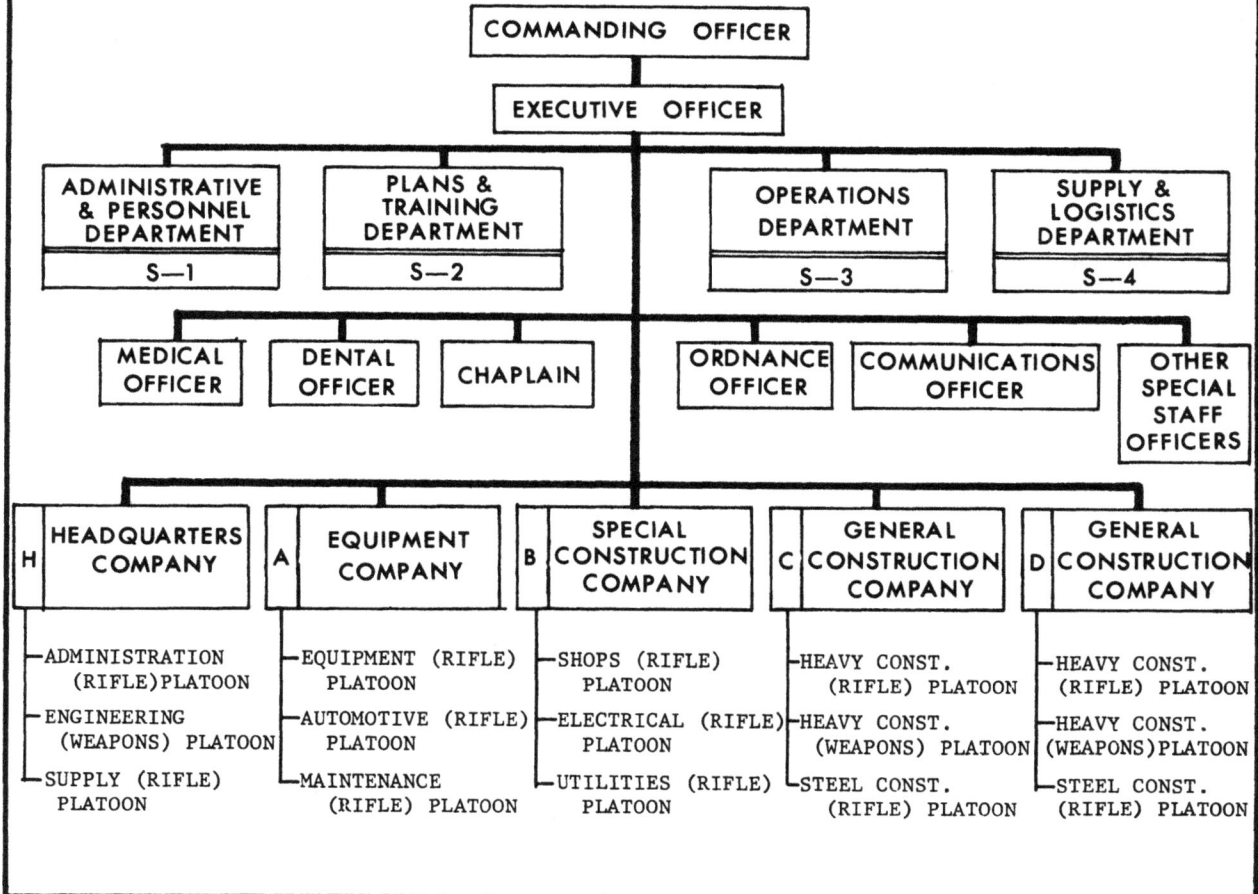

```
                            COMMANDING OFFICER
                                    |
                            EXECUTIVE OFFICER
                                    |
     ┌──────────────┬───────────────┼───────────────┬──────────────┐
ADMINISTRATIVE    PLANS &                        OPERATIONS      SUPPLY &
& PERSONNEL       TRAINING                       DEPARTMENT      LOGISTICS
DEPARTMENT        DEPARTMENT                                     DEPARTMENT
   S—1              S—2                             S—3             S—4

     ┌──────────┬──────────┐            ┌──────────────┬──────────────┐
  MEDICAL     DENTAL     CHAPLAIN     ORDNANCE    COMMUNICATIONS    OTHER
  OFFICER     OFFICER                 OFFICER     OFFICER           SPECIAL
                                                                    STAFF
                                                                    OFFICERS

  ┌──────────────┬──────────────┬──────────────┬──────────────┐
H HEADQUARTERS  A EQUIPMENT     B SPECIAL       C GENERAL       D GENERAL
  COMPANY         COMPANY         CONSTRUCTION    CONSTRUCTION    CONSTRUCTION
                                  COMPANY         COMPANY         COMPANY
```

H HEADQUARTERS COMPANY
- ADMINISTRATION (RIFLE) PLATOON
- ENGINEERING (WEAPONS) PLATOON
- SUPPLY (RIFLE) PLATOON

A EQUIPMENT COMPANY
- EQUIPMENT (RIFLE) PLATOON
- AUTOMOTIVE (RIFLE) PLATOON
- MAINTENANCE (RIFLE) PLATOON

B SPECIAL CONSTRUCTION COMPANY
- SHOPS (RIFLE) PLATOON
- ELECTRICAL (RIFLE) PLATOON
- UTILITIES (RIFLE) PLATOON

C GENERAL CONSTRUCTION COMPANY
- HEAVY CONST. (RIFLE) PLATOON
- HEAVY CONST. (WEAPONS) PLATOON
- STEEL CONST. (RIFLE) PLATOON

D GENERAL CONSTRUCTION COMPANY
- HEAVY CONST. (RIFLE) PLATOON
- HEAVY CONST. (WEAPONS) PLATOON
- STEEL CONST. (RIFLE) PLATOON

their normal letter designation and the platoons retain their letter-number designation to facilitate reference, planning and scheduling.

The basic construction/military support units are the work crew/rifle fire team, the work crew/automatic weapons crew, and the work crew/rocket launcher teams. These units are merged into squads or sections and further consolidated into construction/military support platoons. (See Chapter 1, Section 3, Landing Party Manual). The standard military support platoon consists of three squads (See Figure 4). In a rifle platoon, each squad normally consists of a squad leader and three four-man fire teams. The squad of an automatic weapons platoon is normally composed of a section leader and two four-man teams. This platoon breakdown is included as a basic guide only (See Figure 5). More important is the compatibility between the construction and military support phases. That is, a basic builder crew will form fire teams according to the number of personnel in the crew. The normal fire team consists of four men, but teams of more than four men are possible when required by the size of the work crew. Two or more fire teams are organized as a squad.

In many cases, the MCB is required to conduct operations in two widely separated locations. Figure 6 shows a typical application of the basic concept of divisibility: an MCB separated into a main body and a self-sustaining detachment of approximately equal size to perform separate missions. The same principles may apply to detachments of any size, however, the percentage of overhead labor stays fairly constant while the manpower available for direct labor is reduced as the detachment becomes smaller.

The assignment of a construction program to an MCB which consists primarily of maintenance, repair or improvement work may necessitate a slight deviation from the basic organization indicated in Figures 3, 4 and 6.

FIGURE 4
CONSTRUCTION/RIFLE PLATOON

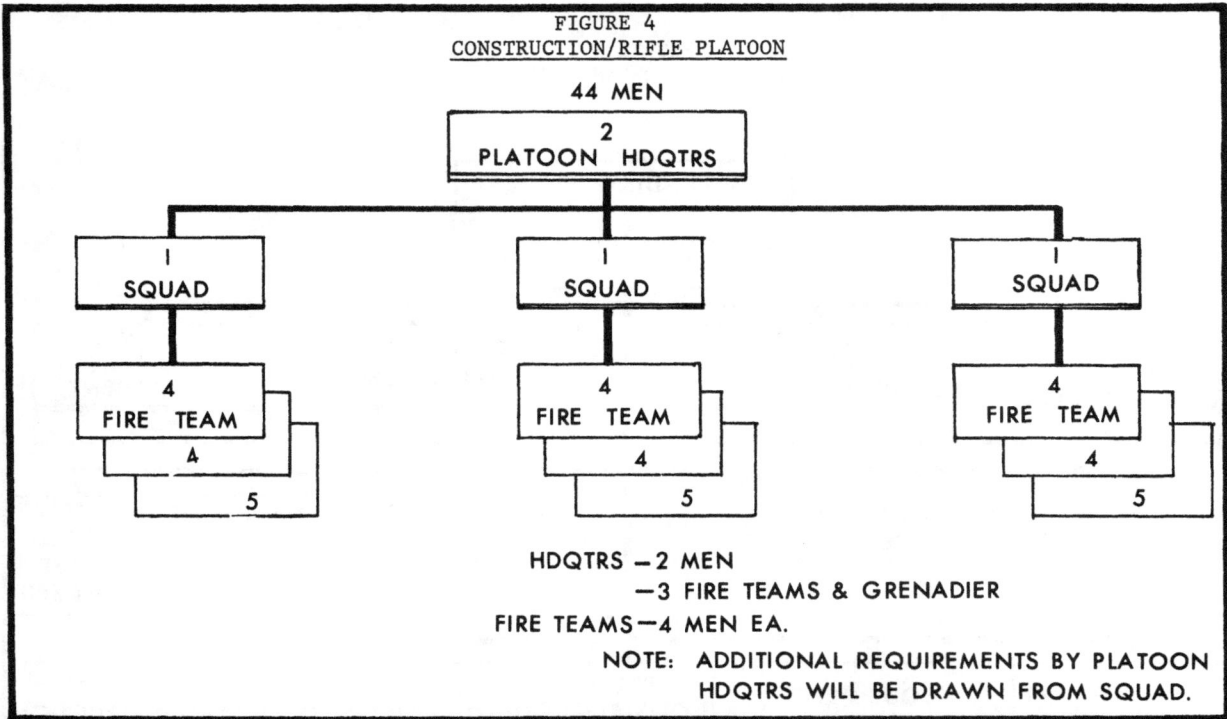

44 MEN

2
PLATOON HDQTRS

1		1		1
SQUAD		SQUAD		SQUAD

4		4		4
FIRE TEAM		FIRE TEAM		FIRE TEAM
4		4		4
5		5		5

HDQTRS —2 MEN
  —3 FIRE TEAMS & GRENADIER
FIRE TEAMS—4 MEN EA.

NOTE: ADDITIONAL REQUIREMENTS BY PLATOON
HDQTRS WILL BE DRAWN FROM SQUAD.

FIGURE 5
CONSTRUCTION/WEAPONS PLATOON

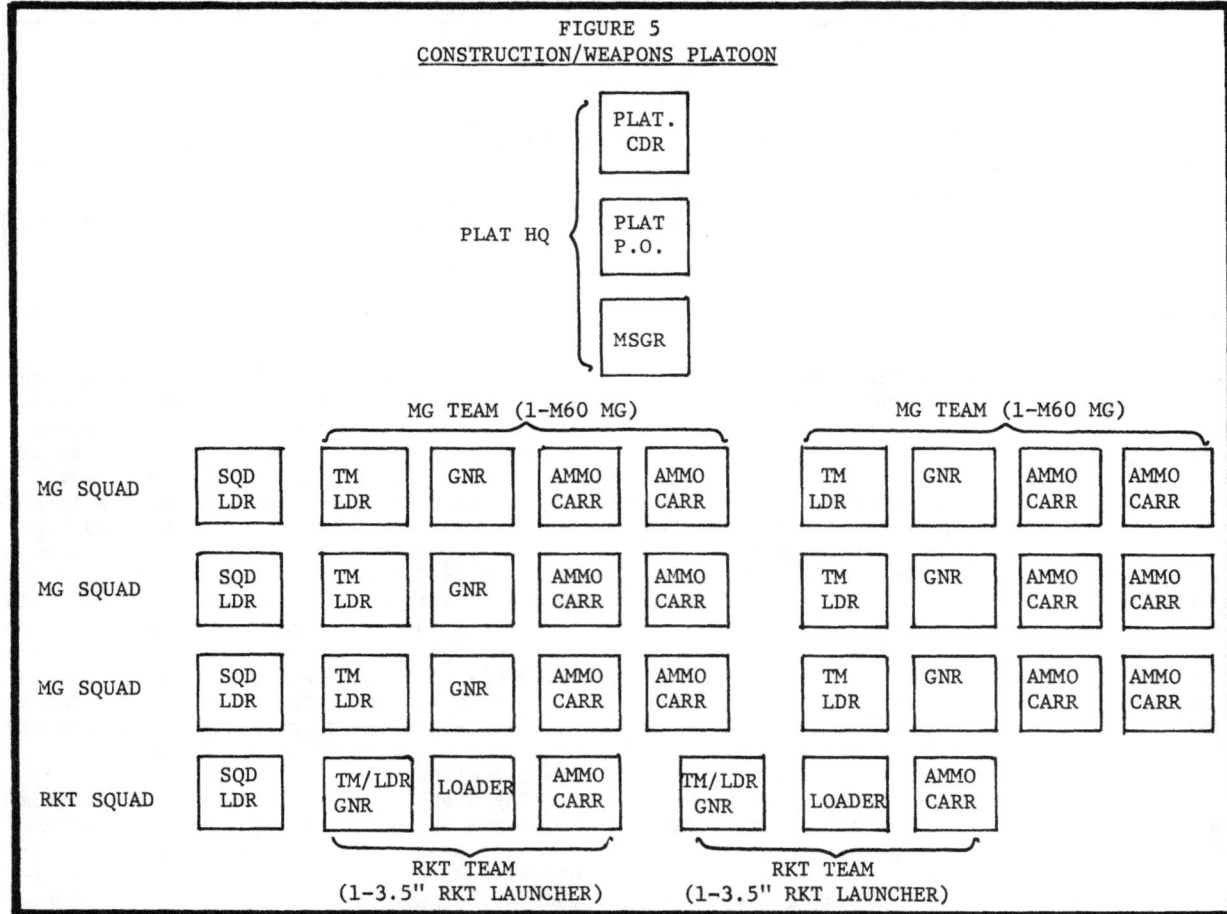

PLAT HQ
- PLAT. CDR
- PLAT P.O.
- MSGR

MG TEAM (1-M60 MG) · MG TEAM (1-M60 MG)

		MG TEAM (1-M60 MG)				MG TEAM (1-M60 MG)		
MG SQUAD	SQD LDR	TM LDR	GNR	AMMO CARR	AMMO CARR	TM LDR	GNR	AMMO CARR	AMMO CARR
MG SQUAD	SQD LDR	TM LDR	GNR	AMMO CARR	AMMO CARR	TM LDR	GNR	AMMO CARR	AMMO CARR
MG SQUAD	SQD LDR	TM LDR	GNR	AMMO CARR	AMMO CARR	TM LDR	GNR	AMMO CARR	AMMO CARR
RKT SQUAD	SQD LDR	TM/LDR GNR	LOADER	AMMO CARR		TM/LDR GNR	LOADER	AMMO CARR	

RKT TEAM
(1-3.5" RKT LAUNCHER)

RKT TEAM
(1-3.5" RKT LAUNCHER)

FIGURE 6
THE NMCB DIVIDED INTO TWO EQUAL DETACHMENTS

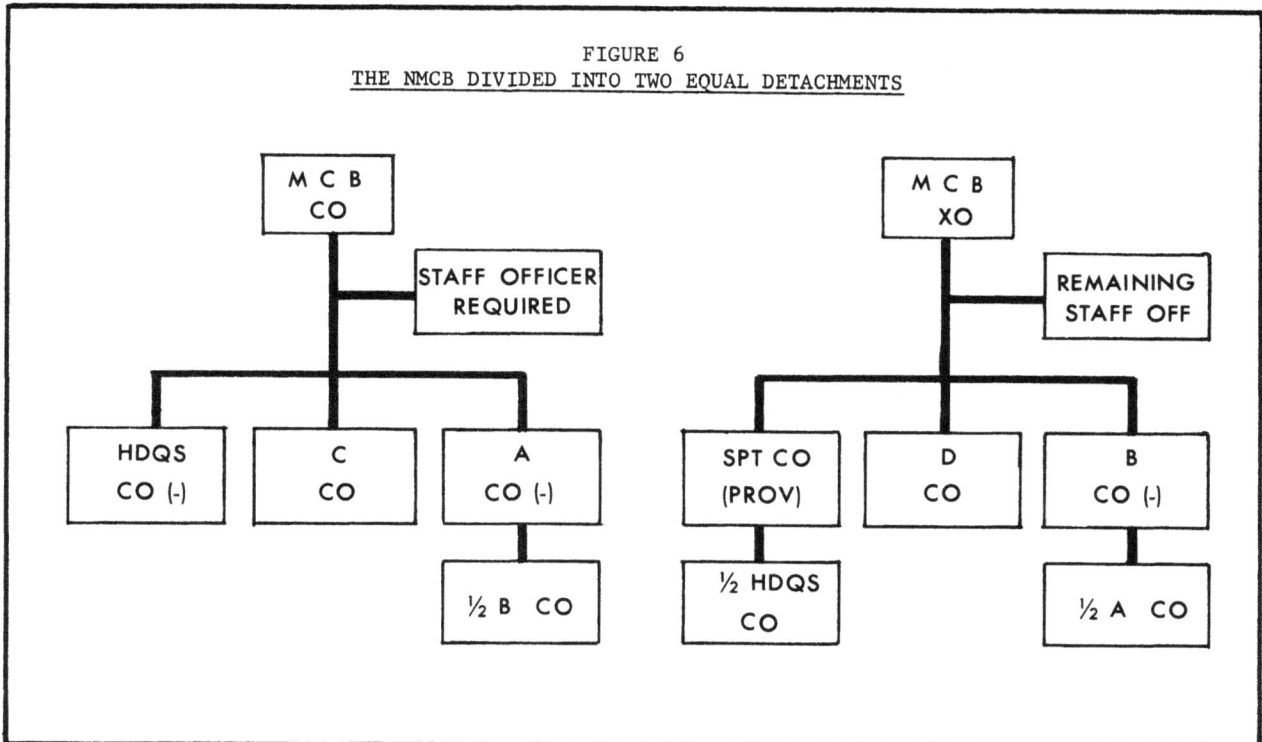

```
        ┌─────────┐                              ┌─────────┐
        │  M C B  │                              │  M C B  │
        │   CO    │                              │   XO    │
        └────┬────┘                              └────┬────┘
             │     ┌──────────────┐                   │     ┌──────────────┐
             ├─────│STAFF OFFICER │                   ├─────│  REMAINING   │
             │     │  REQUIRED    │                   │     │  STAFF OFF   │
             │     └──────────────┘                   │     └──────────────┘
   ┌─────────┼─────────┐                    ┌─────────┼─────────┐
┌──┴───┐ ┌───┴──┐ ┌────┴──┐           ┌─────┴───┐ ┌───┴──┐ ┌────┴──┐
│ HDQS │ │  C   │ │   A   │           │ SPT CO  │ │  D   │ │   B   │
│CO (-)│ │  CO  │ │CO (-) │           │ (PROV)  │ │  CO  │ │CO (-) │
└──────┘ └──────┘ └───┬───┘           └────┬────┘ └──────┘ └───┬───┘
                   ┌──┴───┐              ┌──┴────┐          ┌──┴───┐
                   │ ½ B  │              │½ HDQS │          │ ½ A  │
                   │  CO  │              │  CO   │          │  CO  │
                   └──────┘              └───────┘          └──────┘
```

5. **Commander Naval Construction Battalions U.S. Atlantic/Pacific Fleet.** COMCBLANT/COMCBPAC have been established as representatives of the Service Force Commander to exercise administrative control over assigned MCBs and operational control when the battalions are homeported. The staff performs the routine functions related to coordination of administration, training, project selection and logistic support for assigned units. The commander is responsible for the state of readiness of the battalions and their ability to carry out assigned tasks. He conducts the necessary inspections to evaluate the effectiveness of battalion training programs. These inspections usually include administrative, operational and material inspections made simultaneously, once a year while the battalion is operating overseas. When a workload has been agreed upon for a deployment or when directed by the Service Force Commander COMCBLANT/COMCBPAC releases the battalion to the operational control of the area commander, base commander or expeditionary force commander. Upon completion of the construction tasks assigned, the commander either reassigns the MCBs or returns them to his operational control at the homeport.

6. **Brigades and Regiments.** The Chief of Naval Operations may establish Naval Construction Regiments and Brigades to meet certain command requirements in particular geographical areas or situations, usually upon recommendations of the Commander, Naval Construction Battalion. These larger units are composed primarily of basic MCBs supplemented in the planning stage. Regiment and Brigade Headquarters or staffs are echelons of military command, administrative control, and operational control, and are tailored under Navy regulations to meet specific operational requirements. The homeport NCRs have a very different mission from that of the operational regiments described above. The homeport regiments have broad administrative and logistic duties that will be discussed in Chapter X.

7. **Construction Battalion Units.** A Construction Battalion Unit (CBU) may be commissioned by CNO to meet a specific SEABEE mission. The size of the unit can vary from 20 to 400 men and equipage allowance will vary as widely. The unit will be organized and equipped for its special mission and will not have the flexibility of most NCF units. There is one CBU currently in commission. CBU-201 is tasked with construction support of the Antarctic Program.

8. **Construction Battalion Maintenance Unit.** The Construction Battalion Maintenance Unit is a commissioned unit which operates and maintains public works and public utilities at an advanced base after construction has been completed. Its functions resembles that of a public works department at a naval activity (See Figure 7). In the early phases of con-

```
┌─────────────────────────────────────────────────────────────────────────┐
│                              FIGURE 7                                     │
│          CONSTRUCTION BATTALION MAINTENANCE UNIT ORGANIZATION             │
│                                                                           │
│                    ┌─────────────────────────┐                           │
│                    │   COMMANDING  OFFICER    │                           │
│                    └────────────┬────────────┘                           │
│                    ┌────────────┴────────┐    ┌────────────────┐          │
│                    │   EXECUTIVE  OFFICER │────│     STAFF      │          │
│                    └────────────┬────────┘    └────────────────┘          │
│      ┌───────────────┬──────────┴────────┬───────────────────┐           │
│  ┌───┴────────┐ ┌────┴───────────┐ ┌──────┴─────────┐ ┌───────┴──────────┐│
│  │  HDQS CO   │ │  A-CO (TRANS)  │ │  B-CO (SHOPS)  │ │ C-CO (FIELD MAINT)││
│  ├────────────┤ ├────────┬───────┤ ├────────┬───────┤ ├──────┬──────┬────┤│
│  │ ADMIN/PERS │ │EQUIPMENT│EQUIP- │ │UTILITIES│OPERA- │ │BUILD-│BUILD-│UT/ ││
│  │  SUPPLY    │ │OPERATION│MENT   │ │ SHOPS  │TIONS  │ │ ERS  │ ERS  │SW  ││
│  │            │ │         │MAINT  │ │        │       │ │      │      │    ││
│  └────────────┘ └─────────┴───────┘ └────────┴───────┘ └──────┴──────┴────┘│
└─────────────────────────────────────────────────────────────────────────┘
```

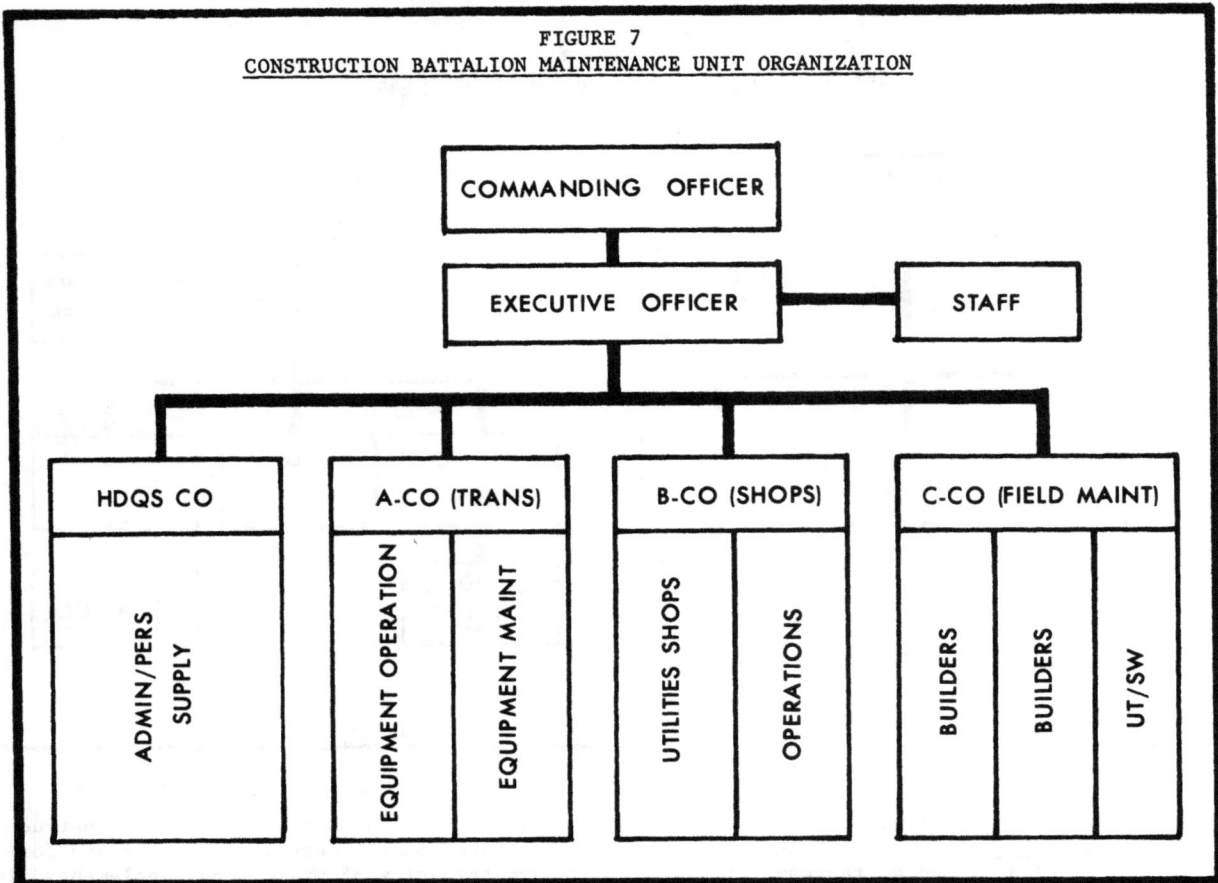

struction, it is attached to the Advanced Base Construction Force to help built the facilities it will later maintain. The CBMU is eventually transferred to the Naval Base Command, the CEC officer in charge often then becomes the Public Works Officer and his unit operates as a public works department.

9. SEABEE Teams. A small highly mobile, air transportable construction unit which can provide disaster relief or technical assistance by working, supervising or conducting classroom instruction is called a SEABEE Team. These teams, which may be tailored to fit any size task, normally consist of one CEC officer and about twelve enlisted SEABEES. They are extremely effective in this nation's counterinsurgency effort.

A typical personnel allowance is indicated in Figure 8 and would include:

    3 Equipment Operators
    2 Construction Mechanics
    2 Builders
    1 Construction Electrician
    1 Steelworker
    1 Utilitiesman
    1 Engineering Aid
    1 Hospital Corpsman

Each team carries sufficient housekeeping supplies, tool kits and automotive and construction equipment that it is able to be self-sufficient in the field and can do a variety of construction tasks. These teams have been employed as engineers to the Army's Special Forces, as technical instructors for the Agency for International Development and as construction advisors under various military assistance programs.

10. Special Detachments. Many other special detachments may be formed from these basic organizations to meet a specific need. These detachments may range from a small lighterage barge crew to one-half of an MCB. Usually,

8

when a project is assigned to the battalion at a site far from the battalion's main location, a regular construction unit of the battalion is selected for the job. This may be a platoon, two platoons, or even a company. The basic unit selected is the nucleus of the detachment. The officer of the unit is designated officer in charge of the detachment. Equipment and crews of specialists are assigned on a subcontract basis, including subsistence and service personnel as needed. In forming detachments, every effort must be made to preserve military integrity. This can be accomplished by detaching the largest possible basic unit, the minimum of which is the work crew/fire team. The exact number, of course, is relegated to a common sense formula governed by practicality. The establishment of commissioned detachments is governed by directives of the fleet commanders.

11. Air Detachments. All active MCBs are required to maintain a contingency plan for the rapid deployment of an Air Detachment of 2 officers and 87 men. The detachment must be organized into two equal strength units. One unit (Sierra) is the horizontal construction detachment (equipment) and the other (Tango) is specialized in vertical construction. (See Figure 9)

12. Naval Facilities Engineering Command. The Naval Facilities Engineering Command has the responsibility of ensuring that the units of the NCF are capable of meeting fleet support missions. The command is responsible for the allowance, funding and engineering design and for advising BUPERS on staffing and training requirements (See Chapter X)

13. Construction Battalion Center. Construction Battalion Centers are permanent shore stations equipped and staffed to support Naval Construction Forces. They are also the major source of logistical support for MCBs as well as other construction units. Each CBC has a Supply and Fiscal Department and a Construction Equipment Department (CED) which furnishes depot maintenance services on automotive and construction equipment. Naval Construction Schools (NAVSCOLCONST) or a Construction Training Unit (CTU) are tenant commands at the CBC and provide training to MCB personnel. The Centers receive, preserve, store, account for, and issue advanced base material and equipment (See Chapter X). Newly commissioned MCBs are usually outfitted at the CBC which also provides homeport facilities. CBCs are under the management and technical control of NAVFAC. They are presently located at Davisville, Rhode Island; Port Hueneme, California; and Gulfport, Mississippi.

14. Advanced Base Construction Depots. ABCDs are functional components located in contingency areas. The depots stock tactical support functional components, construction-consumables and perform fourth echelon equipment mainte-

FIGURE 8
CAPABILITY OF 13-MAN SEABEE TEAM (OFFICER NOT SHOWN)

```
FIGURE 9
NMCB AIR DETACHMENT WITH SIERRA/TANGO DIVISION SHOWN


                          ┌──────────────┐
                          │    O I C     │
                          │  LTJG CEC    │
                          └──────────────┘

   ┌──────────────────────┐                      ┌──────────────────────┐
   │    SIERRA HDQS        │                      │    TANGO HDQS         │
   │  1 SK    1 EOC        │  ┌─────────────┐     │  1 HM    1 BUC        │
   │  1 HM    LEADING      │  │ AOIC ENS CEC│     │  1 CS    LEADING      │
   │  1 CS    CHIEF        │  └─────────────┘     │  1 PN    CHIEF        │
   │  2 EA                 │                      │                       │
   └──────────────────────┘                      └──────────────────────┘


┌──────────────┐  ┌──────────────┐  ┌──────────────┐  ┌──────────────┐
│  EQUIP PLT   │  │ SIERRA SPT   │  │  TANGO SPT   │  │  CONST PLT   │
│  1 EO1       │  │    SECT      │  │    SECT      │  │  2 BU1       │
│  1 CM1       │  │  1 SW1       │  │  1 SW2       │  │  24 BU       │
│  16 EO       │  │  6 SW        │  │  2 SW        │  └──────────────┘
│  6 CM        │  │  2 UT        │  │  1 CE2/UT2   │
└──────────────┘  │  2 CE        │  │  2 UT        │
                  │  2 BU        │  │  2 CE        │
                  └──────────────┘  │  3 EO        │
                                    │  2 CM        │
                                    └──────────────┘

         THE TWO HDQS AND THE TWO SUPPORT SECTIONS
         ARE COMBINED FOR FULL DETACHMENT DEPLOYMENT
```

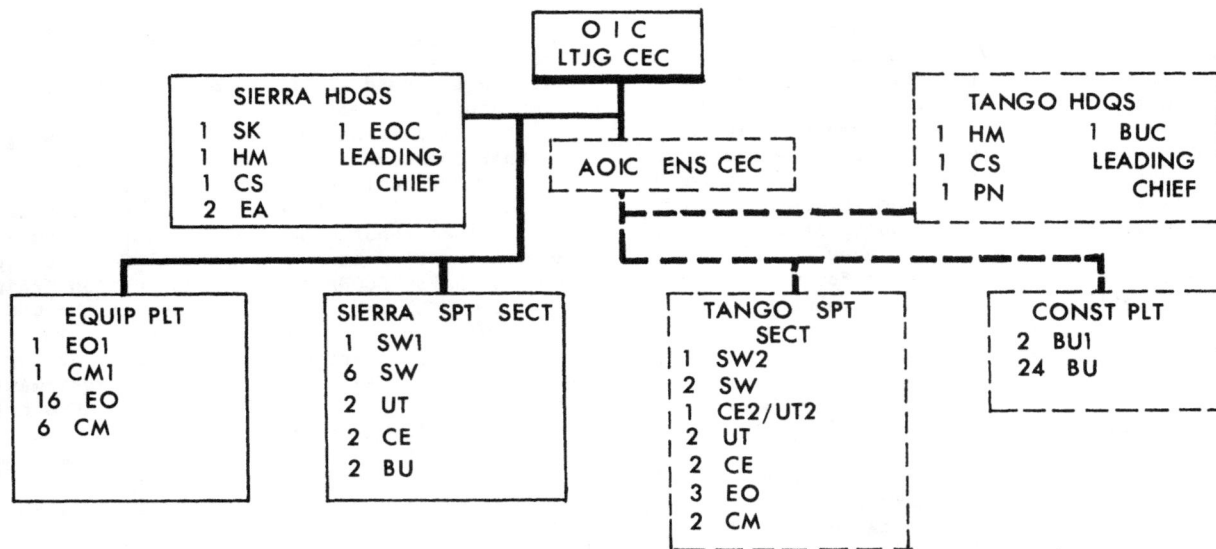

nance. The Officer-in-Charge reports to the contingency Naval Construction Brigade or Regiment for operational control. ABCDs differ from CONUS Construction Depots which are located with the homeport regiments and do perform fifth echelon maintenance.

15. Naval Construction Force Reserve. The active duty units of the NCF are not sufficient to support the Navy and Marine Corps under mobilization conditions. They constitute merely a skeleton force which could be greatly expanded when necessary. In order to facilitate the transferral of Naval Construction Force Reserve units from a reserve to a mobilized status, a six-year training program is conducted at the Construction Battalion Centers. The Reserves, during the six-year cycle, are brought to peak personnel strength by assignment of officers and enlisted men from each of the divisions which normally form a battalion. The RMCBs are also augmented by non-SEABEE personnel (storekeepers, commissarymen, etc.) who are placed in SEABEE divisions instead of other activities when they enter the Reserves. The Reservist's training cycle is composed of six yearly stages. Principal responsibility for training reserve battalions lies with the Construction Training Unit at Gulfport, Mississippi. The six-year program gives the Naval Construction Force Reservists a comprehensive knowledge of SEABEE functions which will enable the various NCF Reserve units to attain a maximum state of operational readiness for mobilization.

SECTION C - DOCTRINE

1. Mission of the Naval Construction Forces. The defense establishment must be efficient and capable of adjusting to new requirements imposed by national security measures. Naval Construction Forces are designed to support the operating goals of the defense effort by maintaining a continuing state of operational readiness. The active NCF operates under the direction of the Department of Defense, Joint Chiefs of Staff, and the Chief of Naval Operations, in compliance with the doctrines and policies outlined in OPNAV INST. 5450.46, "U.S. Naval Construction Forces; doctrine and policy governing." In wartime they support Naval task forces and fleet Marine forces by the construction and maintenance of bases and facilities of a temporary or semi-permanent nature that are needed in pursuing the war. In peacetime they exist to train a nucleus of active duty personnel around which full-scale mobilization may be affected in wartime if required, and to provide a limited force-in-being for military construction required in support of possible contingency/emergency

operations of a smaller scale where rapidity of response is a major consideration.

2. ACB Mission and Doctrine. The primary mission of the ACB is the support of the Naval Beach Group during the initial assault and during the landing operations. Secondly, the ACB assists the Shore Party operations when possible without interfering with its primary duty. When not engaged in emergency action or training maneuvers, the ACB may be designated to test and evaluate experimental amphibious equipment or to perform other tasks specially assigned by NAVFAC or Fleet Amphibious Forces. During amphibious operations the ACB is specifically responsible for:

a. Assembly of pontoon causeways at the base of departure and subsequent installation, operation and maintenance of the causeways at the beachhead for the offloading of LSTs over shallow beaches.

b. Installation and operation of ship-to-shore bulk fuel delivery systems (buoyant hose or bottom laid pipe).

c. Rigging and operating self-propelled pontoon barges for lighterage and transfer operations.

d. Assembly and operation of warping tugs in conjunction with causeways, fuel systems and salvage work.

e. Assembly and operation of pontoon floating drydocks.

f. Developing and improving beach facilities and providing for beach salvage.

The ACB is also capable of undertaking limited construction and special projects, especially those requiring surf, open sea and heavy rigging experience, and those employing floating equipment and pontoons. OPNAV INST. 5440.62 elaborates on the functions and use of the ACB. It is reprinted as Annex B of this manual.

3. MCB Mission and Doctrine. In March 1961, the MCB mission was redefined and expanded by CNO. OPNAV NOTICE 5440 stated the mission in these terms:

a. To maintain a state of operational readiness to provide rapid and effective construction support to Naval, Marine Corps and other forces as necessary.

b. To be prepared to conduct ground defense operations when required by the circumstances of the deployed situation.

c. To be prepared to conduct disaster control operations, including public works functions as directed.

This Notice was followed by OPNAV INST. 5450.46D, of 12 March 1962 which provides the basic doctrine for operation, training and readiness of the active Mobile Construction Battalions. This instruction is reprinted as Annex A.

4. Operational and Material Readiness. In order to meet the requirements outlined in OPNAV INST. 5450.46D for the support of combat forces a minimum of two-thirds of all Naval Construction Forces assigned to each fleet must be capable of redeployment within ten days. After sixty days in their homeports, NCF units will be capable of redeployment within ten days. During the first forty days, they will be capable of redeploying within thirty days. While enroute to or from deployment sites, units will be prepared for immediate diversion to emergency, contingency, or mobilization assignments.

Existing functional components for outfitting Naval Construction Forces units will be tailored or adapted to sustain, without augmentation for ninety days, operations planned or anticipated for emergencies and for either limited or general war assignment. Allowance will be predicated on combat consumption rates based on two ten-hour shifts per day, seven days a week, a total of 1,800 construction hours. Equipment and material will be maintained in a ten-day operational and material readiness state. The basic tables of allowances for the active Naval Construction Forces along with any other functional components developed and maintained to comply with OPNAV INST. 5450.46D, or required for contingency operations are allotted to the cognizant Fleet Commanders for reassignment. Fleet Commanders are responsible for assuring that these components are maintained in a continuing state of readiness for contingency assignments within a period of ten days. Portions of these functional components not required on peacetime deployments may be positioned in other areas ready for immediate movement to the battalion.

5. Command Responsibilities. The Chief of Naval Operations (CNO) approves annual employment plans and all foreign technical assistance programs. COMCBLANT/COMCBPAC and the Commanding Officers of ACBs have the following responsibilities:

a. Determination of deficiencies in equipment, material and personnel which must be remedied in order to meet contingency, limited and general war plans. Deficiencies will be reported to the CNO with recommendations for appropriate action and requests for release of required components or materials.

b. Determination and coordination of transportation requirements to implement policies covering both peacetime and emergency employments.

c. Coordination of routine deployment schedules and assigned projects and training with the policies set forth in OPNAV INST. 5440.62 and 5450.46.

d. Determination of requirements and development of procedures for resupply of the battalions under normal and contingency operations.

6. Deployment Policies.

a. Battalions will normally be assigned construction projects contributing to training and readiness for fulfillment of contingency assignments.

b. Battalions will be deployed to, and

employed in, various environments to enable these forces to meet emergency requirements which may develop in similar areas of the world. Deployment sites will be chosen wherever possible to ensure proximity of forces to possible areas of contingency assignment.

c. Battalions will participate in fleet training exercises in order to coordinate planning and operational procedures with the forces supported.

d. Under conditions of emergency, disaster, or catastrophe caused by enemy action or natural causes, Naval Construction Force units may be utilized as directed by cognizant authority. This includes furnishing assistance to civilian agencies.

e. For special situations, specific requests may be submitted to the Chief of Naval Operations for consideration.

The battalions maintain the maximum possible state of readiness through the performance of such diversified projects as:

a. Assembly of nuclear power plants.

b. Construction of floating drydocks.

c. Counterinsurgency assignments throughout the world.

d. Operation of nuclear reactors.

e. Construction of causeway piers and ship-to-shore pipeline systems.

f. Construction of advanced bases.

g. Construction of Short Airfields for Tactical Support (SATS).

h. Construction of communications centers.

i. Construction of range facilities for space programs.

j. Construction of water front facilities, bridges and roads.

k. Construction of semi-permanent and temporary camps.

SECTION D - CONCEPT OF OPERATIONS

1. General. The NCF units may be called upon to provide engineer support during periods of General War, Limited War or Cold War. When not engaged in one of these conditions, the NCF units will normally be engaged in what have been called "normal peacetime operations." These normal peacetime operations involve training. Not only formal training, but on-the-job instruction and practice. For the ACBs, this will usually take the form of participating in fleet exercises. For the MCBs it may involve a deployment to a U.S. overseas base to do new construction or to perform repair and improvement projects, or it may involve practice exercises with the Marine Corps or other fleet units. During these periods, the NCF units are building a pool of SEABEES trained to perform their wartime tasks, testing allowance for wartime adequacy, gaining experience in operations in environments similar to anticipated wartime climates and generally attempting to develop detailed doctrine and standard operating procedures which will enable them to effectively accomplish tasks assigned under emergency or wartime conditions. SEABEE operations may be placed into four categories for discussion purposes. However, in the field these distinctions are not so sharp and a battalion may be engaged in more than one at a time. These categories are:

Support of Expeditionary Forces
Base Development and Maintenance
Disaster Control Operations
Support of Foreign Assistance Programs
(See Figure 10)

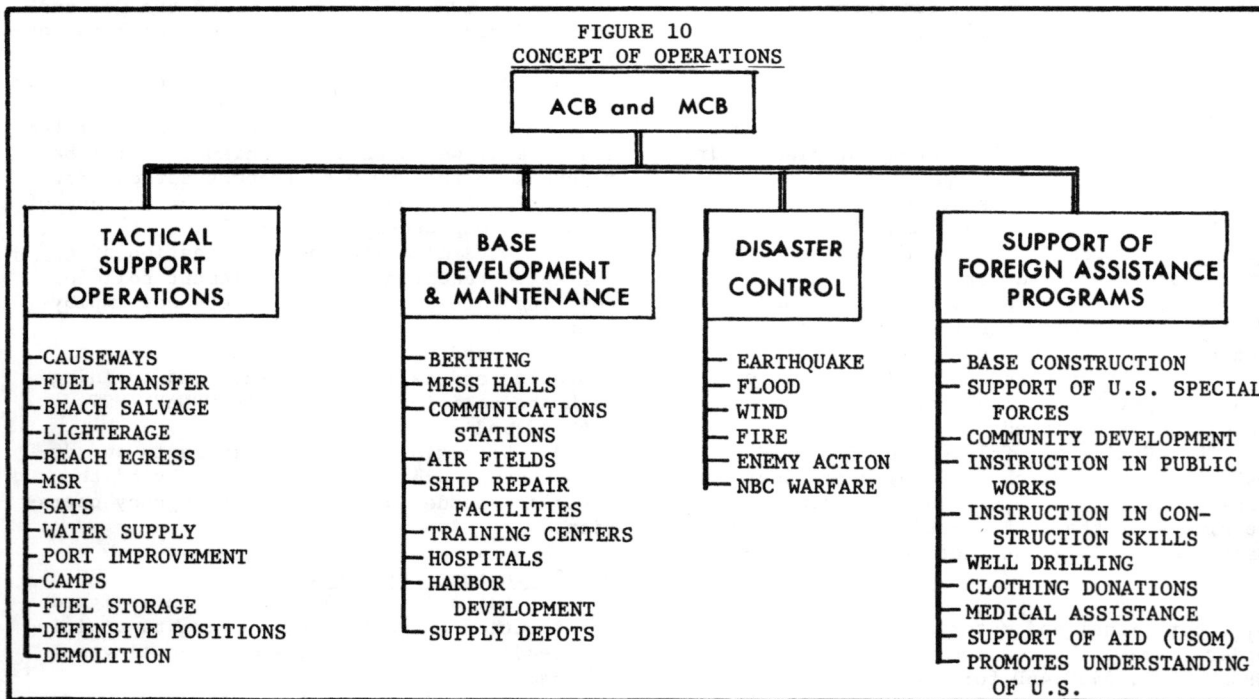

FIGURE 10
CONCEPT OF OPERATIONS

ACB and MCB

TACTICAL SUPPORT OPERATIONS
- CAUSEWAYS
- FUEL TRANSFER
- BEACH SALVAGE
- LIGHTERAGE
- BEACH EGRESS
- MSR
- SATS
- WATER SUPPLY
- PORT IMPROVEMENT
- CAMPS
- FUEL STORAGE
- DEFENSIVE POSITIONS
- DEMOLITION

BASE DEVELOPMENT & MAINTENANCE
- BERTHING
- MESS HALLS
- COMMUNICATIONS STATIONS
- AIR FIELDS
- SHIP REPAIR FACILITIES
- TRAINING CENTERS
- HOSPITALS
- HARBOR DEVELOPMENT
- SUPPLY DEPOTS

DISASTER CONTROL
- EARTHQUAKE
- FLOOD
- WIND
- FIRE
- ENEMY ACTION
- NBC WARFARE

SUPPORT OF FOREIGN ASSISTANCE PROGRAMS
- BASE CONSTRUCTION
- SUPPORT OF U.S. SPECIAL FORCES
- COMMUNITY DEVELOPMENT
- INSTRUCTION IN PUBLIC WORKS
- INSTRUCTION IN CONSTRUCTION SKILLS
- WELL DRILLING
- CLOTHING DONATIONS
- MEDICAL ASSISTANCE
- SUPPORT OF AID (USOM)
- PROMOTES UNDERSTANDING OF U.S.

12

2. Preparation for Support of Expeditionary Forces. SEABEES are designed to built advanced bases in support of the fleet and to provide engineering support for FMF units rather than to serve as combat engineers. However, the battalions are responsible for military engineering forces ashore. Although they are not committed to infantry operations except in emergencies, the battalions are an integral part of the beach and base defense system and constitute a potential reserve force. Military support tactics for the battalion, company, platoon, squad and fire team are explained in the Landing Party Manual.

In the initial planning for an amphibious assault, the Marine Corps recognizes that they do not have enough engineer capability to support a combined Division-Wind landing. They have therefore requested from the fleet commander and have received authority to utilize SEABEES in support of their amphibious operations. The SEABEES have been told that they are expected to provide the bridge from ship-to-shore (ACB); construct airfields, storage areas, main supply routes, etc., (MCB); provide jobsite or beach security against guerrilla-type activity, sabotage of equipment and material, sniper harassment and well-organized enemy patrol action; provide blocking companies to be utilized if the enemy breaks through the attacking forces and provide disaster recovery services from the effects of tactical nuclear weapons, biological or chemical warfare, storm, flood or fire.

The ACB would task organize into the tactical elements that they provide to the various beach party teams (See Figure 11). These ACB elements would then embark on many ships of the task force. The MCB, although tailored to some extent for each operation, will maintain its basic organization and embark as a unit to the extent permitted by the tactical situation

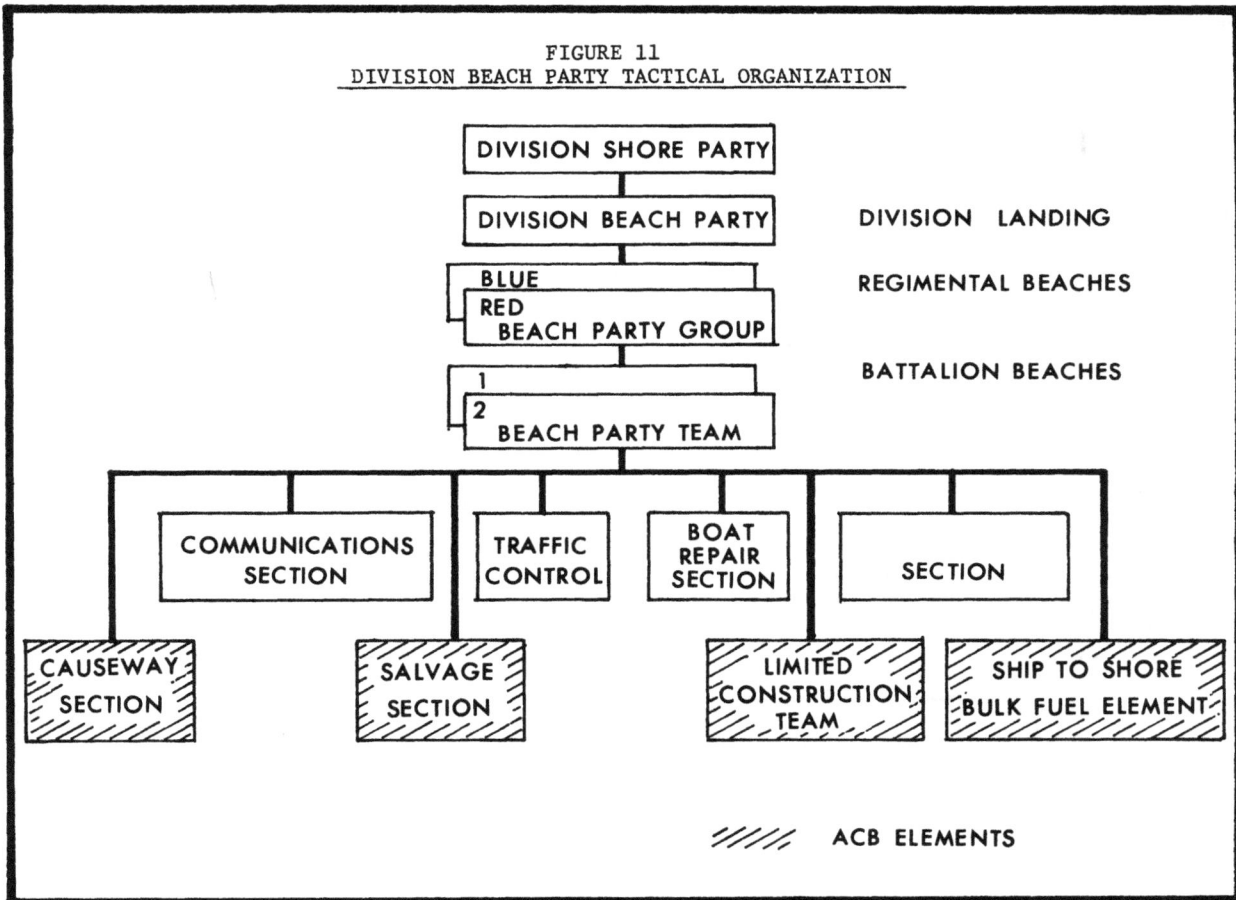

FIGURE 11
DIVISION BEACH PARTY TACTICAL ORGANIZATION

3. Organization of the Landing Force. The Marine Expeditionary Force (MEF) will consist of a Marine Division (MARDIV) and a Marine Air Wing (MAW) and supporting units drawn from force troops. These force troops may be formed into a Force Logistic Group and a Force Engineer Group. Smaller landings can be accomplished by elements of these units. For example, a Marine Expeditionary Brigade (MEB) would consist of an infantry regiment from the division, an air group from the wing, and subordinate supporting units.

The engineer capabilities available to the Commanding General of the MEF are shown in Figure 12. The Division Engineers are organic to the Division and perform combat engineer functions such as clearing minefields, demolition, stringing barbed wire, etc. They are assigned directly to front line troops in as small as squad size units and frequently must operate under fire. Since they do not operate as a battalion, their construction potential is rather low.

The Marine Air Wing has no engineer designated as such. However, with the Marine Air Base Squadron (MABS), there are many items of equipment that could be utilized in the construction of the airfield in addition to its maintenance. Some Marines assigned to the MABS have experience in the construction trades.

Within Force Troops in both the Atlantic and Pacific, there are a Force Engineer Battalion and a Bridge Company. The topographic company is homeported at Camp Lejeune. The Force Engineer Battalion, consisting of about 45 officers and 1060 men has the equipment and talent to perform engineer support tasks for the expeditionary force. It can be assigned as a battalion or more frequently as several separated companies in support of either the division or the wing. Tasks include development of a water supply preparation of a barrier plan, including tank traps and fortified positions; construction of supply dumps; and construction of main supply routes (MSR). Many of the tasks of the Force Engineers are done by assigned MCBs when the quantity of work exceeds their capability to get it done in the required time frame.

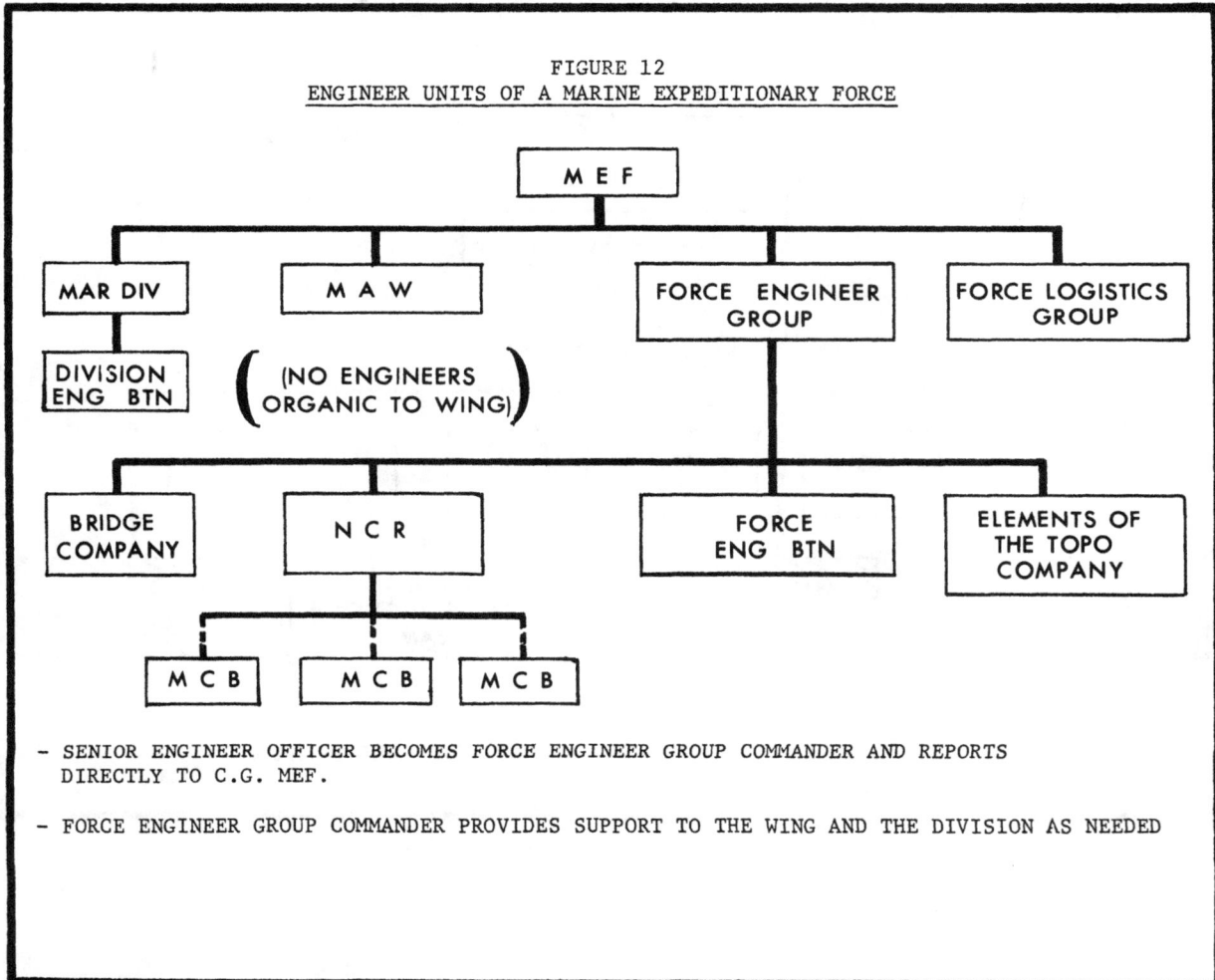

FIGURE 12
ENGINEER UNITS OF A MARINE EXPEDITIONARY FORCE

- SENIOR ENGINEER OFFICER BECOMES FORCE ENGINEER GROUP COMMANDER AND REPORTS DIRECTLY TO C.G. MEF.

- FORCE ENGINEER GROUP COMMANDER PROVIDES SUPPORT TO THE WING AND THE DIVISION AS NEEDED

The Bridge Company consists of several prefabricated structures. The units requiring a bridge supplies the bulk of the labor needed to erect it. Bridge Company personnel supervise and control its construction.

The MCB represents the heaviest construction capability in the assault area. However, this is a relative term and can be misunderstood. The SEABEES have tailored the MCB allowance so that it can perform most general construction tasks. In order to perform any one large job, the allowance must be re-enforced by additional tools, equipment, and perhaps men. These are grouped into Supplemental Functional Components designed for a specific task. For example, in order to conduct an extensive quarry operation, a rock crushing and screening plant, a large power shovel and additional dump trucks, air drills and blasting material would be needed. In many cases, the MCB is expected to provide support to both the division and the wing (See Figure 12). The nature of the projects assigned by the FEG Commander will determine the exact allocations of personnel and equipment made by the MCB Commanding Officer.

The senior engineer will assume the duties of Force Engineer Group Commander. The Force Engineer Battalion has sufficient officers to provide a staff to the FEG Commander. When several MCBs are assigned, they are formed into a Naval Construction Regiment. The Commander and staff of the NCR generally then become the Force Engineer Command element suitably augmented by Marine Engineer Officers and NCRs.

4. Engineer Task Assignments. Once the operation commences, the assault forces move from ship to shore either by crossing the beach or by landing to the rear of the enemy's beach defenses by helicopter. The initial objective of these forces is to gain control of the beachhead by occupying the critical terrain features (hills, bridges, crossroads, etc.). This strong point defense may cause wide separation of forces and the bypassing of unimportant areas. The threat of NCB warfare will also tend to spread friendly forces over many miles. This causes the engineer forces to operate in areas that have not been cleared or occupied by the assault forces. Therefore, the SEABEES are provided with weapons to protect their jobsites from harassment from small pockets or bypassed troops or partisan activities (See Figure 13).

On signal, the ACB causeway team will land the causeway and anchor it in position. This will permit rapid offloading of mobile equipment embarked in LSTs. Engineer units of the FEG can be landed and can start to construct beach egress roads, fuel farms, etc. As the build-up ashore continues, the ACB fuel element will construct ship-to-shore pipelines for motor fuel and aviation gasoline and hook them to the fuel farms constructed by the MCBs or Force Engineers. Other engineer tasks are depicted in Figure 14.

5. Initial Battalion Action Ashore. If the situation permits, the main body of the battalion proceeds to the assigned bivouac area. If possible, the battalion establishes a tactical bivouac near the eventual site of its first camp and immediately establishes an adequate defense and basic sanitary and housekeeping facilities. Communications and liaison are set up with adjacent units and with the next higher headquarters in the chain of command. During the early part of the landing, elements of the battalion may be assigned to

FIGURE 13
ORDNANCE ALLOWANCE LIST (MAJOR ITEMS)

	MCB	ACB
RIFLE, M-14, 7.62-MM (or M-16E1 5.56-MM)	472	600
MACHINE GUN, M-60, 7.62-MM	16	---
LAUNCHERS, 3.5 INCH ROCKET	8	8
PISTOL, .45 CALIBER	126	109
BAYONET, M-6	520	660
GRENADE LAUNCHER	39	39
COMMAND POST TENTS	4	6
BINOCULARS	15	15
INDIVIDUAL FIELD INFANTRY EQUIPMENT	600	700

FIGURE 14
TYPICAL ENGINEER TASKS IN AOA

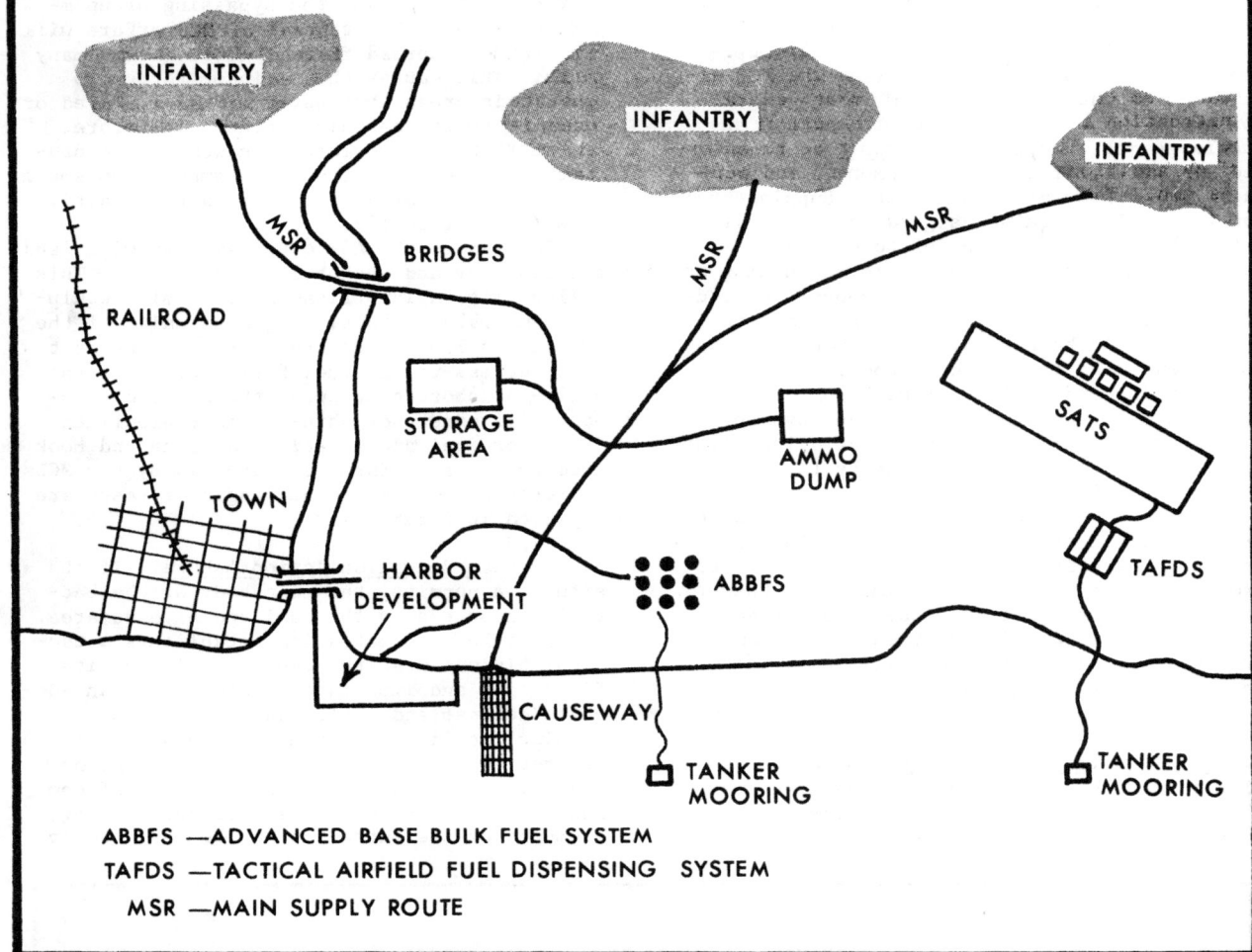

INFANTRY

INFANTRY

INFANTRY

MSR

BRIDGES

MSR

MSR

RAILROAD

STORAGE
AREA

AMMO
DUMP

SATS

TOWN

HARBOR
DEVELOPMENT

ABBFS

TAFDS

CAUSEWAY

TANKER
MOORING

TANKER
MOORING

ABBFS —ADVANCED BASE BULK FUEL SYSTEM

TAFDS —TACTICAL AIRFIELD FUEL DISPENSING SYSTEM

MSR —MAIN SUPPLY ROUTE

unloading operations for infantry support.
Care must be taken in this matter to determine
relative priorities because this unloading effort may detract from the construction effort.
The battalion will commit all other manpower
to consolidating its camp and preparing for
construction operations. Food, water, ammunition, fuel and repair parts supply are established as soon as possible. Shops are set up
as quickly as conditions permit. Engineering
reconnaissance begins immediately, using all
available facilities. Local sources of building material are located, evaluated, and if
required, obtained through the civil affairs
section of the Expeditionary Force.

When the bivouac has been established, the
battalion sets its perimeter defense. The
perimeter defense provides the maximum security against infiltration and, along with other
local ground defense tactics, is described in
the Landing Party Manual. Provisions for the

perimeter defense of a job site must be reviewed daily, obeying the basic rule that any
defense must be continually improved. Enemy
infiltration can only be prevented by 360 degree protection of every unit, large or small.
Perimeter defense should be organized by all
units during any halt, by isolated units in a
sustained defense and by units in a specific
situation where normal security measures are
inadequate.

During daylight observers are posted within each unit area or on critical terrain nearby to keep the ground between these areas under
constant surveillance. At night, listening
posts are established to cover all avenues of
approach to the defense areas. Dense jungle
or woods limit observation and fields of fire
to a few yards, necessitating special security
measures such as the construction of tactical
and protective wire obstacles, and mine laying.

When the MCB is operating a separate camp (Command Post) the Headquarters Company Commander is responsible for the Command Post security and defense. The MCB is usually assigned specific sector defense and disaster control responsibilities by the military commander. Although usually assigned to a ground defense sector which includes camp and jobsites, the battalion may be required to assume additional defense responsibilities as considered necessary by the commander. Guides for the organization and conduct of ground defense for each element of the battalion are described in the Landing Party Manual and appropriate Army and Marine Corps Field Manuals. The senior officer present at the job site, usually the battalion project officer in charge of construction, has command responsibility for the job defense. He prepares a job defense plan and assigns defense responsibilities to each company. The companies in turn assign duties to the platoons; platoons to the squads; squads to fire teams, and fire teams to individuals. Whenever crews, squads, platoons or companies report for sub-contract work, they report as integral units, and are assigned defensive responsibilities at the time of reporting for duty. In the case of a surprise attack, all attached troops are automatically under the tactical control of the senior officer present. The principles of job defense are also applicable to the defense of isolated work details.

As the assault forces move inland, it may be necessary to construct additional support facilities far from the beach. If the operation will be a prolonged one where the invasion forces plan to occupy captured terrain for a length of time, some base development will be necessary for the fleet units operating in the area or as a logistic rear area for the combat forces. This determination will have been made by the Joint Chiefs of Staff (JCS) prior to the invasion and an operation plan for construction would have been developed (See Figure 15).

FIGURE 15
ADVANCED BASE PLANNING PROCESS

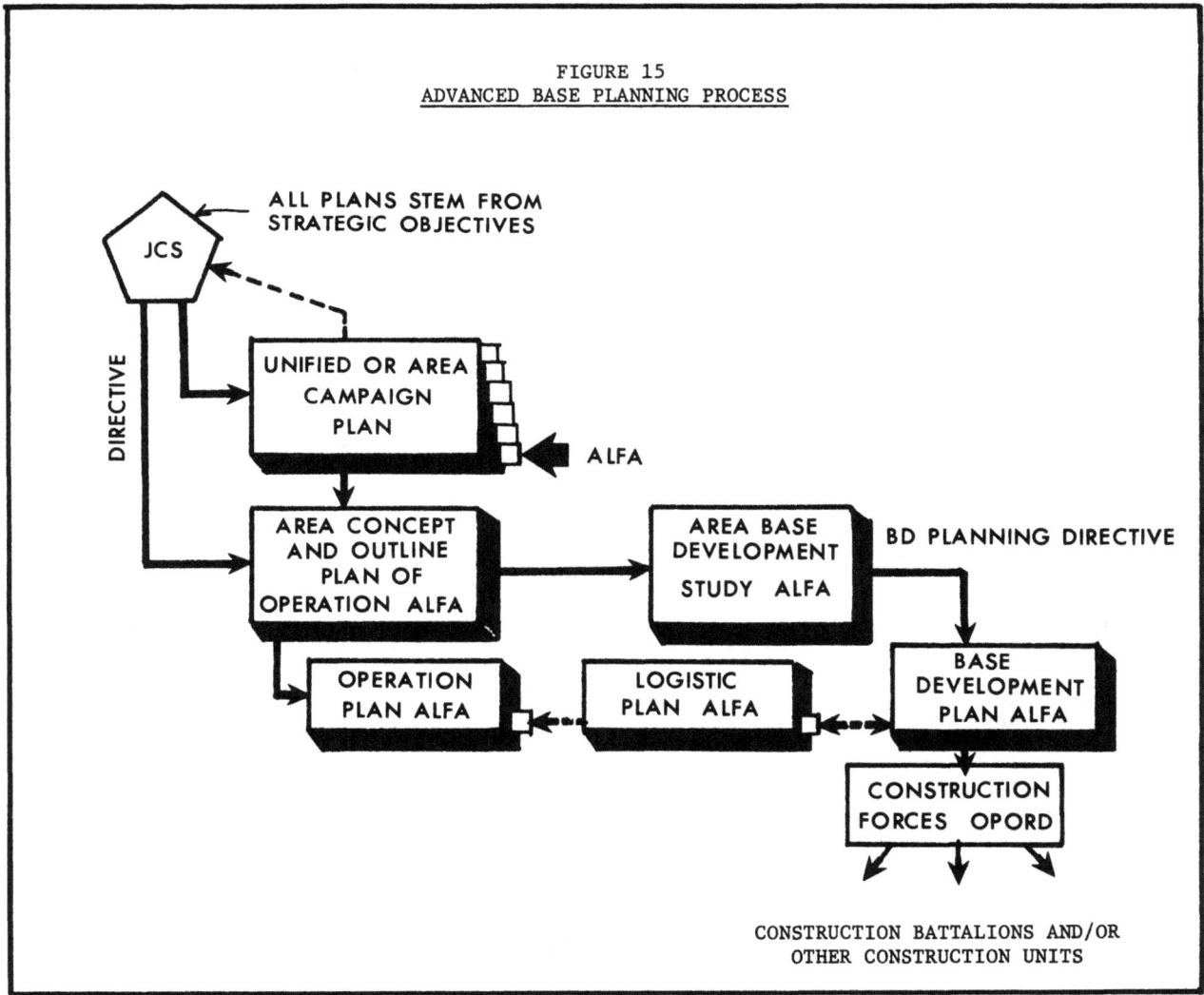

6. <u>Advanced Base Planning</u>. The bulk of the work planned for the NCF in wartime involves base repair and development in forward areas. This base development will be similar to that accomplished during World War II, except that the emphasis will be on repair and expansion of existing facilities by temporary and limited construction rather than large scale new construction.

An advanced base is usually activated as an emergency temporary facility, organized in or near a forward combat area to support operations of the United States or allied military forces. An MCB is capable of design adaptation and the construction of advanced base complete with all necessary facilities. These include, but are not limited to, electric power requirements; water supply and purification; hospital facilities; airfields and their auxiliary components such as parking stands, taxiways, control towers; housing areas; stream and river crossing facilities; roadways and general lines of communications. An advanced base may be developed to support a variety of missions including the defense of strategic areas, the support or protection of a communication or supply line, and particularly the mounting or support of direct offensive action. Advanced bases may, in justifiable instances, be developed as permanent installations. Usually, however, they are temporary facilities which will be deactivated after their mission is served. The total development plan is predicated on the immediate situation involving the phase of hostilities, current weapons, capability and potential of the enemy, location, and logistical and strategic plans.

MCBs are designed to build, repair, improve, or dismantle advanced base facilities and lines of communications during general or limited war or contingency operations, and MCB activities are designed to develop this capability. MCBs train in peacetime by building outlying base facilities for the Navy. In emergency situations, MCBs build or rebuild any facilities required for advanced bases for the United States or allied military operations. They are trained to participate in the ground defense and disaster control activity of the base to which they are assigned. The construction workload will usually be large. Therefore, one can expect that several battalions operating as a regiment will be employed in the repair of facilities damaged by the assault or the extension of existing facilities, or the construction of new facilities utilizing functional components. (See Annex C for a brief description of the Functional Component System). In these situations an MCB will come under the direct operational control of a Naval Construction regiment or brigade, a Naval Base or station, a joint construction command, an island, area or base command or other designated command in the forward area.

Preparation for battalion operations involves long range planning at every level of the Naval Construction Force organization. These preparations are developed months or years before the need arises in order to maintain the ability to erect advanced base structures with extraordinary speed. The Joint Chiefs of Staff (JCS) evaluate all potential military threats to security, develop a comprehensive network of defense, and plan a full range of strategic and tactical counter-measures. Top level missions and general tasks for the Navy which emanate from the National Security Council through the JCS are implemented by the Chief of Naval Operations by means of the Navy Planning System (OPNAV INST. 5000.19). Base Development (NWIP 11-23) is the planning document used in translating assigned support missions into facility requirements for advanced bases. In an emergency, the JCS issues directives to unified, specified, or area commanders. These directives state the strategic goals and tentatively allocate the necessary forces. A JCS directive may be supplemented by a Strategic Study, A Base Development Feasibility Study or both. Advanced base planning in the area of operations proceeds from the JCS statement of strategic goals to actual working plans through the following stages:

a. An area campaign plan.

b. An outline plan for each operation.

c. An Area Base Development Study. This document determines the site that will best support the planned naval operations. It is in short, a feasibility study.

d. Base Development Planning Directive. This indicates the base site, its mission and sets the operational target date desired by the Base Commander.

e. The Base Development Plan. This plan is usually published as an annex to a logistics plan and furnishes full information to all concerned units as to the intent of the fleet or area commander in the construction and operation of the base. The plan includes anticipated base loading, proposed command structures, a schematic site plan showing areas allocated to specific functions and the estimated construction effort involved.

f. Operation Order (OPORD). This document directs specific construction units to undertake specific projects and assigns these construction units to the Base Commander.

Planning will be specific to the needs of an operation or readiness situation. It is based on intelligence, maps, and other valuable information obtained from the appropriate service group or furnished by higher command. No two operations are completely parallel. However, the MCB is equipped and trained so that standard allowances and special items of equipment, with trained men to use them, are always available to carry out specific missions and meet all contingencies. In planning an advanced base, the following factors are considered: economy of time and manpower, simplicity of design, economy of materials, location, terrain and climate. Military assistance such as air defense measures, location and protection of existing facilities, and similar factors are especially important during emergencies.

The Engineers Handbook for Planning Advanced Bases, NAVFAC TP-PL-5 (Confidential) is a guide for CEC officers on the advanced base planning level. It describes the planning background at various levels, sets the construction and shipping requirements of functional components, describes the computations for determining MCB capabilities, reviews site selection requirements and tabulates pertinent logistic data. Other guides are the NAVFAC Detailed Catalog of Equipment and Material Requirements for Advanced Base Functional Components, NAVFAC P-103 and Advanced Base Drawings, NAVFAC P-140. The Army manuals that are most helpful in the area are the Staff Officers Organization, Technical and Logistical Data, FM-101-10; Engineer Field Data, FM 5-34; and Army Tables, Drawings and Bills of Material of Engineer Functional Components, TM 5-301, 302, and 303.

An adequate investigation of the site and careful study of the design details are essential for the greatest economy in construction time and effort. After information on the site is collected and the type of construction is determined such details as establishing grade lines, the location of work areas for equipment and the procedure for material compaction must be resolved. Embarkation planning, particularly the stowage of material and equipment, is a predominant factor in the logistics of engineer forces. Personnel, with their tools and equipment, must be loaded for disembarkation in a sequence planned in the order of utilization. Without synchronized unloading of men with the equipment they are to use the battalion cannot perform its tasks with the efficiency inherent in its organization. Scheduling and timing must be coordinated to meet the military plan as well as to meet normal construction phasing. When the unit to do the work is selected, the OPORD is issued by COMCBLANT/COMCBPAC. Through this OPORD, COMCBLANT/COMCBPAC delegates the projects to his subordinate units. He designates their individual objectives and defines the limits of their responsibilities and authority in carrying out their task assignments. The relative priorities for the components of each project, as well as target dates for initial occupancy and final completion are indicated. Schedules are based on man-day estimates, adjusted for anticipated conditions. The OPORD is distributed to all elements of the future advanced base command and to tactical construction units, to assure that subordinate commands understand the duties, responsibilities and command relationship of the construction units. The standard organization of an OPORD is described in Annex D.

7. Advanced Base Organization for Construction. Advanced base construction is directed by the Commander, Advanced Base Construction, who reports to the Advanced Base Commander. The Advanced Base Commander may have jurisdiction over an area including the civilian population, or only over the base itself. He and his staff, when appointed early enough, assist the Area Command in the preparation of the Base Development Plan. In a joint operation, the Advanced Base Commander may be an officer of the Navy, Army, or Air Force.

The Commander, Advanced Base Construction Force, commands all construction units assigned to develop a given base. In a joint operation, he may be an officer of any of the three services. In a Naval Operation, he may be a commander of a naval construction brigade or regiment, or at a small advanced base he may be the Commanding Officer of an MCB. Except in an emergency, the Advanced Base Construction Force does not perform tactical or combat engineering tasks, concentrating instead on base construction and engineering assignments. Usually, the Maintenance Units (CBMUs) are transferred to the base upon completion of base facilities. Occasionally, however, they may be used for construction of the base they will later maintain. On his arrival at the site, the Commander, Advanced Base Construction Force, concentrates on coordinating the effort, expediting production, and changing basic precepts or task assignments to meet unexpected changes required by local conditions or higher command decisions.

The Base Civil Engineer (Public Works Officer) serves on the staff of the Advanced Base Commander as a technical expert and advisor, having no direct command or construction responsibilities. He usually participates in the Base Development Plan preparation and maintains liaison between the Advanced Base Commander and the Commander, Advanced Base Construction Force during base construction. After the construction phase, he will act for the Base Commander in matters of maintenance and minor construction.

8. Phases of Development. Advanced base development usually moves through four general phases, Assault and Consolidation, Exploitation, Base Development and Base Operation. Although no two operations are identical, they usually follow this pattern. During the assault and consolidation of the landing forces ashore, the Expeditionary Force Commander establishes headquarters ashore and relieves the Shore Party Commander of responsibility for base development which is already underway. The most critical phase from the standpoint of base development occurs during the exploitation of the beachhead. Any uncertainty as to construction, handling of shipping, unloading or clearing of cargo may result in lost time and material and serious military reversals. During this phase, the pre-invasion plans for base construction are modified, if necessary, after a swift reconnaissance and analysis of the area situation. The MCB must be ready to

regroup and improvise plans in response to any unexpected situation. When the Amphibious operation is secured, Shore Party operations over the beaches continue under the Expeditionary Force Commander until relieved by appropriate authority. The Advanced Base Commander develops and improves unloading facilities and storage areas in order to expedite over-the-beach operations.

When the area has been secured, the Advanced Base Commander proceeds with full scale development until all facilities are fully operational. All construction and weight-handling equipment is operated on two ten-hour shifts per day on highest priority work and mechanics must be expeditiously supplied with repair parts to keep every piece of equipment in action. The rear boundary for the tactical forces is established and the Advanced Base Commander assumes full control of the rear areas specified for the development of the base. The Shore Party is attached to the base command, which absorbs the Shore Party's functions. As the base reaches full operation, the functions of the Shore Party and other tactical support units are delegated to other units operating under the Advanced Base Commander. Construction required to complete or expand base facilities is pressed during this period and, when completed, the construction units are relieved for restaging and outfitting and another operation. CBMUs remain to operate as a public works unit for the base.

9. <u>Disaster Control</u>. The term "Disaster Control" used throughout this manual in the broad and inclusive sense means the basic actions of protective and recovery measures taken in cases of natural disaster or unfriendly military action. Disaster Control may be defined as the sum total of all action taken (other than military action) to prevent, reduce, or repair damage from natural disasters or military action without employing weapons or initiating offensive action against an enemy. Certain measures of protection or recovery are directly associated with an enemy attack while others are related to acts of natural disaster. However, some measures such as emergency communications, shelters and advance stockpiling of critical materials, are directly related to both enemy action and natural disaster.

Each battalion is responsible for disaster control measures to protect its own personnel, equipment, camp and job sites and may be assigned the responsibility for participation in the defense of other activities. Additionally, the standard organization also makes it an effective Disaster Control Unit (DCU) ready to give direct assistance to any military or civilian installation or community during an emergency. The NCF units by nature of their missions and complements have most of the capabilities of a Disaster Control Unit.

10. <u>Disaster Control Plan</u>. The functions performed by a disaster control organization fall generally into the areas of control and communications, personnel welfare, security, engineer, fire, ordnance, medical, transportation, supply and NBC defense operations. A battalion, by the nature of its mission and complement, has many of the capabilities required of a disaster control unit. The battalion disaster control plan is based upon these assumptions:

a. Work will be done using the standard battalion organization when possible.

b. Where specialized training is necessary, teams will be identified and trained.

c. Within these teams, fire team and squad organizations will be maintained.

The Disaster Control Plan will include provisions for the following situations:

a. Nuclear Warfare.

b. Biological and Chemical Warfare.

c. Natural disaster (storm, earthquakes, etc.).

d. Air raid.

e. Fire.

f. Civil disturbance.

The plan must be specific in:

a. Assignment of personnel to emergency recovery teams.

b. Assignment of specific equipment and material to recovery teams.

c. Establishment and equipping of Control Centers.

d. Indication of equipment dispersal areas.

e. Preparation of and assignment to shelter areas.

f. Provision for emergency communications

g. Care of survivors.

h. Plan for salvaging materials.

i. Training of disaster control teams.

j. Individual training in self-protection and aid.

k. Conduct of readiness drills.

As indicated in the Disaster Control Plan, each work crew/fire team is trained and equipped to operate as a disaster control team. Thus, the battalion is prepared to operate as the independent Disaster Control Unit (DCU) or as part of a Disaster Control Group (DCG). This transition provides the minimum variation from the normal organization. After specialized teams have cleared the damaged area and rendered it safe for normal repair operations and procedures, the construction companies commence the repair and rebuilding phases. Their activities and responsibilities are essentially similar to those of the battalion's daily routine. Officers normally assigned as prime or sub-contractors are utilized in their normal capacity. In the event that SEABEE forces are required to assist in areas other than their own, the local control commander would rely on the construction battalion to function as a self-contained unit. Complete conformance with the disaster control concept as set forth in NAVFAC INST. 3440.12 is not considered applicable to SEABEE operations.

11. Disaster Control Teams. The following teams, formed from battalion personnel, must be organized and trained. Normal battalion operations do not provide any training in these areas.

NBC FUNCTION

```
AA Teams (2) - NBC Survey
AB Teams (2) - Personnel
              Decontamination
AC Teams (6) - Facilities and Area
              Decontamination
AE Teams (2) - Dosimetry
AF Teams (2) - Clothing Decontamination
```

CONTROL AND COMMUNICATIONS FUNCTION

```
CA Teams (1) - Control Services
CB Teams (2) - Command Post Services
```

Team assignments within the battalion are shown on the Personnel Control Boards in the battalion operations office and in individual company offices.

12. SEABEE Support of Foreign Assistance Programs. Foreign aid-type projects may be undertaken upon specific approval by the Chief of Naval Operations. Each individual project will be evaluated in terms of overall worth to the nation and the Navy considering all of the political and military factors involved. Normally, projects will not be approved without concurrence of the cognizant Fleet Commander, unless overriding factors of the urgency of the requirement dictate otherwise. Generally, foreign assistance will be either direct assistance to foreign governments or engineer support to other U.S. agencies that are providing foreign assistance. For example, in Vietnam SEABEES constructed camp facilities and defensive structures for the U.S. Army Special Forces teams. This is an example of support to U.S. Agencies, whereas the USOM-sponsored construction of hamlets for Vietnamese civilians constituted direct assistance.

Usually, employment of the entire battalion would be limited to supporting U.S. agencies. The construction of U.S. bases overseas not only brings new sources of money to foreign countries, but also demonstrates in steel and concrete, the U.S. pledge to protect her allies. A task that is always assigned to every SEABEE who deploys overseas is that of showing the American way of life and our devotion to freedom. The individual American overseas represents our national policy to the foreign people with whom he comes in contact. SEABEES, through constructing school playgrounds, distributing donated clothing, to providing medical assistance have always been participants in civic action. They have traditionally been admired by others because they do not shirk from working with their hands. SEABEES have avoided becoming involved in local politics

which would cause hard feelings, even though they are deployed to overseas locations for the majority of every year.

13. SEABEE Team Operations. From these battalions, a few trained petty officers were selected to provide direct technical assistance to friendly nations throughout the world. They were formed into SEABEE teams which furnish assistance on engineering problems, supervise a wide range of construction work, instruct and assist local personnel in the construction and mechanical trades, and assist in disaster control operations. Their operations aid these nations to achieve a more rapid economic and social development and to foster self-reliant improvement programs.

The Chief of Naval Operations directed the Commanders in Chief, U.S. Atlantic and Pacific Fleets, to organize two SEABEE Teams within each of the Construction Battalions. Enlisted personnel serving with the teams have received construction trade and military training as members of battalions and they are given a wide range of specialized training.

Every team member must be experienced at the journeyman level in at least one construction trade and have had additional practical experience in one other construction trade. A team represents all the skills available in a battalion and is limited only by manpower and equipment. Teams have been trained to concentrate on projects which provide on-the-job training or classroom instruction to in-country personnel. For example: construction of roads, bridges, airstrips, earth dams, water wells, sanitation systems, and schools. Team training programs are highly specialized with emphasis on technical material that augments conventional military skills. Team personnel must acquire the skill to undertake general construction, specialized construction and recovery operations. Specific training topics are listed:

```
Civil Action
Communications
Counterinsurgency
Country Familiarization
Defensive Tactics
Demolitions
Disaster Recovery and Control
First Aid
Instructor Capability
Intelligence
Leadership
Logistics
Soil Testing
Surveying
Weapons Employment
Survival Evasion Resistance & Escape (SERE)
Inservice Training (instruction in other
   SEABEE ratings)
```

Responsibility for SEABEE team operations in each foreign country lies with the United States Ambassador. These operations may be undertaken by any of the following agencies: Agency for International Development; Peace Corps; Military Assistance Advisory Group; and the United States Information Agency. The Ambassador has the authority to assign task forces to programs which offer maximum benefit to the country involved. SEABEES can be used as a part of the Military Assistance Program to provide construction advice, field inspection, design assistance, and on-the-job training. Through interdepartmental agreements (i.e., between DOD and DOS) SEABEE teams can also act as contractors for the Agency for International Development to provide project planning assistance and technical instruction. In addition they can aid the Peace Corps by training Peace Corpsmen and by providing engineering advice (See Figure 16).

FIGURE 16
SEABEE TEAM USE IN ECONOMIC AND MILITARY ASSISTANCE PROGRAMS

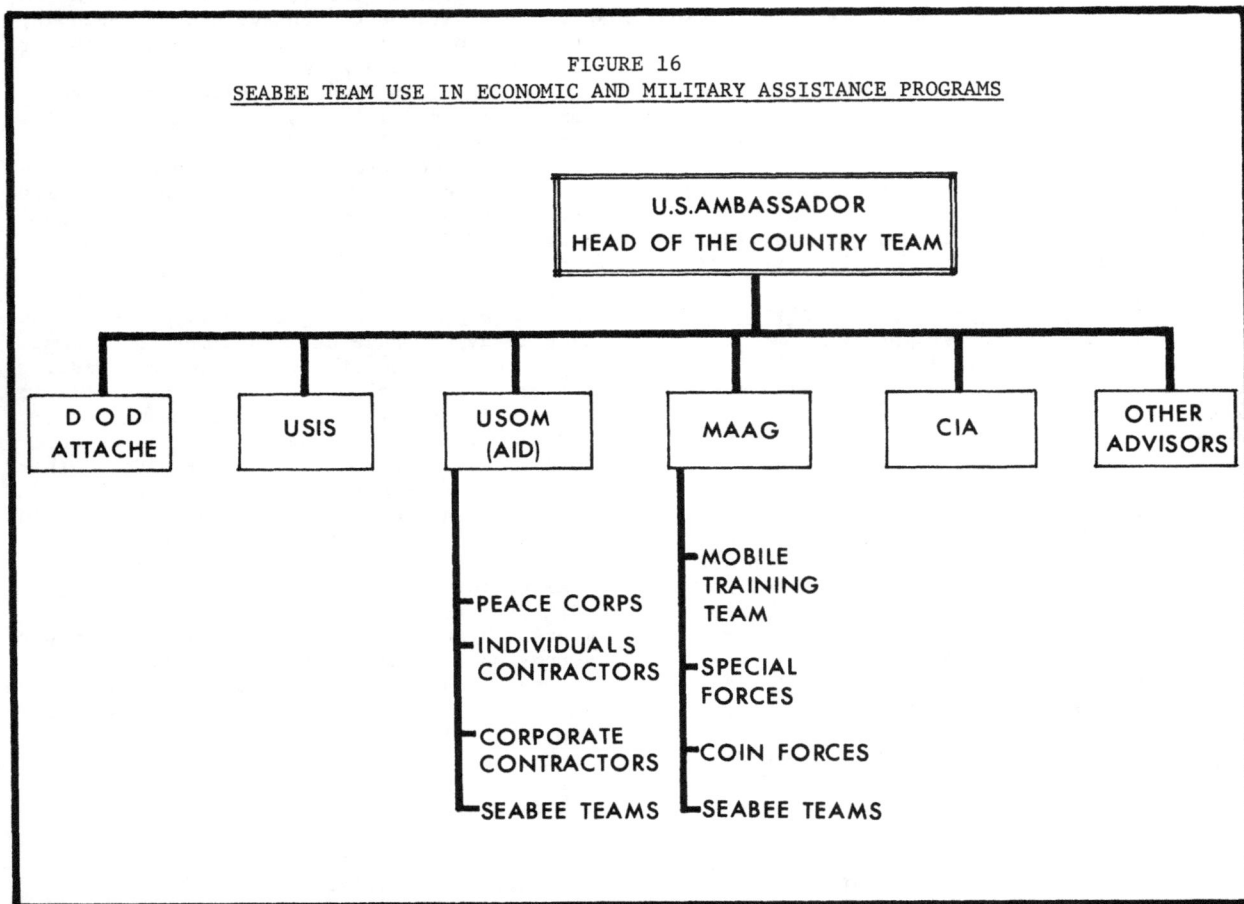

U.S. AMBASSADOR
HEAD OF THE COUNTRY TEAM

DOD ATTACHE   USIS   USOM (AID)   MAAG   CIA   OTHER ADVISORS

USOM (AID)
- PEACE CORPS
- INDIVIDUALS CONTRACTORS
- CORPORATE CONTRACTORS
- SEABEE TEAMS

MAAG
- MOBILE TRAINING TEAM
- SPECIAL FORCES
- COIN FORCES
- SEABEE TEAMS

Chapter II
COMMAND AND STAFF

SECTION A - THE COMMANDING OFFICER

1. **Duties.** The Commanding Officer of an ACB or MCB is an officer of the Civil Engineer Corps. He is responsible for organizing the battalion so that the work is accomplished in a most expeditious manner. As Commanding Officer of a Commissioned Naval Unit, he has complete responsibility for all phases of battalion performance and may delegate authority to his subordinates (when not contrary to law), but this does not relieve him of utimate responsibility for the security, efficiency, and well being of his command. The chain of command passes from the Commanding Officer through the Executive Officer to the Company Commanders and down through the Platoon Leaders to the small unit leaders. In addition the Commanding Officer has the battalion staff to assist him in formulating command policy and making command decisions.

It is a command responsibility to expand and supplement the guidance from higher authority to ensure the ability of the command to promptly and aptly respond to assigned mission tasks. This includes a continuing evaluation of commitments and assets. The former have been indicated in a general way in Chapter I, the latter consists of the officers and men assigned (including their skills and level of training), and the equipment and material allowance. General responsibilities of the Commanding Officer are stated in U.S. Navy Regulations, Articles 0701 through 0739, some of which are reprinted as Annex E. The Commanding Officer is also governed by area and type commander regulations, fleet regulations and applicable directives of higher authority.

When the NCF units are committed to support an amphibious assault, the tactical organization of the MCB remains essentially the same, but the ACB forms tactical teams described in Chapter V. The Commanding Officer of the MCB may become the Force Engineer Group Commander of a Marine Expeditionary Force reporting directly to the Commanding General of the MEF. The Executive Officer would essentially assume responsibility for directing the day-to-day effort of the battalion. The Commanding Officer of the ACB serves on the Staff of Commander Amphibious Force or with the Commander Naval Beach Group during an amphibious operation. When required to provide technical assistance, an ACB representative may be assigned to the Transport Group Commanders staff, and the Commanding Officer of the ACB involved may be directed to serve the Division Beach Party Commander ashore in a similar capacity.

The Commanding Officer is involved with every facet of battalion administration and operation, and he may delegate as much authority as he desires. Therefore no discussion of his duties would be complete unless one included the duties of all the officers under his command. (See Annex J) However there are certain duties that require command action. The Commanding Officer is required to:

   a. Organize the battalion to perform its construction, combat defense and disaster control missions.

   b. Ensure that officers and men of the battalion are well trained to perform their peacetime and wartime tasks.

   c. Ensure his command is ready to perform their wartime tasks through conducting inspections and drills.

   d. Report to his appropriate senior any deficiency which lessens the effectiveness of his command.

   e. Use all proper means to promote the morale and preserve the moral and spiritual well being of the personnel under his command.

   f. Award punishment under Article 15 of the UCMJ and convene Courts-Martial and investigations.

   g. Afford the opportunity for persons under his command to make requests, reports, or statements to him.

   h. Ensure that noteworthy performance of personnel receive timely and appropriate recognition.

   i. Preside as president of the wardroom mess when the battalion is operating its own officers mess.

In the event of death, incapacity, or absence of the Commanding Officer, command devolves upon the senior CEC officer present who is regularly assigned to the battalion. This officer exercises the full authority of the Commanding Officer (See Article 1322, U.S. Navy Regulations). Unless otherwise directed by the Secretary of the Navy, the normal line of succession after the Commanding Officer would be the Executive Officer followed by the next senior unrestricted CEC officer.

2. **Inspections by the Commanding Officer.** The Commanding Officer can best acquaint himself with the actual conditions in his command through the use of planned, regularly scheduled and unscheduled inspections of each aspect of battalion operations. While the ultimate responsibility for inspection rests with the Commanding Officer, it is his duty to delegate this responsibility to appropriate subordinates

Navy Regulations emphasize the importance of inspections. The Commanding Officer shall, when circumstances permit, hold an inspection of the personnel and material of the command on each day of the week, except Sunday, as may be most expedient. (See Article 0708, U.S. Navy Regulations). Inspections, when well planned and executed, serve to improve morale by promoting pride in the appearance and efficiency of the equipment and personnel assigned to the battalion.

The Commanding Officer must be able to look at the battalion from a unique viewpoint without being encumbered by day-to-day tasks. He must maintain a clear picture of the battalion and battalion operations not only as they are at the moment, but as they may be. The flexibility of the battalion is such that it can be deployed on very short notice. Therefore, consideration must be given to the operation and maintenance of all equipment under any unusual climatic conditions. This includes, deserts with the extremes of heat and sand, the tropics where the problems are moisture and humidity, the polar regions where cold is the extreme and beach operations where salt water and sand cause deterioration. Command inspections are intended specifically to keep battalion personnel, equipment, tools and material in constant readiness for continuous operation. They represent an important phase in keeping units in condition to carry out missions in the quickest and most efficient manner.

The Commanding Officer conducts formal personnel inspections, bag inspections and locker inspections periodically, often in conjunction with a pass in review. A notice is issued in advance, designating the time, place, uniform and other pertinent information for these inspections. After the inspection, the Commanding Officer may present decorations, awards or commissions. Chapter 3 of the Landing Party Manual prescribes procedures for inspections, reviews and other ceremonies.

The Commanding Officer may desire to conduct a weekly materiel or zone inspection of the battalion spaces, equipment and supplies to ensure that they are maintained in a satisfactory state of cleanliness and preservation. Such flaws as fire hazards, unsafe or unsatisfactory working conditions are noted and corrected.

Much of the inspection authority is delegated by the Commanding Officer. He may participate by inspecting selected segments, or he may leave an inspection entirely to Company Commanders and Staff Officers. Some examples are:

a. Daily inspection of the barracks and the general mess is normally performed by the Officer of the Day.

b. At morning quarters daily working personnel inspections are performed by company commanders or unit leaders.

c. Tool box inspections are conducted monthly by various battalion officers.

d. Formal and informal safety inspections are conducted at frequent irregular intervals by Company Commanders, the Safety Officer and Safety Chief and project petty officers.

e. Inspection of construction progress for correctness in accordance with the plans and for good engineering practices is performed by the Commanding Officer and Executive Officer on occasion and regularly by the Operations Officer and his staff.

FIGURE 17
COMPANY COMMANDER'S INSPECTION

The condition of each man's 782 gear is carefully checked by company officers during equipment inspections.

3. Discipline. While discipline is a responsibility of every command element in the battalion, the Commanding Officer alone has punishment authority. He may impose certain punishments under Article 15 of the UCMJ, usually called Captain's Mast in the Navy. Captain's Mast is the method of dispensing non-judicial punishment in the Navy and most minor offenses are handled under this procedure. Captain's Mast is not a form of courts-martial, but it is a means for the Commanding Officer to correct minor deficiencies in the conduct of his subordinates. The authority to administer non-judicial punishment appears in Article 15, UCMJ. This authority is explained in Chapter XXVI and Appendix 3, MCM, and in NAVJAG Manual, JAG INST. P5800.7.

The Commanding Officer may impose non-judicial punishment on either officer or enlisted personnel. Authority to impose such punishment may not be delegated. Effective use of Captain's Mast contributes greatly to maintaining discipline with a minimum of time and red tape. Captain's Mast is scheduled at the discretion of the Commanding Officer. It usually requires the presence of many individuals: the Commanding Officer, the Executive Officer, the legal yeoman, the MAA, the accused's petty officer supervisor and Company Commander or Platoon Leader and the witnesses. An attempt should be made to dispose of cases within a day or two after the incident, while testimony is still fresh and when punishment (if warranted) will be more effective. Speedy disposal of cases is especially important if the accused has been placed in a restricted status. When evidence clearly points to trial by courts-martial, much time is gained by completely by-passing a hearing at Captain's Mast. The MAA notifies all persons concerned of the time and place of Captain's Mast and escorts the accused to and from the proceedings. Captain's Mast is carried out with the solemnity that prevails during a courts-

FIGURE 18
THE BATTALION COMMAND POST

Under wartime conditions, pertinent information about the construction projects, enemy situation and location of friendly units is displayed in the battalion command post by members of the battalion staff working for the Executive Officer.

martial (See Annex F). During the conduct of mast, situations will arise where it becomes apparent that referral of the case to trial by courts-martial is indicated. In these situations the Commanding Officer must take care not to become so involved that he becomes the accuser, thereby precluding himself from being the convening authority. To accomplish this, the Commanding Officer must always conduct Mast on a highly professional and objective level. If non-judicial punishment is inadequate for the offense charged, the accused should be tried at a single trial by the lowest court that has the power to adjudge appropriate and adequate punishment for the alleged offense.

The Commanding Officer of a construction battalion may convene summary courts-martial, convene special courts-martial, order pretrial investigation of general courts-martial charges, or recommend general courts-martial. A flag officer may authorize the officer in charge of a detachment to convene summary courts-martial. The battalion Commanding Officer usually issues standing appointing orders for summary and special courts-martial. By appointing more than one summary and more than one special courts-martial, the personnel appointed can prepare in advance for their duties; in this way the battalion will be ready to deal promptly and surely with any cases which might arise.

4. <u>Request and Meritorious Mast</u>. There are two types of Captain's Mast other than those for disciplinary reasons. These are Request Mast and Meritorious Mast. Any man in the battalion may ask to talk to his Commanding Officer in private. To do this he would submit a request in writing up the chain of command. No one in the chain may disapprove the request, but they may make comments on it and must forward it to the Commanding Officer. He will schedule a time and place to talk with the individual concerned. This procedure is known as request mast.

Meritorious Mast is utilized by the Commanding Officer to recognize the superior performance of members of his command. Usually a man is recommended by his supervisor and his Company Commander prepares a citation in writing and submits it to the Commanding Officer. If approved, the Commanding Officer sets a time and place for the mast. It is usually held in front of the entire battalion, perhaps in conjunction with a formal personnel inspection. The citation is then placed in the man's service record.

SECTION B - THE EXECUTIVE OFFICER

1. <u>Duties</u>. The Executive Officer is the direct representative of and principal assistant to the Commanding Officer. He executes the policies and instructions of the Commanding Officer. He acts for the Commanding Officer in the event of sickness, leave or other absence.

The Executive Officer is detailed by the Chief of Naval Personnel and is the Officer who is eligible to succeed to command, being the next CEC officer in rank to the Commanding Officer. If no officer has been detailed by the Chief of Naval Operations, or if the officer detailed is absent or incapacitated, the Commanding Officer acts in accordance with Article 1353, U.S. Navy Regulations.

While in the execution of his duties, the Executive Officer takes precedence over all persons under the command of the Commanding Officer. All orders issued by the Executive Officer have the same force and effect as though issued by the Commanding Officer.

The general responsibilities of the Executive Officer are stated in Chapter 8, U.S. Navy Regulations. The specific duties and the authority of the Executive Officer are assigned by the Commanding Officer. The Executive Officer is usually required to attend to the following matters:

a. Ensure that good order and discipline are maintained throughout the command.
b. Supervises company commanders and department heads.
c. Coordinate and supervise the battalion staff.
d. Supervise the administration and personnel departments.
e. Supervise training programs.
f. Execute the policy of the Commanding Officer.

2. <u>Discipline</u>. Like the Company Commander, the Executive Officer has no punishment authority. He handles minor infractions of battalion regulations by conducting classes involving extra-military instructions. He screens more serious cases by conducting premast hearings and recommends action to the Commanding Officer.

In fulfilling his duty to ensure that good order and discipline are maintained he must:

a. Set the example.
b. Demand performance in accordance with the rules.
c. Be consistent.
d. Follow up.
e. Communicate.

He is assisted by a Chief Master-at-Arms who reports to him directly. The MAA force consists of a CMAA and subordinates as required. Battalion MAAs, usually boatswain's mates, are assigned on a full-time basis and have complete authority over enlisted men regarding the maintenance of order and discipline and barracks routine. Members of the Master-at-Arms force:

a. Enforce battalion directives, station

regulations and naval law.

b. Report rule, regulation or law infraction to superior authority.

c. Indoctrinate incoming enlisted men in layout and routine of the base and the battalion and maintain current listing of bunk and locker assignments.

d. Instruct personnel in the performance of shore patrol duties.

e. Instruct and supervise compartment cleaners and inspect the company areas.

f. Take inventory of personal effects of deserters, absentees and deceased personnel with disposition according to regulations.

g. Notify all persons concerned of the time and place of Captain's Mast and assist in arrangements for courts-martial.

h. Take charge of muster and escort prisoners and restricted men.

i. Direct prisoners in cleaning, maintenance and other work outside the detention area and in the performance of extra duty and hard labor.

j. Raise and lower the colors at prescribed times and sound prescribed calls via the battalion public address system.

3. <u>Chief Staff Officer</u>. The Executive Officer must constantly prepare himself to assume command of a large detachment from the battalion. This detachment may constitute approximately one-half of the entire command. It would be formed when battalion operations are at widely separated locations. For this reason and because he is the next officer in the chain of command he must keep current on everything that the battalion does. He does much of this acting in his capacity as Battalion Chief Staff Officer.

The battalion staff has been formed to assist the Commanding Officer in establishing command policy and making command decisions. The staff performs certain work that the Commanding Officer would do for himself if he had unlimited time, knowledge and capacity. The Commanding Officer and his staff should be considered a single entity since the staff acts solely in the name of the Commanding Officer. All policies, plans, decisions and orders must be authorized or approved by him before they are put into effect. The principal assistants to the Commanding Officer in determining policy and assisting the Executive Officer in the execution of programs to carry out this policy are two groupings of staff officers closely parallel to the staff organization described in the U.S. Marine Corps Staff Manual (NAVMC-1110-A03F). These groupings are the Executive Staff and the Special Staff shown in Figure 19. The Executive Staff consists of:

a. The Administrative and Personnel Officer, called an S-1, who is responsible for preparation, routing and filing of reports, correspondence and directives; transfer and receipt of personnel preparation of the personnel diaries; and maintenance of service records.

b. The Plans and Training Officer, called an S-2, who is responsible for contingency planning, intelligence collection, and the scheduling and monitoring of technical and military training in the MCB. In the ACB, however, contingency planning and intelligence collection are the responsibility of the Operations Officer.

c. The Operations Officer, called an S-3, who is responsible for the planning and scheduling of current battalion construction, combat and disaster control operations. He is responsible for construction quality and safety and for providing engineering support to the Company Commanders. In the ACB, the Operations Officer is also responsible for contingency planning and intelligence collection.

d. The Logistics Officer, called an S-4, who is responsible for procuring, receiving, storing, issuing and accounting for all equipage, repair parts and construction materials. He disburses government funds for battalion purchases and military pay, and he operates the general mess and the ships store.

This constitutes the Executive Staff of a SEABEE unit. Additional information concerning the duties, operation and organization of these staff officers and their departments is contained in Chapters III through VI.

The Special Staff consists of a narrower grouping of staff specialists. Some are assigned to the staff functions on a full-time basis, such as the Medical Officer, while others serve in this capacity as a collateral or part-time duty, such as the battalion newspaper advisor or the battalion historian. The Chaplain, Medical and Dental Officers are special staff billets which are filled by officers holding 4100, 2100 and 2200 designators respectively. Other special staff functions of the battalion may be performed by Civil Engineer Corps Officers.

A battalion officer may serve in several of these staff billets simultaneously. Because of the similarity of proximity of some staff duties to primary billets or because of special qualifications of certain officers, some combinations of assignments are usually routinely made. For example, only the Medical Officer has the qualifications to function as Sanitation Officer, while the physical grouping of the Administrative, Personnel, and Postal sections make it expedient that one officer be designated to supervise all three. The assignment of staff duties is at the discretion of the Commanding Officer, who may also decide that some of these functions can be performed better by each Company Commander as he instructs and counsels his men.

Senior officers are assigned to certain posts in keeping with the authority their rank and Naval regulations or customs demand. Medical and Dental Officers or Chaplains and other specialists are not given duties incompatible with their non-combatant status. Senior officers and specialists should not be

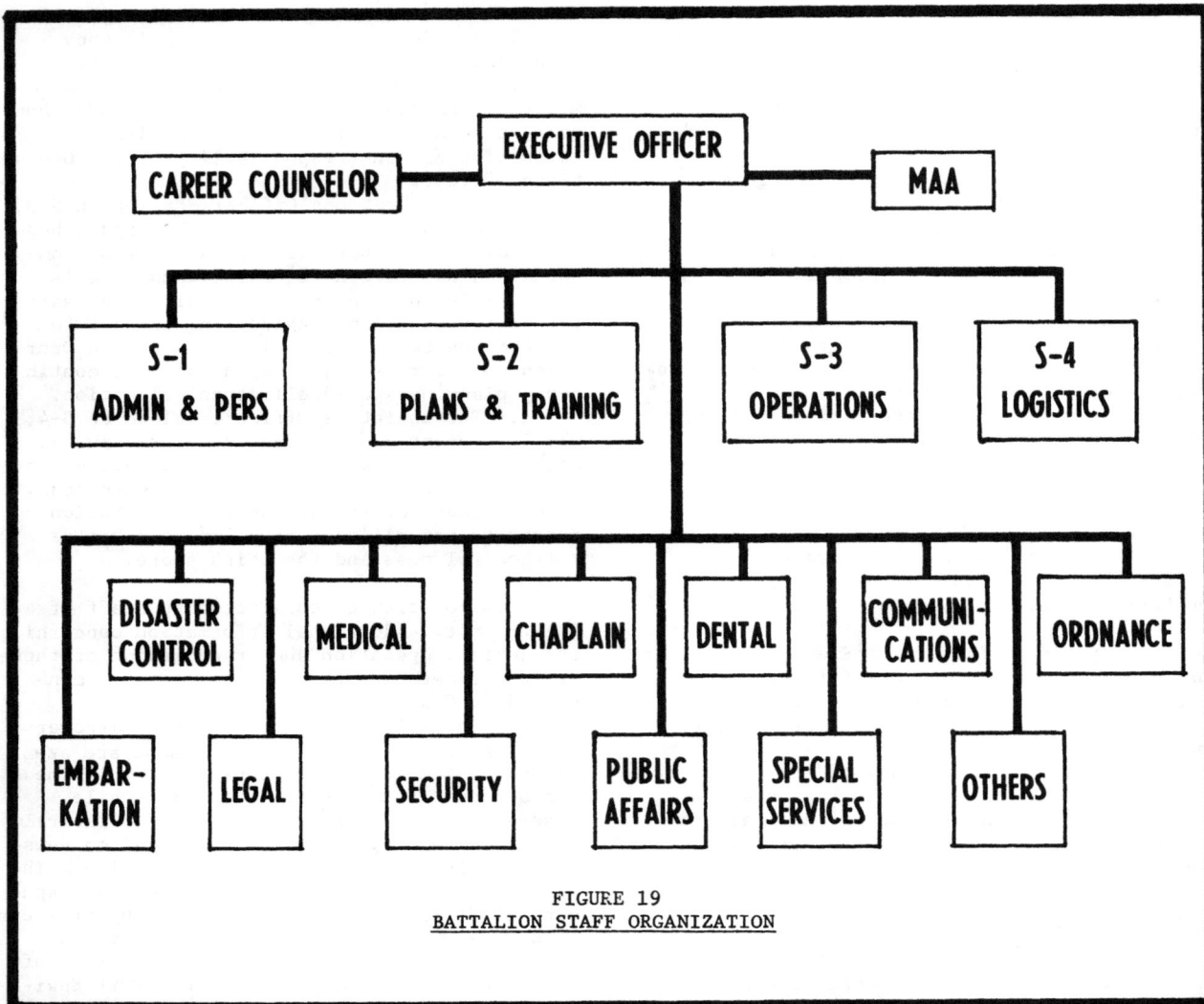

FIGURE 19
BATTALION STAFF ORGANIZATION

burdened with excessive duties which will interfere with their primary responsibilities. However, staff duty assignments for junior officers gives them an opportunity to acquire a full range of experience. Many of these duties also require the assistance of qualified enlisted personnel. Each staff member in carrying out his assignment performs the following duties as standard procedures:

a. Keeps informed on directives and policies issued by the Commanding Officer and advises him on the effectiveness of directives and on situations under the cognizance of the staff office.

b. Drafts directives and correspondence for the approval and/or signature of the Commanding Officer.

c. Maintains liaison with staffs of other commands.

The Executive Officer must coordinate the programs of all staff officers and departments and resolve conflicts as to resource allocation. He must balance training requirements with project requirements and overhead requirements and recommend a course of action to the Commanding Officer. The result of this staff work should be approved instructions, notices and operation orders signed by the Commanding Officer. Through these and other documents, the policy of the Commanding Officer is set forth and explained to every member of the command. The Executive Officer is the channel through which this information flows. The Executive Officer may also use informal means of communications to members of the battalion such as articles in the battalion newspaper, information pamphlets concerning deployment areas or dependents benefits, meetings with his officers, notices in the plan-of-the-day or announcements made at morning quarters.

SECTION C - THE SPECIAL STAFF

1. **Chaplain**. An officer of the Chaplain Corps, responsible for the performance of all duties relating to religious activities is normally assigned to each MCB. In the ACB there is no Chaplain billet. (See Article 0807, U.S. Navy Regulations). In addition to general religious and moral guidance, the Chaplain performs the following functions:

    a. Conducts divine service and forms voluntary classes for religious instruction as directed by the Commanding Officer.

    b. Visits the sick, and in combat, attends the wounded.

    c. Visits the homes of the men of the battalion residing in the vicinity in cases of sickness and for religious guidance.

    d. Visits confined men.

    e. Maintains liaison with the American Red Cross and the Navy Relief Society in solicitation of services for battalion personnel.

    f. Investigates requests for emergency leave, humanitarian shore duty and hardship discharge, and advises the Commanding Officer in regard to them. A detailed list of duties performed by the Chaplain Corps is contained in the Chaplain's Manual, NAVPERS 15664.

Under Article 9 of the Geneva Convention, Chaplains are accorded the status of non-combatants. He may not be assigned the staff function of special services officer nor may he be made custodian of any funds except the Chapel Fund. This fund may be established by the Commanding Officer to accept funds voluntarily offered by individuals and civilian activities for the use of the Chaplain in matters pertaining to morale. Expenditures may be made by the Chaplain upon the approval of the Commanding Officer.

FIGURE 20
THE BATTALION CHAPLAIN CONDUCTING SERVICES

Wherever and whenever he can, the Chaplain conducts divine services for battalion personnel.

FIGURE 21
SICK BAY IN A FORWARD AREA

The Dispensary will normally be a strongbacked 16 x 32 tent. Note that the emergency power supply is sandbagged to protect the generator from sniper fire or the effects of a mortar blast.

2. The Medical Officer. The battalion Medical Officer is responsible for maintaining the health of the personnel of the command. He is an officer of the Medical Corps who is the head of the battalion medical department and a member of the battalion staff (See Article 0969 through 0972, U.S. Navy Regulations). He provides medical care and treatment to personnel, and when directed by the Commanding Officer or authorized by Naval law and regulations or the Manual of the Medical Department, to other people in the Armed Forces, dependents and civilians as necessary. In addition he is responsible for:

a. Conducting physical examinations, diagnosing ailments, caring for the sick, and advising the Commanding Officer regarding the physical fitness of the battalion.

b. Maintaining and inspecting the health standards of the command in matters having a direct bearing on health such as camp sites, mess halls, galleys, berthing facilities, heat and ventilation, working conditions, insect and pest control, water supply and waste disposal, and advising the Commanding Officer on these matters.

c. Collaborating with the Safety Officer in discovering and countering safety hazards and devising standards in this regard.

d. Providing continuing educational programs for all battalion personnel in first aid, hygiene, sanitation and veneral disease.

e. Issuing instructions to prepare the men for environmental changes, special health measures and safety hazards.

f. Maintaining health records of personnel, preparing required reports and drafting battalion directives on medical and health matters and medical provisions of the Disaster Control Plan.

g. Procuring, inspecting, storing, issuing transferring and accounting for medical stores and equipment (except dental supplies) and assuring that the Medical Department is at all times prepared to meet emergency situations.

h. Training and directing appropriate recovery teams.

i. Operating the Battalion Aid Station in combat and training litter bearers from each company.

30

j. Performing other duties prescribed by U.S. Navy Regulations, the Bureau of Medicine and Surgery instructions and other pertinent directives as well as collateral duties assigned by the Commanding Officer.

k. Assisting local health authorities in quarantine inspections, advising the Commanding Officer regarding the medical aspects of all pertinent quarantine regulations and reporting any condition within the command or community which may endanger health.

l. Identifying and caring for the deceased.

When operating alone, a construction battalion will maintain sufficient medical equipment and facilities to treat its personnel. Sick call is held at hours prescribed by the Medical Officer and approved by the Commanding Officer. If facilities are lacking to treat a patient who is seriously ill, the Medical Officer arranges to have him evacuated to an area where facilities are available. When operating at a base with complete hospital facilities, serious cases are dispatched there for treatment. Morning sick calls are normally held in the battalion area. Items of reference include the Guide for Naval Medical Department Officers (NAVMED P5049), the Handbook of the Hospital Corps (NAVMED P5004), the Manual of Naval Preventive Medicine (NAVMED P5010) and the BUMED Instructions of the 6820.4 series which list other authorized reference books.

Because of the diverse nature of the tasks assigned the ACB and the relatively small team assignments which divide the battalion geographically, medical treatment requiring the services of a doctor is administered by the base medical department when the troops are in a stationary situation and by the ship's doctor or task force doctor when deployed on a mission. However, the battalion maintains a medical section headed by a chief Hospital Corpsman who is in charge of administering sick bay, sick call and first aid not requiring a doctor. He assigns corpsmen to operational detachments as specified by orders and rotates his staff to assure that a corpsman is on duty during all hours. His responsibilities include the maintenance of battalion health records, the preparation of reports as required and arrangement of personnel and provision of equipment for first-aid training of battalion personnel.

3. The Dental Officer. An officer of the Dental Corps assigned to the battalion is a member of the staff and is head of the battalion Dental Department (See Article 0975-0979, U.S. Navy Regulations). In combat and disaster control situations, the Dental Officer and members of his staff assist the Medical Officer. Responsibilities of the Dental Officer are:

a. Preventing and controlling dental diseases and deficiencies, supervising dental hygiene within the battalion and advising the Commanding Officer on all related matters.

b. Conducting dental examinations and providing dental care and treatment to battalion personnel and, when directed by the Commanding Officer, to other persons in the U.S. Military Service.

c. Furnishing other dental services as authorized by law, Navy Regulations, BUMED Instructions and other pertinent directives.

d. Procuring, inspecting, storing, transferring, issuing and accounting for dental stores and equipment.

e. Cooperating with local authorities in regard to inspecting and controlling any condition within the command or community which may impair oral health.

f. Drafting battalion directives on dental matters and carrying out assigned collateral duties.

When a construction battalion is operating alone, the Dental Department is fully equipped to care for the needs of the personnel (with the exception of prosthetic treatment). When the battalion is at or near an established base, the Dental Officer and members of the Dental Department may be assigned to temporary duty at the base Dental Department. Personnel then receive treatment at the base facilities.

4. Communications Officer. The ability of a battalion to carry out its mission is strongly dependent on communications; therefore, rapid, efficient and reliable communications are of vital importance to each battalion. In its broadest sense communications refer to all means, radio, wire, sound, visual messenger and liaison personnel and no means should suffer through neglect or deemphasis in training or maintenance. The Communications Officer is responsible for:

a. Providing wire and radio communication from the battalion Commanding Officer to subordinate units and staff officers when required for construction operations, disaster control work or battalion defense.

b. Maintaining assigned equipment in peak condition and stocking sufficient quantities of consumables such as batteries and wire to support extended operations in the extreme conditions imposed by combat or disasters.

c. Training selected personnel to install, operate and maintain field radio equipment, field telephones, switchboards and related signal equipment, operate the message center and act as messengers.

d. Arrange for communications with higher headquarters.

e. Prepare the communications plan for the battalion and directives of the 2300 series. DNC 5C Naval Communications, NWIP 16(A) Basic Operational Communications Doctrine and appropriate Field Manuals and Technical Manuals should be consulted for assistance. A detailed list of these manuals can be found on pages 3 and 4 of enclosure (2) of COMCBPAC INST. 5070.1.

5. Disaster Control Officer. Staff responsibility for Disaster Control requires that the officer keep informed of the latest Disaster Control techniques and maintain liaison with local and regional commands. He drafts the battalion disaster control plans and, with the cooperation of other staff officers, coordinates them with the watch and training plans and the overall operation of the base. He makes appropriate inspections and formulates and conducts disaster control drills. The Disaster Control Officer should initiate procurement of all special equipment and materials, act as advisor for personnel training and suggest precautionary design measures for incorporation in the layout of job sites and installations in contingency situations and protection for battalion personnel and equipage from natural or nuclear disasters. The U.S. Navy Disaster Control Manual, NAVFAC INST. 3440.12 and local disaster control plans should be referenced.

6. Embarkation Officer. The Embarkation Officer is responsible for loading and unloading of personnel, equipment and supplies in movements by ship, aircraft and railroad. He compiles and maintains statistical data pertaining to the movement of the command and determines requirements for the allocation of ships, planes and railroad cars. He maintains liaison with movement agencies, and may supervise actual loading and unloading operations of the battalion or its detachments.

His tasks are so exacting that he should have attended embarkation courses prior to assuming responsibility for loading. The base or station usually plans for and loads the battalion in normal peacetime operations but under emergency mount-out or during fleet training exercises, the entire task would be in the hands of the battalion embarkation officer. His specific duties and responsibilities are described in Chapter VI.

7. Legal Officer. The Legal Officer advises the Commanding Officer on matters of military justice and is available to assist all battalion personnel regarding step-by-step procedures of any military judicial action. Although, not necessarily trained in civilian law, he should be a graduate of the Naval Justice School (Long Course) and must have a thorough and up-to-date knowledge of Naval Law. He drafts battalion directives of the 1640 and 5800 series, prepares appointing orders for investigations and courts-martial, trial guides and any judicial action to be taken by the Commanding Officer. If he reviews charge sheets, courts-martial records, and investigation reports for the convening authority, the legal officer should not be appointed as a member, trial counsel, assistant trial counsel, defense counsel, assistant defense counsel, or investigative officer in accordance with Article 6(c) of the UCMJ. His additional duties consists of:

a. Maintaining liaison with legal staffs and assisting in preparation of cases for Captain's Mast.
b. Keeping the Unit Punishment Book and other legal records.
c. Handling claims against the government brought by, or arising from, actions by battalion personnel. His main sources of information are Manual for Courts-Martial, Manual of the Judge Advocate General, Courts-Martial Reports, and Digest of Opinions of the Judge Advocate General of the Armed Forces.

8. Ordnance Officer. The Ordnance Officer is responsible for the operation of the armory, handling of ammunition, range firing and safety along with issue and control of field infantry equipment. If he has not been specifically trained in ordnance, Navy courses are available and must be completed before assuming the responsibility for this assignment. He must be familiar with all current technical publications, NAVORD and Ordnance Supply Office (OSO) ordnance allowances and training directives and bulletins that effect the battalion. In particular those indicating faulty lots of ammunition or modifications to battalion weapons. He is responsible for the requisitioning, inspection and issue of weapons, repair parts, ammunition and any authorized equipment or supplies and maintenance of related records of receipt, issue and inventory as required by NAVORD and OSO. He also advices on care, maintenance, storage, security and use of ordnance material and on the preparation of training programs. He has equipage accountability to the battalion Supply Officer for NAVORD equipment that he reissues on subcustody. He must provide for proper stowage and registry of private firearms and report the loss of any firearm to both local authorities and the cognizant ONI office.

9. Public Affairs Officer. In administering the public information program for the battalion, this officer keeps informed on all policies and directives at the battalion-level and navy-wide. He makes and maintains contacts with press representatives, prepares or reviews all releases to be published outside the battalion in service newspapers, military periodicals or the local press. On occasions of promotions, awards, reporting for duty or other appropriate times, the Public Affairs Officer prepares the letters to next-of-kins and articles to the Fleet Home Town News Center. In addition, he suggests topics for release and assists in their preparation. He formulates plans for handling any major news item, requests photographs as required and keeps a scrapbook on battalion activities. The Public Affairs Manual, NAVEXOS P1035 is the main reference source for guidance.

10. Security Officer. The Security Officer assists the Executive Officer in matters relating to the battalion watch, internal and

external security, and physical protection. In addition it is his duty to study the security requirements of the battalion, review every battalion activity from the standpoint of security and prepare recommendations for effective safeguards against theft, sabotage and lawlessness. He drafts the battalion watch plan, directives of the 1601 series and assists in preparation of the OOD Instruction Book and special orders for each watch billet. The Security Officer also works with the classified Material Control Officer in safeguarding classified material and cooperates with the Base Security Officer. The Landing Party Manual (OPNAV P34-03), Army Manuals entitled Barriers and Denial Operations (FM 31-10), Physical Security (FM 19-30), and Operations against Irregular Forces (FM 31-15), and the Marine Corps publication, The Marine Rifle Squad (FMFM 6-5) would be helpful in devising an appropriate security system.

11. Special Services Officer. The Special Services Officer administers the battalion's Special Service Program. The Commanding Officer usually designates other officers or petty officers to assist him on a collateral duty basis. For example, one officer may be assigned as Athletic Program Officer, other officers and petty officers may be delegated to coach

or manage various teams or leagues. The Special Services Officer administers the Special Services Program as a staff assistant to the Commanding Officer. He:

a. Arranges a well-rounded program of athletics, entertainment, motion pictures, theater, music, hobby crafts and other entertainment and social activities to appeal to all hands.

b. Organizes and supervises the operation of the Special Services Library, and battalion cruise books.

c. Performs liaison on matters related to Special Services with higher command and with local military and community organizations.

d. Serves as Custodian of the Recreation Fund.

e. Serves as a member of the Recreation Council.

f. Attends meetings of the Enlisted Recreation Committee.

g. Represents the battalion on the Base Recreation Council.

h. Obtains and maintains Special Services gear.

i. Controls the Recreation Gear Locker and provides for the issue and return of Special Services gear.

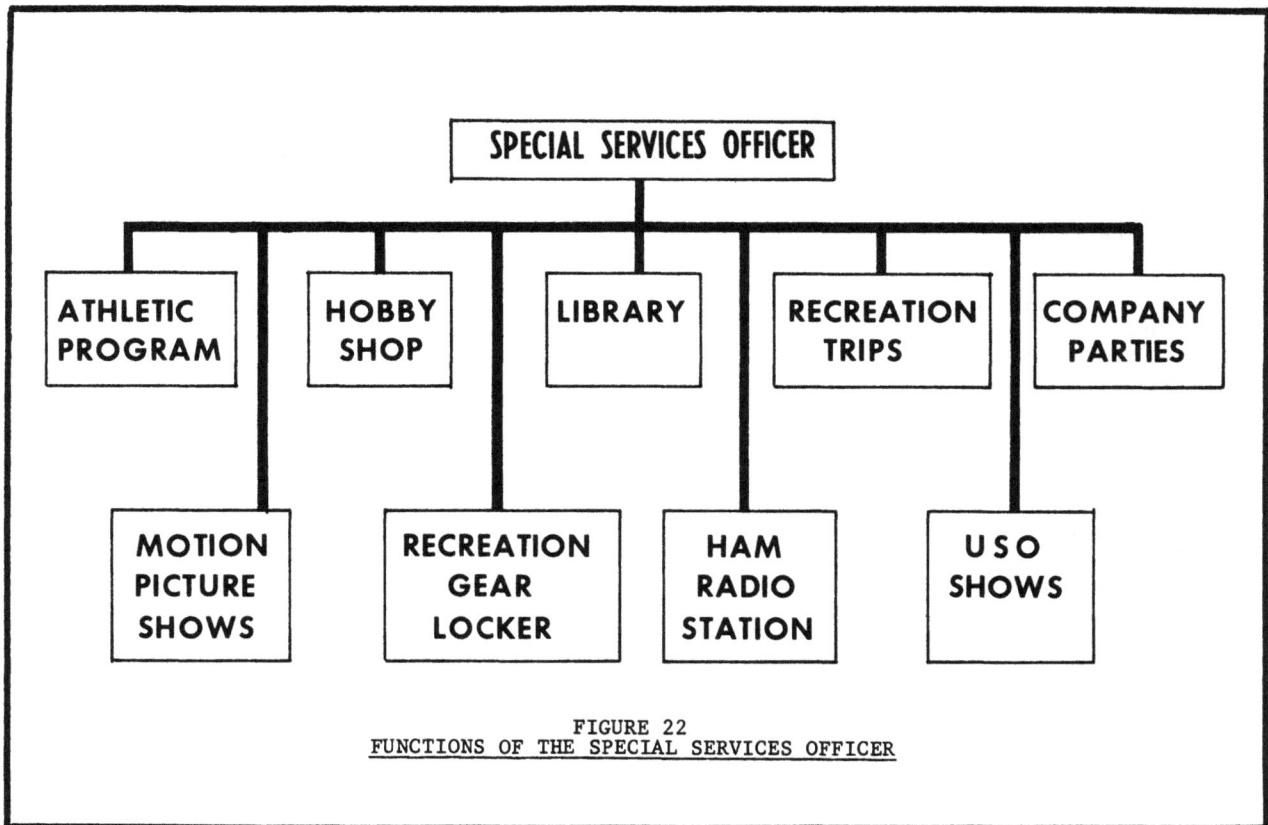

FIGURE 22
FUNCTIONS OF THE SPECIAL SERVICES OFFICER

Members of the Special Services Department assist the Special Services Officer in the performance of his duties, namely, in the planning, developing, organizing, directing, producing, publicizing and supervising entertainments, picnics, dances, parties, concerts, dramatics, game rooms, libraries, arts and crafts, tournaments and other Special Services activities. They help to select, distribute and maintain recreation equipment and facilities. They keep Special Services records and accounts and operate the Recreation Gear Locker (See Figure 22). Peacetime personnel limitations frequently prevent the inclusion of trained specialists to assist the Special Services Officer in carrying out Special Services programs. Thus, he must be alert to spot and train volunteer help. The ACB Special Services Officer must consider the operational differences of his battalion and those of an MCB. Shipboard limitations, for instance, dictate the type of activities in which his men may participate. However, he must keep in mind that some type of physical activity is important for the well being of the battalion and should be available to the men except in times of emergency. Such games as volleyball and badminton, for example, may be set up in a relatively small area, even on a beach after a manuever. It is the duty of this officer to maintain as broad a program of diversified activities and interests as possible in order to maintain physical fitness and a high standard of morale of all enlisted personnel.

A standard program is neither possible nor desirable. The program must be adapted to the needs of the personnel and to the facilities available. The success of the program is proportionate to the interest of the Commanding Officer, the Special Services Officer, the Recreation Council, the Enlisted Recreation Committee, and others who assist in the program. The ideal program should be established with one thought in mind -- to care for all wholesome desires insofar as practicable. A variety of activity to meet varied tastes and situations are essential.

Motion pictures are the most important recreation facility in a battalion. As part of its allowance equipment, every battalion is assigned 16-mm projectors. A qualified operator must be available to run the projector before films may be shown. Films are distributed through Navy Picture Exchanges to ships and activities within a specified area. The films are then passed along the chain of activities according to a specified schedule. A record of exhibition, transfer and inventory is prepared each month for the signature of the Commanding Officer. The Special Services Officer instructs and supervises motion picture operators, arranges for pick-up, delivery, stowage and transportation of films, plans for repair of equipment, arranges for publicity for the movie schedule, and strives to obtain

good screening, good sound and comfortable seating.

The Hobby Craft Program is intended to provide facilities for the pursuit of hobbies during leisure time. Leatherworking, painting, sculpture, photography and woodworking are a few of the hobbies that may be carried on in the Hobby Shop. Funds necessary to operate the Hobby Shop are handled as part of the battalion's Recreation Fund. Operational costs may include the purchasing of material, equipment and occasionally the payment of salaries to personnel who act as supervisors or instructors in the shop. In many cases, a qualified enlisted man is assigned to the shop on a part-time or service watch basis at no extra compensation. An inventory of consumable supplies such as model kits, leather, lumber and lacquer may be carried in stock for resale to users of the Hobby Shop. Items that are normally obtainable through the Ships Store or Navy Exchange may not be carried as part of the Hobby Shop's supplies. The Hobby Craft Program must necessarily vary according to the funds and space available, the location of the battalion and the interest of personnel

SECTION D - COLLATERAL DUTIES

1. Athletic Program Officer. The Athletic Program Officer assists the Special Services Officer with the battalion athletic and physical fitness programs. He is generally assisted by other officers or petty officers in the management of various teams and leagues. The Athletic Program Officer assists in the preparation of directives of the 1710 and 6100 series. He supervises physical fitness testing and reporting in accordance with BUPERS INST. 6100.2 series. The battalion athletic program consists primarily of competitive sports with both team and individual sports being offered. Basketball, softball, volleyball, swimming, fishing, horseshoes and table-tennis are typical sports activities.

Intramural activities are the heart of the athletic program. But in addition to intramural sports, teams may be formed to represent the battalion. Leagues or other forms of organized competition often exist in the area where a battalion is operating. Under the Navy sports program, every effort is made to stimulate interest in athletic competition by conducting play-offs with leading service teams in a specified area. A team of champion ship caliber that successfully survives elimination tournaments in service competition is permitted to participate in All-Eastern or All-Western Navy championships. A team representing a naval activity, such as a battalion, may not have more than 50 percent officer personnel participating at any one time. This regulation does not apply to individual-type events (such as golf, tennis, and track) or to athletic events conducted within the battalion. Officers are also prohibited from participating in competitive boxing and wrestling events.

Most athletic equipment is purchased with money from the battalion recreation fund. There are two limitations to the expenditure of funds for the purchase of such equipment:

a. Disbursement may not be made for equipment that is available from appropriated funds or is a part of the battalion's allowance.

b. Expenditures may not be made for equipment or upkeep of equipment for the sole benefit of individuals or groups.

Athletic and other recreation equipment (such as fishing rods, checker sets and darts) is kept in a special gear locker. One man is designated as official custodian of equipment. A custody card is prepared for every man in the battalion to record the issue and return of recreation gear. Racquets, bats, golf clubs, boxing gloves and similar gear are stored in the locker when not in use.

2. <u>Barracks Officer</u>. The Barracks Officer is responsible to the Executive Officer for the general supervision of battalion personnel residing in barracks. He represents the command at base material inspections and acts as liaison officer between the Commanding Officer and enlisted personnel in matters pertaining to barracks residence. His duties include assurance that all battalion personnel comply with base and battalion barracks directives. He performs periodic inspections of lockers, clothing and bedding for properly marked and complete issue of clothing, gear and effects.

3. <u>Benefits and Insurance Officer</u>. The Benefits and Insurance Officer assists battalion personnel, their families or survivors in obtaining information regarding benefits and insurance. He keeps informed of all related laws, directives, and policies, publicizing and explaining as required. NAVPERS 15885-B of May 1964 entitled "Rights and Benefits of Navymen and their Dependents" is a collection of various articles that have appeared in ALL HANDS and can be of great assistance in counseling battalion personnel.

4. <u>Brig Officer</u>. A construction battalion does not usually operate a brig but makes use of the nearest correctional facilities. If the battalion is operating alone, however, the S-1 may serve as Brig Officer and assume responsibility for confined men of the battalion and prisoners of war. Regulations for operating a brig are contained in the Corrections Manual, NAVPERS 15825 rev. Instructions for prisoners of war are normally contained in the administration plan or annex to the OPORD.

5. <u>Camp Maintenance Officer</u>. In the ACB the

A Company Commander is usually designated as Maintenance Officer. He has charge of the maintenance shops and within the capabilities of his men and equipment is responsible for the physical upkeep of the buildings and areas assigned to the battalion. When the work involved exceeds these limitations, he initiates work requests channeled to the basic Public Works Department or other activities. He conducts the necessary liaison and has large projects approved by the Operations Officer. He also supervises the preparation of any required designs, layouts or drawings. In the MCB, the B Company Commander is usually charged with camp maintenance. Details of his responsibilities are listed in Section D of Chapter VII.

6. <u>Career Appraisal Counselor</u>. A mature career Chief Petty Officer or Senior Petty Officer is designated as career counselor. Serving in this capacity, he provides information for the personnel department regarding military and civilian vocations. He advises battalion personnel in opportunities for advancement within their chosen specialties. In order to operate effectively, he must continually keep abreast of policies and directives on reenlistment, informing the Commanding Officer about the reenlistment rate and factors affecting it. The Navy publishes information to assist in this effort of keeping personnel informed and has assembled a career counselor's kit, NAVPERS 15959 containing brief descriptions of most Navy programs. The Officer Fact Book, NAVPERS 15898, and the Career Counseling Guide, NAVPERS 15878 are also valuable sources of information on a Navy career.

7. <u>Censor</u>. The battalion officer with this duty drafts directives of the 5530 series and keeps informed on censorship policies, laws and regulations. When it becomes necessary because of the nature of battalion operations or location to censor personal mail, a board is organized and supervised by this officer. His duty includes the examination of all unofficial photographs taken which may be considered breaches of security and the proper disposal of such prints and negatives (See Article 0735 and 0736 of U.S. Navy Regulations and the Armed Forces Censorship Manual, OPNAV INST. 5530.10).

8. <u>Classified Material Control Officer</u>. The Classified Material Control Officer assists the Commanding Officer in fulfilling his responsibility for the security of classified matter and serves as his advisor and representative. The specific responsibilities of a Classified Material Control Officer are:

a. To assure himself that all persons who are to handle classified matter are instructed and cleared in accordance with the U.S. Navy Security Manual for Classified

Matter, OPNAV INST. 5510.1.

  b. To formulate and coordinate security control measures within the command and to exercise security control over visits to and from the command in accordance with OPNAV INST. 5510.49.

  c. To maintain a program of declassification and downgrading of matters and to check all classification assigned by others in the command for the correct level of classification and downgrading in accordance with OPNAV INST. 5500.40.

  d. To maintain records of current classification of matter for which the command is responsible.

  e. To prepare recommendations regarding release of classified matter to foreign governments.

  f. To review proposed press releases for the purpose of indicating any classified information which must be deleted.

  g. To serve as Top Secret Control Officer if none has been designated.

9. Custodian of the Recreation Fund. Recreation funds are utilized to promote recreation, amusement and the general welfare of battalion personnel. Dependents of battalion personnel may be permitted to participate in the benefits afforded from the fund with the exception of specific activities designated by the Chief of Naval Personnel.

  When a construction battalion is commissioned, BUPERS makes a grant of appropriated funds for special services. Thereafter, most special services funds come from revenue-producing activities such as ship's stores and Navy Exchanges. This revenue is maintained and accounted for in:

  a. A composite recreation fund.
  b. A unit recreation fund.
  c. A command recreation fund, or
  d. The BUPERS Central Recreation Fund.

  The Special Services Officer of a battalion is normally designated as Custodian of the Recreation Fund.

  ACBs and MCBs may participate in the benefits of a composite Recreation Fund when working at a base or station. The fund is used to pay all expenses of the recreation program of general benefit to all participating units including the battalion. An allocation from the composite recreation fund is usually made to the Battalion Recreation Fund after allowing for financing the common recreation program in which all activities are participating. This allowance is based on the ratio of the number of ACB and MCB men using the revenue-producing facilities to the total number of personnel entitled to benefits of the fund.

  The administrator of a command Recreation Fund can levy assessments on any unit recreation fund within his command. The command recreation fund makes loans or grants to ships or activities within the command whose ship's store does not generate sufficient revenue to provide for recreation needs.

The BUPERS Central Recreation Fund supports, equalizes and administers the recreation program of the entire Navy. The Central Recreation Fund receives revenue from donations, various assessments and balance in recreation funds of decommissioned activities. It may grant or loan funds to command, composite or unit recreation funds.

  The Accounting Handbook for Non-Appropriated Funds, NAVEXOS P2409 and the Special Services Manual, NAVPERS 15869 should be used for guidance by the custodian of the battalion's recreation fund.

10. Educational Services Officer. The Educational Services Officer is responsible for the administration of all the details of the Navy Education Program. He publicizes educational opportunities available and administers General Educational Development Tests (GED). His specific duties are included under the duties of the Plans and Training Officer in Chapter IV.

11. Fire Marshal. The Fire Marshal is responsible for initiating fire prevention plans and drives and conducting periodic inspections of battalion areas for assurance that fire hazards are minimal. He is responsible for fire trucks assigned and for all fire fighting equipment and inspection and readiness training of fire-fighting personnel. He assigns the fire watch detail and requires one battalion member, trained in fire fighting techniques and familiar with local fire fighting resources, to be on call and ready at all times. OPNAV INST. P11320.15, U.S. Navy Structural Fire Fighting Manual contains additional instructions.

12. Historian. The Battalion Historian records the history of the construction battalion describing events of enduring significance but containing enough detail to capture the units individuality. He records encountered difficulties and major problems along with techniques used in their solutions. He prepares a weekly illustrated summary of operations of the battalion's prime mission in perspective which becomes an integral part of the Historical Log. This summary consists of an edited composite of the weekly and monthly operations reports, the weekly battalion newspaper, the monthly battalion familygram and other battalion reports and publications. He may review project folders at the completion of each job for historical adequacy, content and progress photographs for possible use in the Historical Log, requesting reprints, maps or drawings as required. He prepares drafts of directives of the 5750 series and possibly current history articles for military and general publications. He reviews the reports and articles written by others for historical accuracy.

13. Legal Assistance Referral Officer. The Legal Assistance Referral Officer is not a licensed attorney offering legal service but, when so requested, helps battalion personnel

obtain competent professional advice for any personal problem including such things as divorce, non-support, financial contracts and damage suits. He keeps informed regarding the Navy legal assistance programs and maintains liaison with the nearest Legal Assistance Officer.

14. The Library Officer. The purpose of the library is to provide battalion personnel with an organized collection of books and articles intended for informational, technical and re-creational reading. The majority of books are provided by the Bureau of Naval Personnel. When a battalion is commissioned, it is given a basic allotment of books. New books are supplied regularly to supplement the collection. Requests for specific books may be forwarded to BUPERS. Books are not charged to any battalion allotment. Magazines, newspaper, and other periodicals however are not furnished by BUPERS and must be purchased by the battalion. The Recreation Fund usually supplies the necessary money. The Library Officer plans for and supervises the Library. In addition to the regular enlisted librarian, a few men may be trained to stand watch as duty librarians. Running the library requires much more than merely applying a set of rules concerning hours of operation and checking books in and out. Programs intended to keep habitual readers satisfied and to attract non-readers to the library such as making the physical surroundings attractive and comfortable, making the collection varied enough to attract all types of men, and advertising the library programs are effective methods that help to make men read.

15. Motor Vehicle Accident Investigator. An officer, usually the "A" company commander, because of his responsibility for battalion automotive and construction equipment is appointed to investigate motor vehicle accidents involving construction battalion personnel. Every accident involving Navy motor vehicles or construction equipment must be reported by the operator on Standard Form 91, Operator's Report of Motor Vehicle Accident. A pencil and a copy of this form must be carried in each Navy vehicle at all times. Instructions for completing this report are given in the Navy Driver's Handbook, NAVFAC MO-403 and in NCPI 5100. The operator involved in an accident must deliver the accident report or ensure its immediate delivery to his supervisor who will forward the report to the Motor Vehicle Accident Investigation Officer. He prepares a case study of each accident for use by the Safety Officer and a report for the Commanding Officer on a Standard Form 91A.

16. The Newspaper Adviser. Most battalions publish some type of newspaper. It may be published daily, weekly, or monthly. A battalion officer always serves as advisory editor. One or two men may be assigned newspaper duty on a part time basis. Insofar as practical, men with newspaper experience and knowledge of the Navy are selected. These men have the responsibility of editing, producing and perhaps writing and illustrating. It is usually possible to obtain volunteers to write sports commentaries, book reviews or some type of company column. A newspaper should carry general, local and battalion news. Where commercial newspapers are available, emphasis is placed on Navy activities. The Armed Forces News Bureau supplies news that is considered appropriate for all military units.

Some battalions prepare a cruise book to commemorate the activities of the battalion during a given assignment. The Recreation Committee and Council determines the wishes of the battalion personnel before undertaking such an enterprise. The usual procedure is to arrange with a commercial firm to print the cruise book after it has been prepared by battalion personnel. The book is then sold, approximately at cost, to battalion members. Copies may be sold to outsiders at a profit. In form and content, the cruise book is similar to a college yearbook. It gives a brief description of the battalion, contains photographs of battalion members and shows action shots of various battalion activities. If the battalion has a Historian, much of the material in the cruise book will be his account of noteworthy events. The Cruise Book Officer coordinates the activities, articles, illustrations and photographs and may even suggest an appropriate layout to the enlisted personnel who do the work of arranging the book. Further guidance is given by the Navy Public Affairs Manual, NAVSO P-1035, and the Navy Publications and Printing Manual, NAVEXOS P-35.

17. Photography Officer. The Photography Officer supervises the battalion's photography section and is responsible for maintaining the battalion photographic equipment and files. He requests procurement and controls the use of photographic equipment and supplies in addition to supervising the operation of the photographic laboratory. He is responsible for taking and processing photographs of construction operations and photographs of news events and Navy subjects in accordance with the policies of the Commanding Officer. These photographs are used for the reporting of project status, for public information releases and for historical records. The Photographer's Mate (PH) assigned to the Photography Section operates motion-picture and still cameras and processes film and prints in the photography laboratory.

18. Postal Officer. Navy post offices are operated by the Commanding Officer through a designated Navy Postal Clerk under the supervision of a Postal Officer. The Postal Officer is responsible for:

a. Keeping abreast of the policies and directives of the U.S. Post Office Department, the Department of the Navy, and of other higher authority on all matters affecting postal operations.

b. Supervising operation of the battalion Post Office.

c. Instructing post office personnel and company mail orderlies in proper postal procedures.

d. Preparing required postal reports, including the quarterly statistical report on postal operations for submission to CNO.

e. Advising the Commanding Officer on postal matters and drafting directives for his approval.

f. Keeping battalion personnel informed on office hours and postal procedures.

One Navy Postal Clerk, responsible to the Postal Officer and a suitable number of assistant clerks, responsible to the Postal Clerk are designated to operate the Post Office. The requirements for assistant clerks is usually based on one per every 500 persons served by the facility. The Postal Clerk and his assistants are nominated on NAVPERS Form 2864 and must execute an oath of office before commencing their duties. These assignments are normally filled by rated postal clerks (PC). A Postal Clerk is designated for a specific post office and when he is transferred or removed or when his designation is revoked, this information is entered in his service record and transmitted via the Chief of Naval Personnel to the post office of which the battalion is a branch. Assistants are designated in the same manner in accordance with BUPERS Manual, Article B-1202. In the absence of the Postal Clerk, an assistant may be selected to assume the operational responsibility for the post office for a period up to 90 days, after which, if another Postal Clerk is not nominated, postal operations must cease. The Navy Postal Clerk takes charge of all postal business in accordance with Navy Postal Service Regulations, OPNAV INST. 2700.14 and all instructions and directives issued by the Post Office Department, in particular the Postal Manual. He is also responsible to the Postal Officer for the general condition, financial situation, security, cleanliness and operating efficiency of the post office.

The function of the mail orderlies is to collect mail at the battalion post office and deliver it to men in their companies. Each company normally will have a mail orderly who is nominated and approved on DD Form 285 in accordance with OPNAV 2700.14A, Chapter 8. Company clerks usually serve as mail orderlies.

19. Registered Publications Officer. The Registered Publications Officer is responsible for the security of registered publications in the custody of the battalion. Registered publications are classified publications to which register numbers are assigned for security reasons. Periodic accounting of custody, receipt, transfer and disposition is required (See the Registered Publication Manual).

20. Research, Development, Test and Evaluation Officer. With the assistance of the cognizant Company Commanders, the RDT&E Officer performs evaluation, analyses and tests of equipment and methods when such assignments are given to the command. He originates and recommends special tests and develops training exercises for new equipment and methods. He maintains liaison with the Naval Civil Engineering Laboratory on matters affecting SEABEE Operations.

21. Safety Officer. The Navy spares neither expense nor effort to provide effective safety measures and equipment to protect its men. However, these provisions are worthless unless they are efficiently applied. It is the responsibility of the Commanding Officer and every officer and petty officer not only to see that safety is practiced daily by all members of the battalion but each must learn to foresee and provide against hazards in the daily activities of his unit. Since safety precautions normally apply only to usual conditions, Commanding Officers or others in authority may find it necessary to issue special precautions to their commands to cover local conditions and unusual circumstances. It is the responsibility of each man to use every available means to prevent accidents. Violations of safety directives and failure to exercise good judgment are subjects for disciplinary action. Each individual must report to his immediate superior in the chain of command any condition or equipment which he considers to be unsafe. He must exercise reasonable caution in the event of an unforeseen hazard and in addition, warn others whom he believes to be endangered by known hazards. The U.S. Navy Safety Precautions, OPNAV 34P1 and COMCBPAC Safety Manual, COMCBPAC INST. 5100.1 are excellent references.

The Commanding Officer organizes the safety program of the battalion which consists of six points:

a. A safety policy statement by the Commanding Officer.

b. Expert Staff direction.

c. Active participation by all officers and petty officers.

d. A program of safety discipline, training, and recognition.

e. Inspection.

f. Accident records and program evaluation.

The duties of the safety officers are involved with a continuing staff responsibility for the safety and efficiency of battalion operations. The safety officer acts in an advisory capacity to company commanders, department heads and supervisors and represents the Commanding Officer at management conferences and meetings on matters of accident prevention. He has the responsibility to:

a. Prepare a safety program, using demonstrations, lectures, movies and posters, to be submitted to the safety council for review and to the Commanding Officer for approval. Reference materials and minimum safety libraries are listed in COMCBLANT/COMCBPAC instructions.

b. Review new projects and correct unsafe work practices and conditions before they cause accidents.

c. Instruct personnel in the selection, use and maintenance of safety equipment; supervise safety inspections; and consult with the medical officer in the control of health hazards.

d. Organize a complete program for motor-vehicle operator education and accident prevention.

e. Organize and monitor the battalion program of brief (5-15 minutes) safety meetings conducted by crew leaders on hazards and precautions related to current work.

f. Maintain adequate stocks of safety equipment and materials and submit requests for non-allowance articles when required by the nature of the assigned projects.

g. In conjunction with the educational services officer, prepare a program for off-duty safety including instructions on leave, liberty, swimming, boating, hunting and operation of private automobiles.

h. Provide adequate bulletin boards, posters, safety publications, educational materials and conspicuous warning signs at danger points.

i. Maintain and encourage the use of an industrial and motor vehicle safety library.

j. Verify and when necessary investigate accident reports and prepare summaries and analyses to be forwarded to the Commanding Officer.

k. Review corrective action proposed by supervisors and advise the command as to those measures that should be implemented.

A Chief Petty Officer is assigned, usually on a full-time basis, as Safety Inspector to assist the Safety Officer in executing the details of the battalion safety program. He regularly visits each job site, the shops, galley and camp, consulting with supervisors and helping to develop and encourage safe practices. He should be prepared for this assignment by special indoctrination and courses of instruction.

The Safety Policy Committee meets periodically to recommend safe work practices and procedures, to analyze the causes of accidents and to formulate a safety program.

Each officer and petty officers is responsible for the safety of personnel under his charge and must continually look for and eliminate situations potentially hazardous to them. It is the officer's duty to instruct and supervise personnel in the performance of their jobs. Each supervisor:

a. Instructs persons under his charge by describing what is to be done, illustrating the safe way to do the job and requiring his subordinate to demonstrate that he can do the work in a safe manner.

b. Checks each crew frequently to assure that safe methods are being applied and corrects unsafe working conditions at once.

c. Requires the use of proper tools, equipment and safety devices and inspects equipment before and during use, removing defective equipment from service at once.

d. Contacts his superior in the chain of command when unable to neutralize unsafe conditions, and consults the safety officer when in doubt as to safe work practices for a given job.

e. In case of accident, arranges for prompt treatment of the injured, completes Supervisor's Report of Accident, NAVEXOS Form 108, and forwards it to the Safety Officer through channels, notifies the OOD and the Safety Officer and when disciplinary action is indicated, places persons involved in accidents on report.

Various accident records and reports are required as a basis for legal, administrative, disciplinary and corrective action. All such documents must be accurate and complete. The accident reporting procedures are prescribed in COMCBLANT/COMCBPAC Instructions of the 5100 series. The investigative report includes findings as to whether the accident occurred in the line of duty and whether negligence or misconduct were involved.

In all cases, the investigation aims at revealing the causes and fixing the responsibility for each accident. As soon as the cause of an accident is revealed, action is taken to prevent reoccurrence.

The effectiveness of the battalion safety program requires periodic evaluation. No means have yet been devised to determine precisely how many and what kind of accidents were prevented or precisely what prevented them. However, certain comparisons reveal trends and the effectiveness of the safety program. The frequency and severity of lost-time accidents in the battalion can be compared with parallel data for various industries and organizations (See Figure 23).

Ten percent of the organizations in the construction industry are members of the National Safety Council. This segment of the construction industry has an accident frequency rate almost half that of the industry as a whole.

Construction firms under contract to the Army Corps of Engineers are required to comply with safety standards specified in their contracts. The NAVFAC figure includes civilian contractors and construction battalions.

22. _Sanitation Officer_. The battalion medical officer usually performs the functions of the sanitation officer. He assumes the initiative in maintaining health standards in fields under the cognizant of other departments; such as food handling, food preparation, lighting, heating, ventilation, housing, insect, pest and rodent control, water supply, waste disposal and clothing -- all of which have a direct bearing on the health of battalion personnel. In addition to BUMED Instructions, U.S. Army Field Manuals are helpful, particularly FM 21-10, Military Sanitation.

23. Senior Watch Officer. The senior officer eligible to serve as Officer of the Day is usually assigned Senior Watch Officer. As Senior Watch Officer, he prepares the OOD duty roster and makes certain that assigned officers understand and comply with all general and special orders, instructing them as necessary to acquaint them with specific orders of the day as well as routine watch matters. He makes periodic examinations of the battalion Administration Log for adequacy and format and recommends remedial training when any relaxation in watch performance standards is noted.

24. Technical Publications Library Officer. The Custodian of the technical publications library is responsible for the stocking, filing and control of issue of all technical publications held by the battalion. These books range from NAVFAC publications to U.S. Army Technical Manuals (TM) and Field Manuals (FM) to manufacturers catalogs. COMCBPAC/LANT INST. 5070.1 series contains a listing of required reference publications.

25. Voting Officer. To assist the battalion personnel in obtaining absentee ballots and to inform them of election schedules and voting requirements in the various states, the Voting Officer keeps current on all laws and directives of the 1742 series and delivers post card applications for absentee ballots to all hands.

26. Wardroom Mess Treasurer. At an established base, battalion officers normally make use of existing base facilities. Most bases oper-

ate a Commissioned Officer's Mess (COM) (closed) which provides essential lodging (BOQ) and food service. In such cases, Stewards (SDs) and Stewardsmen (TNs) are placed on TAD with the closed mess. When the battalion is living in a SEABEE camp or is operating in the field, battalion officers operate a Wardroom Mess. In such cases, food is usually purchased from the general mess and is served to the officers by stewards who prepare the food or who bring meals from the general mess at prescribed hours. The mess treasurer is responsible for collecting mess shares from each officer monthly, preparing menus for the Commanding Officers approval and ensuring that the stewards keep the galley and messing space in proper condition for the serving of meals and maintain the officers quarters in a habitable condition. The Manual for Messes Ashore, NAVPERS 15951 and portions of the BUSANDA Manual should be consulted for instructions.

27. Welfare Officer. As far as possible, the Navy attempts to provide for all the welfare needs of its personnel. Nevertheless, when a situation arises in which an outside agency can provide valuable service to naval personnel, the Navy cooperates in every way possible to bring about assistance. The primary duty of the Chaplain is to conduct divine services and provide religious instruction. Because of his professional training and wide experience with personal problems, the Chaplain also frequently serves as an advisor to battalion men with troubles of a personal nature.

FIGURE 23
SAFETY RECORDS

FREQUENCY RATES NUMBER OF LOST TIME ACCIDENTS PER 1,000,000 M/H WORKED			SEVERITY RATES NUMBER OF DAYS LOST PER 1,000,000 M/H WORKED		
	1962	1963		1962	1963
ALL U.S. CONSTRUCTION	32.90	32.50	ALL U.S. CONSTRUCTION	2,400	2,350
NATIONAL SAFETY COUNCIL	19.90	19.50	NATIONAL SAFETY COUNCIL	2,397	2,349
ARMY CONTRACTORS	5.31	6.16	ARMY CONTRACTORS	2,656	2,327
ALL NAVFAC	6.78	7.73	ALL NAVFAC	1,527	2,706

Domestic and economic problems are the most common types. Men who go to the Chaplain receive his counsel and suggestions. He keeps in close contact with public and private welfare agencies (such as the Navy Relief Society and the American Red Cross) and gives advice on what services are available and how they can be utilized.

The Navy Relief Society is a private organization set up to aid Naval personnel in times of urgent need. Although it is not an official organization, the Society works closely with the Navy.

The principal service provided by the Navy Relief Society is financial assistance to Navy personnel and their dependents. This assistance is in the form of a loan without interest, an outright grant, or a combination of the two. In general, a grant is made to dependents who are left in dire straits after a Navy man dies. It is BUPERS policy to inform the Society when Naval personnel die or are killed. A representative of the Society then visits the dependents to determine what assistance is needed. Such assistance might be necessary to help a widow complete a training course for some form of employment, to help the family move to a new location, to keep children in school, or to meet immediate financial needs. Loans may be made to Naval personnel who need funds to provide burial for a member of the family, for transporration of dependents in case of death or critical illness, for living expenses during periods when approval of dependent allowance is pending and for similar cases. BUPERS issues periodic instructions on the procedure to be followed in contacting the Navy Relief Society for assistance. The Chaplain and Company Commanders should be familiar with these instructions.

The American Red Cross is a voluntary organization with a quasi-governmental position. (See 0738, U.S. Navy Regulations, and C-9207 (3), BUPERS Manual). Its funds are derived from contributions. A member of the American Red Cross is assigned at many shore activities. The principal function of Red Cross personnel is to serve as a link between the Navy and the homes of Naval personnel. The Red Cross on request will obtain facts concerning dependents and forward the information to the appropriate commanding officer. The information may determine the advisability of granting emergency leave, approving a request for transfer or recommending a hardship discharge. Most battalion Commanding Officers have a policy of requiring that the Red Cross verify that a serviceman is needed at home before granting emergency leave from overseas locations. Red Cross investigative reports are held in strictest confidence and are not shown to the individual concerned. Consultation service regarding family problems also is available through Red Cross. Sometimes this organization is able to obtain a report concerning the health and welfare of a man's family. Information that is not available to a family of a serviceman by direct means is often transmitted through Red Cross. In emergencies, the Red Cross may provide dependents of Naval personnel with both counsel and financial assistance. Under certain conditions, loans or grants may be provided. The Chaplain generally maintains close liaison with Red Cross representatives. All personnel in the battalion should be made to understand what services are provided by the Red Cross and how its members can be contacted.

28. Wine Mess Treasurer. The Commanding Officer may establish a wine and/or cigar mess. These messes are governed by COMCBLANT/COMCBPAC instructions. The wardroom mess treasurer should not serve as wine mess or cigar mess treasurer.

SECTION E - COURTS AND BOARDS

1. Courts-Martial. Courts-martial are convened by appointing order in accordance with instructions contained in. the Manual for Courts-martial, United States 1951 (MCM).

The battalion Commanding Officer is not empowered to convene a general courts-martial or a court of inquiry but he may convene a summary court and a special court.

A summary courts-martial consists of one officer, usually a lieutenant or above.

A special courts-martial consists of three or more members. The senior member, usually lieutenant or above, is president of the court. Battalion officers may also serve as trial counsel and defense counsel of a special courts-martial.

If a detachment has but one officer and he has been empowered to convene summary courts-martial, he serves as both the convening authority and as the summary court.

The Manual for Courts-Martial, 1951, the Manual of the Judge Advocate General (NAVJAG INST. 5800.7) and the Special Courts-Martial Trial Guide for President and Members (NAVPERS 10096) are important reference books.

2. Investigations. The Commanding Officer may appoint one officer or a board of two or more officers to conduct investigations. An investigation will be conducted in accordance with the Manual of the Judge Advocate General. They may be formal or informal. A formal investigation is convened by written order of the Commanding Officer. Testimony of witnesses is taken under oath and a verbatim record is kept of the proceedings.

An informal investigation may be ordered verbally. An appointing order is needed only when a party to the investigation is designated. Statements of witnesses are not under oath and only an investigative report is required. Under certain conditions, an informal investigation will suffice as the pretrial investigation required by Article 32 of the UCMJ preliminary to trial by general courts-martial.

The Commanding Officer may direct a preliminary inquiry to be conducted informally to determine the facts in a particular matter. The report may be written or oral.

In cases of injury which does not result in the loss of a full day from duty or in possible disability, an investigation may not be necessary. In such cases an informal inquiry is sufficient.

3. Field Boards. Field boards are established to hear the cases of personnel who have conducted themselves in such a manner that the Commanding Officer desires the recommendation of an impartial board as to the character of the discharge he will recommend to BUPERS for the man in question.

4. Surveys. A formal survey of Navy property is made by an officer or board of three men, one of whom must be a commissioned officer, appointed by the Commanding Officer. The board may be standing or appointed as required but must not include the Commanding Officer, the officer on whose records the survey material is carried or the officer charged with the custody of the material. An informal survey, however, may be conducted by the officer having custody of the equipment or material being surveyed. The Commanding Officer determines whether the circumstances warrant a formal or an informal survey and approves or disapproves the action of the survey officer or board. A disapproval requires the appointment of a second board and, if the results are still unsatisfactory, a referral of the matter to the cognizant technical command. Procedures for formal and informal surveys are outlined in the BUSANDA Manual.

5. Audits. The Commanding Officer appoints audit boards in writing as required by higher authority and as considered necessary for the efficient administration of the battalion. In general, the audit board counts cash, checks the accuracy and validity of books, statements, expenditures, receipts and inventories and makes certain that sound business practices as well as all current directives and regulations, have been observed during the period covered by the audit. All findings are reported to the Commanding Officer.

One officer with previous disbursing or accounting experience should be included on the board of two or more assigned to conduct the disbursing officer's cash verification audit. When a battalion has billets of both a Supply Officer and Disbursing Officer, the Supply Officer shall be a member of this board. The verification should be a surprise audit, unannounced and at irregular intervals, but at least once during each quarter.

At least once a month, without notice and on no set date, an officer must audit the post office funds. Although this audit is usually assigned to the Disbursing or the Postal Officer, it may be conducted by any commissioned officer except the one making the daily money order audit in accordance with OPNAV INST. 2700.1. The audit report, prepared on a form prescribed by CNO, lists the official recipients of copies and must contain accounting of stamp and money order funds and a comparison of the duplicate records, one set of which is kept by the postal clerk while the second is turned over to the Disbursing Officer.

If the MCB maintains a closed mess for officers, the funds controlled by the mess treasurer are audited monthly according to procedures prescribed in the Manual for Commissioned Officers' Mess Ashore. The audit is accomplished by a board of three officers appointed by the Commanding Officer.

When a separate mess for COPs exists, it is subject to the same type of audit as a closed officers' mess. However, a chief belonging to the mess also serves as a member of the auditing board.

6. Narcotics Inventory Board. The Narcotics Inventory Board checks the supplies and disbursement records of narcotics and, when directed, the stocks of alcohol and antibiotics, in custody of the Medical Officer in accordance with instructions and provisions in the Guide for Naval Medical Department Officers.

7. Advancement in Rating Examining Board. This board is composed of three officers appointed by the Commanding Officer to administer examinations taken by the enlisted men of the battalion for advancement in rating. The senior member of the board arranges for all facilities, announces location and time and makes certain that examinations are on hand for all eligible candidates. Servicewide examinations, drawn up and distributed by the Naval Examining Center, Great Lakes, Illinois, are held on dates prescribed by BUPERS. The battalion receives a test for each man who was recommended to participate. Strict accountability for each examination paper is maintained and all necessary precautions are taken to prevent premature disclosure of test contents. Completed answer sheets, along with the examinations and related documents are sent to the Naval Examining Center for correction and evaluation. All excess papers are disposed of as directed.

8. Recreation Council. The Recreation Council consists of officers appointed in writing by the Commanding Officer and always includes the Special Services Officer. The specific purpose of the council is to supervise the operation of the recreation program and fund. In fulfilling this responsibility, they make regular inspections and audits to determine if funds and gear are used in accordance with regulations, reporting all findings regarding cash on hand, verified bank balance and

property inventory to the Commanding Officer. Financial reports are made monthly or whenever a fund custodian is relieved of duty while property inventory reports are made either monthly or quarterly. The fund custodian does not participate in the audit of property inventory. Decisions of the council are made by a vote of the members and are based on regulations, recommendations of the Enlisted Recreation Committee, suggestions of the Special Services Officer and experience of the council members. Minutes of the council proceedings are submitted to the Commanding Officer for approval prior to implementation of voted measures.

9. Enlisted Recreation Committee. Enlisted men representing each company are appointed by the respective company commander at the direction of the Commanding Officer. This Committee meets monthly prior to the Council meeting to present the recreation desires and requirements of the men in their companies. A program or series of proposals are agreed upon and set forth in a written report which is sent to the Recreation Council for action.

10. Others. The Commanding Officer may appoint other groups to assist the command perform various functions such as the Enlisted Service Records Verification Board, the Training Policy Board, the Budget and Planning Board and the Leadership and Character Education Committee.

Chapter III
BATTALION ADMINISTRATION

SECTION A –
ADMINISTRATIVE AND PERSONNEL OFFICER

1. **Introduction to the S-1.** The Battalion Administrative and Personnel Officer (S-1) assists the Executive Officer in administrative details and in personnel administration. If the Battalion Administrative Officer and Personnel Officer duties are performed by two separate officers, the Administrative Officer will be the senior and act as S-1. When the Battalion Administrative Officer and Personnel Officer is a single billet, he is the head of the Administrative and Personnel Departments and serves as the S-1 on the battalion staff. The billet is normally filled by a CEC officer but when the battalion is below allowance in officers, the Headquarters Company Commander may act as the S-1 and the senior yeoman may supervise the administrative and personnel sections. A breakdown of functions performed by the S-1 section is shown in Figure 24.

2. **Duties of the S-1.** It is the duty of the S-1 to:

a. Maintain the battalion library of directives and other official publications making all required changes and route these publications to all officers concerned.

b. Prepare administrative reports and maintain a tickler file on all special and recurring reports submitted by the battalion.

c. Draft, reproduce, and distribute battalion directives, plans-of-the-day, and internal forms.

d. Route and file incoming messages and mail.

e. Operate the battalion guard mail service.

f. Provide copies of directives and publications to detachments of the battalion.

g. Provide clerical assistance to other officers as required.

h. Prepare movement reports in accordance with NWIP 10-1.

i. Perform all personnel accounting including preparation and distribution of the personnel diary (NAVPERS 501), consolidate muster reports and substantiate PAMI reports.

j. Prepare liberty and leave papers, TAD and PCS orders, rotation data cards, identification tags, Geneva Convention Cards, meal passes, and embarkation passenger lists with the appropriate service record entries.

k. Prepare reports pertaining to casualties, prisoners of war, and enemy civilians.

l. Control custody, periodic verification, upkeep and authentication of service records.

m. Supervise check-in/check-out procedures and interviewing of battalion personnel, maintain records of changes in billet, transfer or rotation of men, and monitor the information on the PAMI Monthly Verification Report (1080-14m).

SECTION B – ADMINISTRATIVE PROCEDURES

1. **Instructions and Notices.** The S-1 is the coordinator of reports going to higher authority and the distributor of guidance coming from superiors in the chain of command. This task of office management requires detailed knowledge of the Navy Directive System and correspondence practices. Use of the directives system makes available to each company and department, a complete, up-to-date reference file of all battalion-wide policies and procedures.

The Navy Directive System is designed to provide effective communications between many persons or departments (See SECNAV INST. 5215.1 series). Strict adherence to this system will result in a logical grouping of subject matter and at the same time produce a complete up-to-date reference file of all pertinent policies and procedures. Directives in the system take the form of written communications or issuances which (a) prescribe or establish policy, organization, methods, or procedures; (b) require action; or (c) contain information essential to the effective administration or operation of activities concerned. Instructions are directives of a permanent nature and are set forth as standing procedures while notices are directives of a one-time or temporary nature (6 months or less) and contain provisions for self-cancellation.

Directives received in the administrative office which affect battalion operations are filed in the battalion "miscellaneous directives" binder with one copy being routed to cognizant officers within the command. The department head, staff officer, or company commander concerned with the implementation of the directive received from a higher authority prepares a battalion directive containing the pertinent subject matter and any supplemental information directed by the Commanding Officer. After the directive is signed by the Commanding Officer, it is then reproduced and distributed to the appropriate battalion distribution list by the S-1.

FIGURE 24
BATTALION ADMINISTRATION

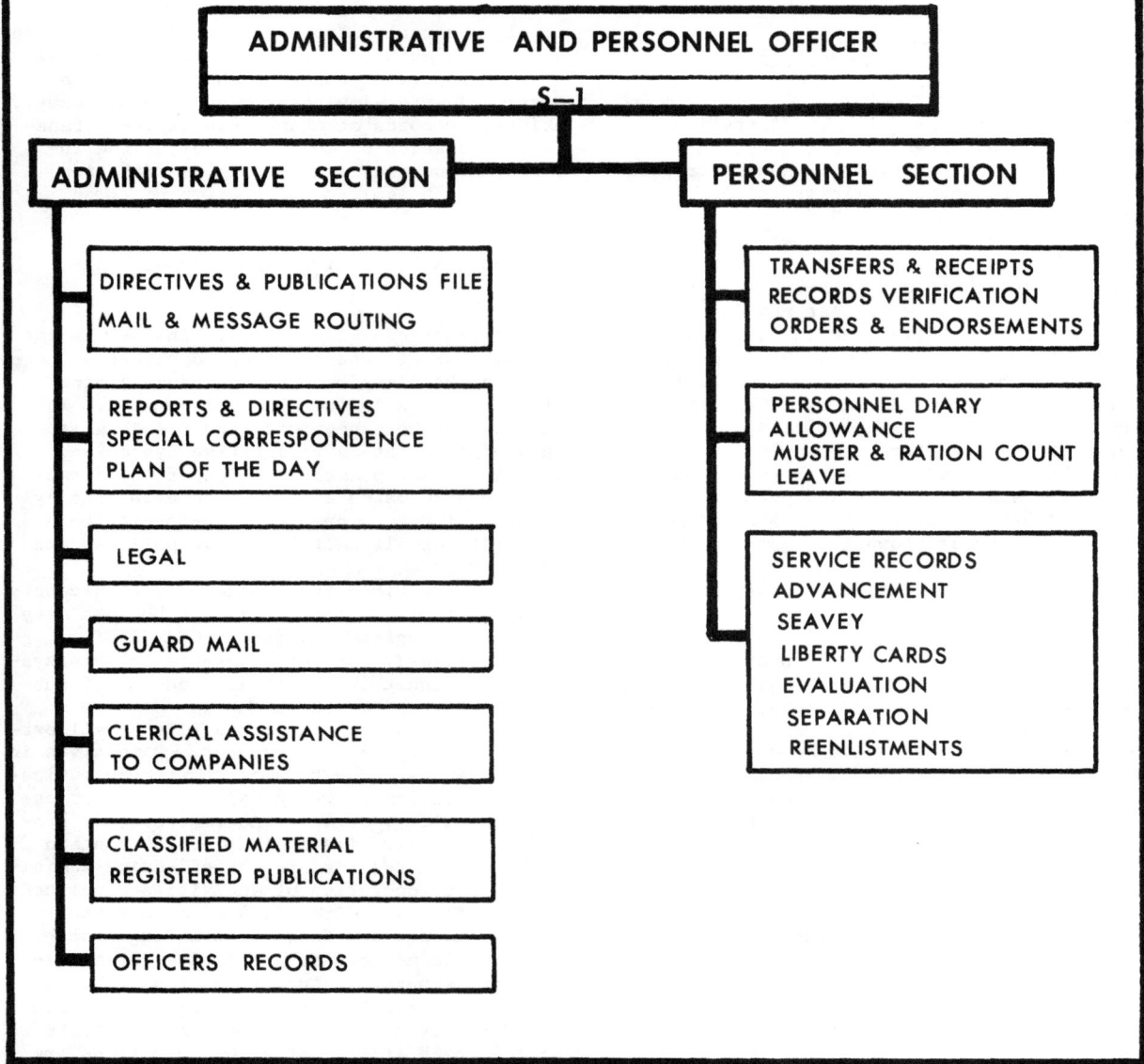

```
                    ┌─────────────────────────────────────────┐
                    │  ADMINISTRATIVE  AND PERSONNEL OFFICER    │
                    │                 S—1                       │
                    └─────────────────────────────────────────┘

┌────────────────────────────┐              ┌────────────────────────────┐
│   ADMINISTRATIVE  SECTION   │              │      PERSONNEL  SECTION     │
└────────────────────────────┘              └────────────────────────────┘

 ┌──────────────────────────────────┐        ┌──────────────────────────────────┐
 │ DIRECTIVES & PUBLICATIONS FILE     │        │ TRANSFERS & RECEIPTS              │
 │ MAIL & MESSAGE ROUTING             │        │ RECORDS VERIFICATION             │
 └──────────────────────────────────┘        │ ORDERS & ENDORSEMENTS            │
                                              └──────────────────────────────────┘
 ┌──────────────────────────────────┐        ┌──────────────────────────────────┐
 │ REPORTS & DIRECTIVES               │        │ PERSONNEL DIARY                   │
 │ SPECIAL CORRESPONDENCE             │        │ ALLOWANCE                         │
 │ PLAN OF THE DAY                    │        │ MUSTER & RATION COUNT             │
 └──────────────────────────────────┘        │ LEAVE                             │
                                              └──────────────────────────────────┘
 ┌──────────────────────────────────┐        ┌──────────────────────────────────┐
 │ LEGAL                              │        │ SERVICE RECORDS                   │
 └──────────────────────────────────┘        │ ADVANCEMENT                       │
                                              │ SEAVEY                            │
 ┌──────────────────────────────────┐        │  LIBERTY CARDS                    │
 │ GUARD MAIL                         │        │  EVALUATION                       │
 └──────────────────────────────────┘        │  SEPARATION                       │
                                              │  REENLISTMENTS                    │
 ┌──────────────────────────────────┐        └──────────────────────────────────┘
 │ CLERICAL ASSISTANCE                │
 │ TO COMPANIES                       │
 └──────────────────────────────────┘

 ┌──────────────────────────────────┐
 │ CLASSIFIED MATERIAL                │
 │ REGISTERED PUBLICATIONS            │
 └──────────────────────────────────┘

 ┌──────────────────────────────────┐
 │ OFFICERS  RECORDS                  │
 └──────────────────────────────────┘
```

The S-1 maintains up-to-date files of all battalion directives and those from COMCBLANT/ COMCBPAC. In addition, basic Navy instructions and notices are kept available for reference along with publications that are not included in the Directives System, such as:

a. U.S. Navy Regulations
b. Navy Department General Orders
c. Uniform Regulations
d. BUPERS Manual
e. Manual for Courts-Martial
f. Manual of the Judge Advocate General
g. Awards Manual (NAVPERS 15790)
h. NMIS Manual (NAVPERS 15642)
i. Transfer Manual (NAVPERS 15909)
j. Manual of Qualification for Advancement in Rating (NAVPERS 18068)
k. Navy Enlisted Classification (NAVPERS 15105)
l. The Public Information Manual (NAVEXOS P1035)
m. The Landing Party Manual (OPNAV P34-03)

Battalion instructions should be prepared so that they are seldom affected by change of station; but all notices should be reviewed every time there is a major change in the battalion's assignment. The Administrative Officer disposes of obsolete instructions and notices in accordance with Section 5215, Article II of SECNAV Instruction P5212.5 series. Current information which is of interest to more than one company or detachment is published by instruction or notice and is distributed with the aid of distribution lists, which, by their very nature, can be used as a guide for posting information on bulletin boards throughout the battalion as well as an aid in routing. Company commanders, shop and field superintendents, and department heads may issue directives which apply only to personnel serving under them. These local directives set forth policies and procedures to be followed in that company or department. Shop regulations and safety precautions are typical examples.

Some battalions issue a book or manual giving the battalion's organization. Such an organization manual may be merely a compilation of effective battalion instructions or it may contain a complete discussion of battalion command control, regulations, daily routine, discipline, watch bills and inspection procedures. An organization manual is intended to clarify battalion policies and procedures and to supplement publications of higher authority.

2. Plan-of-the-Day. A plan of the day is issued to inform all hands of the routine scheduled for the day and to publish special notices. The S-1 supervises the preparation of the plan of the day and submits it to the Executive Officer for approval. The plan of the day fixes the time for reveille, meals, sick call, and other routine and special events. When a standard daily routine (with alternatives for Sundays and holidays) has been published in a battalion directive, the plan-of-the-day merely refers to the appropriate standard routine. The plan-of-the-day also gives the uniform of the day, the name of the OOD, a list of watchstanders, the duty section, assignment and transfer of personnel within the battalion, and any other notices approved by the Executive Officer.

3. Correspondence. Each officer in the battalion who prepares correspondence is responsible for its correctness both in content and form. If security information is involved, he is also responsible for assigning the lowest classification and automatic, time-phased downgrading consistent with proper safeguarding of the information. Subordinates who prepare correspondence requiring the Commanding Officer's signature must initial the proof read copy before submitting it to him for signature. The Commanding Officer shall sign all official correspondence addressed to higher authority relating to the mission or efficiency of his battalion. He shall also sign all official correspondence which is required by law or regulation to be signed by the officer in his own handwriting in the execution of the duties of his office. However, he may authorize other battalion officers to sign routine correspondence "by direction."

All official correspondence must comply with Chapter 16, U.S. Navy Regulations. Preparation is governed by the Navy Correspondence Manual, SECNAV INST. 5216.5. Also useful is the NAVPERS 10009, Writing Guide for Naval Officers.

Incoming correspondence is carefully controlled in order to ensure that prompt action is taken by the proper battalion officer. All official mail to the battalion is addressed to the Commanding Officer. It is opened by the mail yeoman, who stamps it to show the date received and routes it to the companies and departments concerned. Routine mail is routed by means of a simple rubber stamp routing form. Non-routine correspondence is controlled by a routing slip similar to Figure 25. The Administrative Officer usually is the first to see non-routine correspondence and in some battalions, may see all incoming correspondence. He examines the routing and may revise it. He designates a tickler date, which is the date on which action must be completed or a reply prepared.

4. Naval Messages. Naval messages are used when it is determined that letters or speed-letters will not relay the information rapidly enough. They are prepared for release by the Commanding Officer. The Navy Correspondence Manual and U.S. Navy Communication Instructions (DNC-5C) contain guidance on the use and format of naval messages. Normally the battalion uses the facilities of the nearest naval activity for sending and receiving messages. The

FIGURE 25

TYPICAL ROUTING SLIP FOR NON-ROUTINE CORRESPONDENCE

MCB 51 ROUTING SLIP   FROM *COMCBLANT*   FILE NO. *N21* | TICKLER DATE |

SUBJ: *REPLACEMENT OF EQUIPMENT*   DATE RECEIVED *5/28/68*

	ROUTING SEQUENCE	ACTION*	INITIALS	DATE	REMARKS	
COMMANDING OFFICER	2					*SYMBOLS:
EXECUTIVE OFFICER	1					A - Action
S-1						I - Information
S-2						R - Retain Copy
S-3	3	P			*Reply by 6/4/68*	P - Prepare reply
S-4	5	I				
MEDICAL OFFICER						
DENTAL OFFICER						
CHAPLAIN						Do not hold more than two days
Hdqs CO CDR						
A CO CDR	4	A				RETURN TO FILE
B CO CDR						
C CO CDR						
D CO CDR						

message center operated by the S-1 handles intra-battalion communications and the routing of incoming messages. However, if deployed in a remote area, the allowance could be supplemented with the personnel and equipment to maintain message communications with designated activities.

5. Reports. Reports inform those in authority of progress or problems and enable them to take action or to coordinate activities among several units. A consolidated list of periodic, situation and feeder reports required by the commands and offices of the Navy Department is published in OPNAV Instructions of the 5213 series. Recurring reports required by activities in the chain of command are listed in the directives issued by each command. The battalion also lists the reports it must submit in directives of the 5213 series. This list includes the officer responsible for compiling each report, the action addressee, the information addressees, the type of form or format (letter, speed-letter, message) and the Navy directive to be used as reference in preparing the report.

A "tickler" system is used to insure that each report is submitted on time. The "tickler" system usually consists of a set of index cards filed by date and maintained by the administrative yeoman. These reports range from the Daily Personnel Diary (NAVPERS 501) through Monthly Operations Reports, Quarterly Recreation Fund Report to Annual Budget Submissions. Many reports, however, are submitted only to inform higher authority of an event. These situation reports range from Project Completion Reports to Report of Delivery of Personnel to Federal or Civil Authorities.

6. Records. Battalion records are filed according to the subject file classification system specified in SECNAV INST. 5211.3. However, the filing regulations must be adapted and supplemented by each activity to suit its particular needs. Thus each battalion specifies a list of standard file subjects in directives of the 5211 series. These files are disposed of in a manner and at such times as specified in SECNAV Instructions of the P5212.5 series, entltiled Disposal of Navy and Marine Corps Records.

7. Classified Correspondence. Instructions, correspondence, messages or reports may contain classified information which requires protection in the interest of national defense. The Department of the Navy Security Manual for Classified Matter, OPNAV INST. 5510.1 series and OPNAV INST. 5510.49 series contain instructions pertaining to the receipt stowage, issue control, preparation, transmission and destruction of classified correspondence. Procedures for classification and automatic time-phased downgrading are located in the OPNAV INST. 5500.40 series.

8. For Official Use Only Correspondence. The term "For Official Use Only" is assigned to official information which requires protection in accordance with statutory requirements or in the public interest, but which is not within the purview of rules for safeguarding information in the interest of national defense. Its description, use and limitations are set forth in the effective edition of SECNAV INST. 5570.2 series.

FIGURE 26
PERSONNEL PROCEDURES

SECTION C - PERSONNEL PROCEDURES

1. **Programs**. Every officer and petty officer is directly concerned with personnel administration. Each is responsible for the performance, morale and well being of the men under his authority. Therefore all company officers and petty officers must have an understanding of Naval personnel practices, the organization through which they are carried out and the principles that underlie various procedures. This is a large order because Navy personnel administration is complex and dynamic. The officer who takes the time to learn enough about the system so that he can counsel his men, will find that he could not have invested his time in a more worthwhile or productive pursuit. A brief introduction to some of the major programs in presented under the S-1 because he is the primary staff assistant to the command and provides services and instruction to the company commanders.

Personnel procedures are governed primarily by the BUPERS Manual (NAVPERS 15791). Part A deals with personnel organization, allowances and pay while Part B provides guidance on correspondence, records and reports. Part C contains the heart of regulations governing personnel administration, enlistment, classification, transfer, performance, decorations and civil matters. Separations and retirement are also treated in Part C, Part D is set aside for training and education instructions. Other essential tools include some knowledge of personnel accounting, the SEAVEY/SHOREVEY systems, incentive and training programs and evaluation and promotion.

2. **Personnel Accounting**. Since personnel accounting must be accurate in order to match current skills and numbers with future requirements, an integrated Navy-wide Manpower Information System was established under the overall administration of BUPERS. It consists of a Data Processing Center at BUPERS, several Electronic Accounting Machine Units located to provide information on recruits, and three major field activities called Personnel Accounting Machine Installations (PAMI), which record and store information on billets and personnel assigned to the Atlantic area, the Pacific area and the Continental United States. This high speed computer oriented system is tied together by a data transmission system fed by reports from the ships and commands of the fleet and the shore activities.

In the battalion the S-1, acting in his capacity as personnel officer, is responsible for accurate and timely input of information concerning the officers and men of the command. Input has been made standard through the use of the NAVPERS 501 form called the Personnel Diary (See Figure 27). Details concerning format, types of entries and distribu-

tion of copies are contained in the Naval Manpower Information Systems Manual (NMISMAN), NAVPERS 15642. The NAVPERS 501 is submitted by fleet units daily if changes occur. This procedure allows the PAMI to maintain current information on all personnel. Each month the PAMI sends out to all commands the Officer Distribution Control Report (ODCR), NAVPERS 2627 and the Enlisted Verification and Distribution Report (1080-14M). These important documents bring together all the significant factors relating to the distribution of personnel. They show the battalion allowance, onboard strength, changes expected in the next six months and current and projected manning levels. Information is furnished on those personnel currently in each billet and for those expected to arrive on board in the near future. Every month the information on these reports is verified and corrections made via the personnel diary when necessary. Without these reports, personnel planning is impossible.

3. **Personnel Distribution**. The discussion of rotation to sea and shore duty will be limited to enlisted personnel in this manual since officer assignment policy varies with designator. In order to provide an equitable distribution of men to billets while maintaining fleet units in operational readiness condition, the Career Enlisted Rotation Plan was introduced in 1956. The plan called for a survey of men at sea to determine their eligibility to be rotated ashore (SEAVEY), and a similar survey conducted ashore (SHOREVEY) for men who are about to be assigned to sea. The lengths of tours at sea and ashore depend upon the ratio of shore billets to sea billets for each rate. This is further modified to hold the minimum tour ashore to two years to allow men to have time with their families and to hold the minimum tour at sea to three years to provide a degree of crew stability. For example, if there were 220 billets at sea and 108 billets ashore for EO1s, they could expect to spend about four years at sea for every two years ashore.

When a determination by rate of the number of billets ashore that will become vacant during the next period has been made, BUPERS compiles a list of the men who are on sea duty, starting with the one who has been at sea for the longest continual period. Then by counting down the list of those on sea duty until a corresponding number has been reached in each rate, BUPERS establishes and publishes in a 1306 notice the Sea Duty Commencement Cutoff date for each rate. If a man went on sea duty before that time then he knows that he is eligible for rotation ashore. This fact is picked up by the PAMI and a Rotation Data Card is sent to his command. The man will be called into the personnel office to fill out his duty preferences. It is at this point that the S-1

FIGURE 27
PERSONNEL DIARY NAVPERS 501

PERSONNEL DIARY   BUPERS REPORT 1080-01

NAVPERS 501 (REV 7 58)

DATE	CONS NO.	NAME	FILE OR SERVICE NO.	GRADE OR RATE	OFF DESIG OR ENL B/C SERVICE
1	2	3	4	5	6
1 MAY 69		Mobile Const.Batt 11 2045 0011 00 Enlisted 1233			

	040	SMITH Roy P Rec for dut fm CBC PORT HUENEME Calif.	243 76 35	EOC	USN
	041	BROWN Thomas L. Rec for TAD fm MCB Three	569 33 02	CMI	USN
	042	JONES Ray L Tran to NAS Lakehurst NJ for dut. EDA 31 May 69 PCS B15263-018	653 28 42	UT3	USNR-R

15 MAY 65		Mobile Const.Batt.11 2045 0011 00 Enlisted 1233 No changes			
31 MAY 65		Mobile Const. Batt.11 2045 0011 00 Enlisted 1233			
	043	O'TOOLE Clancy A Ch rate to CN. Auth:	313 10 28	CA	USNR-R

Certified to be correct. Approved and forwarded
this 3rd day of JUN 69

a.B. See

A.B.SEE
Commanding Officer
MOBILE CONSTRUCTION BATTALION ELEVEN

NOTE: Due to the fact the original diary is microfilmed, it shall be prepared with a typewriter equipped with a black ribbon capable of producing dark, distinct characters

ORIGINAL

and company commanders have an obligation to ensure first that the man has sufficient obligated service to complete a full 24-month tour ashore when he does receive orders. (Therefore he should have about 30 months of obligated service when he submits the Rotation Data Card. If he does not, he may execute a conditional agreement to extend his enlistment) and second that the man has made his duty choices sufficiently broad that the detailers will have a better chance to find a billet for him rapidly. By limiting duty choices, the man may stay at sea for a long period waiting for his specific billets to become available. Periodically ALL HANDS, a monthly BUPERS information bulletin, publishes an unofficial listing of the shore billets at naval activities. This can be used as a guide in counseling personnel concerning their best chances to move ashore.

When the Rotation Data Card has been filled out it is sent to PAMI. This card, together with a series of cards containing personal and career information which has been stored by the PAMI, is sent to BUPERS for detailing. BUPERS selects those men best qualified to fill what have been called, BUPERS controlled billets. These include, recruiting duty, instructor duty and duty in Washington or with Military Assistance Advisory Groups (MAAG) along with any critical ratings. For the rest of the men of SEAVEY, BUPERS selects the naval district only and leaves the exact billet selection to an Enlisted Personnel Distribution Office (EPDO), a field activity which works closely with the PAMI. There are three EPDOs corresponding to the three PAMIs. But in the rotation of men at sea to shore duty, EPDO CONUS has the task of selecting a specific billet for each man within the district assigned by BUPERS. The EPDO sends to the battalion an assignment card which contains the specific duty station, the month for the transfer and the authority to write the orders. The details of this system may be found in the Transfer Manual, NAVPERS 15909 and BUPERS INST. 1306.20 series.

4. Check-in and Assignment. Each battalion develops a check-in procedure for incoming personnel. These procedures assure prompt completion of administrative details. Men reporting aboard are given a check-in sheet which provides for such matters as:

    a. Temporary billeting by the duty
    b. Verification of service record.
    c. Interviews to evaluate experience and aptitudes and to determine construction/military assignments.
    d. Orientation to battalion facilities and projects.
    e. Issue of special clothing (greens, boots, etc.).
    f. Delivery of pay, health and dental records.

He is then assigned to a specific company,

usually by the S-3, and the company commander determines his exact fire team/work crew assignment. The squad leader will instruct newly arrived men on uniform requirements, bunk and locker standards and the necessary uniform markings. He normally assigns compartment cleaner duties to each man in his squad. The fire team leader will be responsible for seeing that the new men comply with company policy and assist them to overcome any difficulties. On the job, the work crew leader will assign new men to tasks as necessary, ensuring that they are familiar with or are instructed in correct procedures and safe practices.

5. Uniform Regulations. The Battalion Commanding Officer issues instructions concerning the uniform requirements for men at the job site. However, directives on uniforms must conform to policies of the base or area commander. Uniform regulations for Navy personnel are based on U.S. Navy Uniform Regulations, NAVPERS 15665, U.S. Navy Regulations, and directives of local commanders. The uniform of the day must be worn at all times except when otherwise prescribed by competent authority. When working in the field, battalion personnel are permitted to wear working uniforms. Working uniforms are dungarees, or the familiar "greens" which are issued as special clothing to battalion personnel. In many localities officers are required to wear cotton khaki as the working uniform rather than greens.

Local regulations govern the hours that the working uniform may be worn. Sometimes they may be worn only on the job. In other localities, working uniforms are permitted in all but a few areas such as the mess, clubs, and ship's store. Regulations prohibit the wearing of mixed uniforms. Thus shoes, caps, socks, shirts, and other apparel must be worn to conform to the basic working uniform. The principal exceptions to the ban on mixed uniforms are foul weather gear, field shoes, overshoes, and similar items required for comfort or protection.

All naval personnel are required to possess certain items of uniform. The basic list is contained in U.S. Navy Uniform Regulations. Requirements are published from time to time by Commanding Officers. Each company commander is responsible for seeing that men in his company have the required articles of uniform. The company commander or other designated representative may inspect the outfit of each man who reports aboard. Thereafter, the company commander regularly inspects clothing in accordance with battalion directives.

Civilian clothing may be worn only at times prescribed by proper authority. Enlisted men may be permitted to wear civilian clothing at certain times, such as when going on leave or liberty. In some areas, wearing of the uniform may be required at all times.

6. Service Records. A service record is the official history of a person's career in the Navy. It is the property of the government

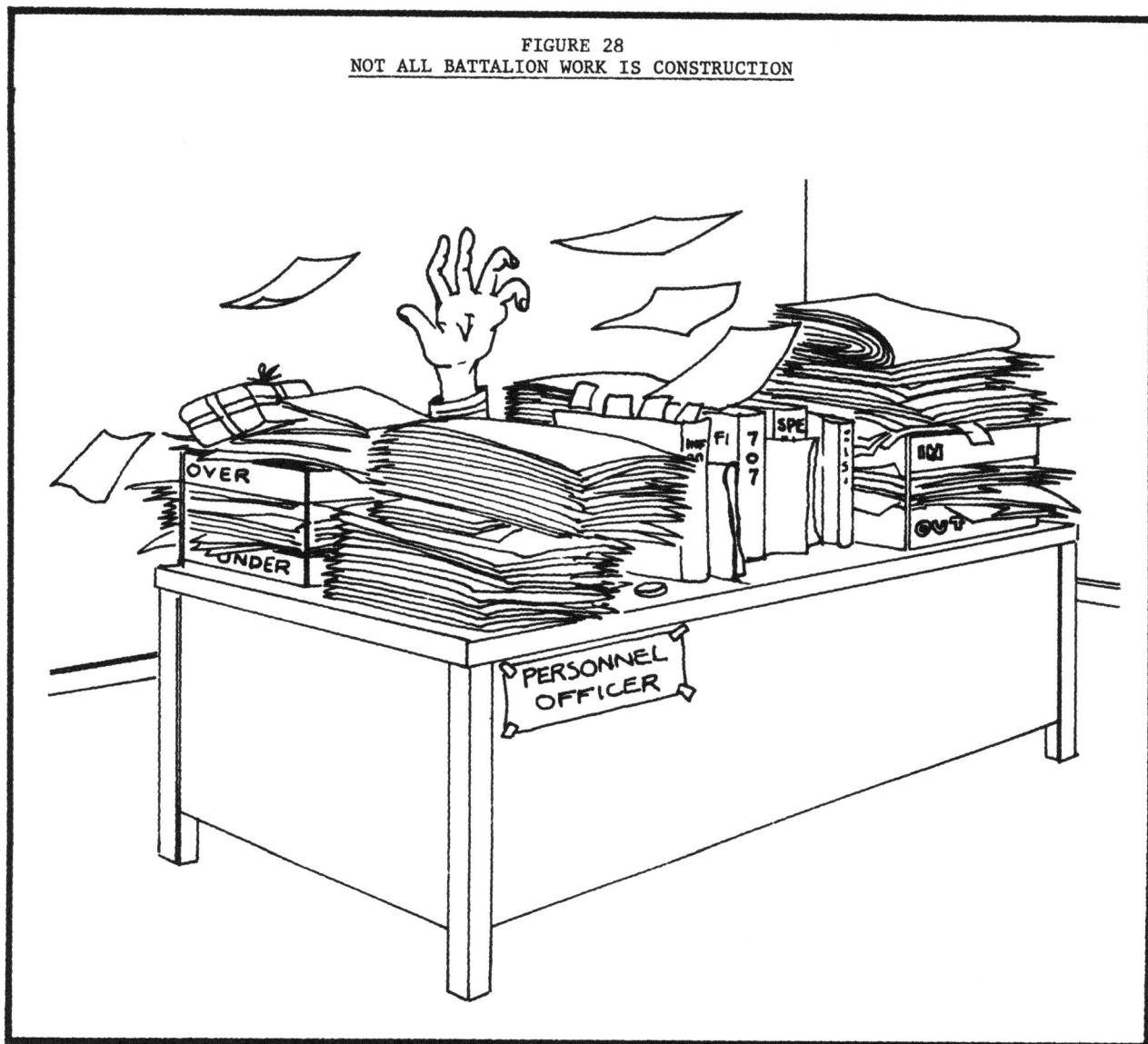

FIGURE 28
NOT ALL BATTALION WORK IS CONSTRUCTION

and its custody and upkeep is one of the more important tasks of the S-1. All information in a service record is classed as official. It may not be divulged except to persons properly authorized and directly concerned. A man is allowed to examine his record but he may not remove it from the personnel office. Altering, making additional entries or removing any portion of the record is prohibited. The enlisted record consists of a flat-type folder containing several loose-leaf pages. These pages are standard forms available through normal supply sources and are filled out to show pertinent information about a man and his enlistment. The right-hand side of the record contains a specified series of pages which must be filled out in strict accordance with Chapter 2 of Part B of the BUPERS Manual. Instructions are

very specific because these entries affect pay, allowances, retirements and veterans benefits. Only an officer designated by the Commanding Officer in writing may make service record entries. When a man checks into the battalion his service record is verified. This consists of interviewing the man and matching other records such as pay, etc. with entries in the service record. The process is repeated every year about 1 September and when the man checks out of the battalion. A record of verification is kept on the inside of the service record cover.

7. Personnel Identification. Personnel are identified by skill and by a series of personal identification cards and tags. In the Navy many skills are grouped into ratings; for

example, carpentry, brick laying, concrete finishing and many others would be found in the Builder Rating. The skill level would be indicated by a man's pay grade; for example, E-6 or first-class Petty Officer. The combination of the rating and the pay grade is called a man's rate and is written BU1. See Annex G for details of the Group VIII Ratings.

There are certain skills not defined by rating that the Navy desires to identify in manpower authorizations and billet allowances and for detailing men to these billets. The Navy Enlisted Classification (NEC) supplements the rating structure and identifies skills within a rating but not common to all personnel in that rating. These codes are divided into trainee, rating, and special series. The trainee codes identify aptitudes and qualifications not discernable from apprentice rates and assist in the planned input into each rating. Skills acquired by petty officers are noted by two letters and four numbers in the case of rating associated codes or by four numbers only in the case of special codes. For example, a certified welder must be a steelworker and his code would be SW-6015, a blaster must be an equipment operator and would be identified by the code EO-5708. However, men with skills such as scuba diver, which has no source rating, would be identified by the code 5345. The trend is toward the identification of as many skills as possible with source ratings. A listing of codes, their relative priority and how the codes are used can be found in the current edition of the NAVPERS 15105, Navy Enlisted Classification.

The Navy has several means of personnel identification. Some must be in possession of the individual at all times while others are required only when prescribed. The most important is the Armed Forces Identification Card (DD Form 2N), popularly called the ID Card. It is issued to all members by authorized activities. Custody and control of these cards is critical and alteration, improper use or possession is punishable under the UCMJ.

Two metal identification tags (often called dog tags) are prepared for each person in the Navy on active duty. During time of war, national emergency, or when traveling in aircraft or reporting to a medical facility under orders or at other times directed by the Chief of Naval Personnel, these tags are worn. They show the name, service number and blood type of the individual.

The Liberty Card (DD Form 345) is issued to all enlisted men of pay grade E-4 and below to cover absences not classified as leave. For this purpose, all branches of the armed services use a card that may be printed in any one of five colors to aid in maintaining control. The battalion is usually divided into four sections. At least one section is on duty at any given time. A different color card is assigned to each duty section. "Section Five" liberty cards are those issued to personnel exempt from duty watches, such as

cooks and bakers, whose unusual work hours do not permit ready assignments to a regular duty section. Enlisted men are required to turn in their liberty cards to the duty petty officer at the expiration of liberty. When a man is detached from the battalion his liberty card is destroyed.

Meal passes may be issued to enlisted personnel who eat in the general mess. The pass identifies those who are not drawing commuted rations and therefore are entitled to their meals without cost. The general mess provides enlisted men with a well-balanced diet through the proper preparation and service of a daily food ration. A ration is one day's legal allowance of provisions for one man. Rations may be either "in kind" or "commuted." Rations "in kind" are foods prepared at Government expense by the command to which a man is attached and issued to him through the general mess, Commuted rations (ComRats) are furnished in the relative money value fixed by law. Personnel drawing ComRats may eat their meals in the general mess but they are charged for each meal. In the deployed battalion, meal passes are seldom used because all men subsist in the general mess. However, many men have their families in the homeport, and so they may draw a commuted ration at the CBC.

The Articles for the Protection of War Victims as authorized by the Geneva Conventions of August 1949 provide for the issue of identification cards in duplicate to persons who may become prisoners of war. One copy is to be turned over to the capturing authority, the other to be retained by the individual at all times. A Geneva Conventions Identification Card (DD Form 528) is prepared for personnel on active duty who are liable to capture. This card is placed in an envelope suitably labeled to show the contents, and filed in the service record. When ordered to duty outside the continental limits of the United States, the card is removed from the record and issued to the individual. In the event of capture by an enemy armed force, he should turn his card over to his captors. The Armed Forces Identification Card (previously described) is treated as the duplicate that is retained by the individual at all times.

Other identification cards include the U.S. Government Motor Vehicle Operator's Identification Card (Standard Form 46), the Immunization Certificate (DD Form 737) commonly called a shot card and the Department of Defense Emergency Instruction Card (DD Form 886-1).

8. Leave and Liberty. Leave is authorized absence from a place of duty chargeable against the individual's leave credit. In general, each person in the Navy on active duty is entitled to leave at the rate of two and one-half calendar days for each month of active duty. A request for leave is made on the Special Request Form. The Personnel Office fills in data from the man's leave record. The Personnel Officer is usually

authorized to approve requests which conform with battalion policy. The OOD may be given authority to grant emergency leave and extension of leave in cases of emergency.

There are five types of leave; earned, advanced, excess, sick and graduation and four ways to use or take leave; regular, transfer, reenlistment and emergency. Instructions concerning leave can be found in Chapter 6 of Part C of the BUPERS Manual.

Liberty is defined by U.S. Navy Regulations as authorized absence of an individual from a places of duty, not chargeable as leave. Liberty may not be taken in conjunction with leave. The Commanding Officer has authority to grant liberty. He may grant liberty at any time for 48 hours or less. A 48-hour liberty may be extended to 72 hours by the Commanding Officer if the period includes a holiday proclaimed by the President or authorized by the Secretary of the Navy. When liberty expires between the end of regular work hours on one day and the beginning of work hours on the next work day, it may be extended to the beginning of work hours. Thus, if a liberty expires at 1600 on Sunday, it may be extended to 0730 or 0800 Monday, if the Commanding Officer so directs. The Chief of Naval Personnel may authorize the Commanding Officer of a unit for which normal liberty is inadequate (because of isolated location or nature of the work performed) to grant liberty for a period of 72 hours or to grant liberty that includes a national holiday for a period not to exceed 96 hours.

During non-working hours, normal liberty may be granted to personnel who are not in a duty status. As previously mentioned, there are generally four duty sections. Men on normal liberty are required to remain in the "general vicinity" of the battalion. General vicinity is variously interpreted to mean from 50 to 150 miles. Notices of instructions always specify the distance that men on liberty are allowed to travel without special permission.

Special liberty is liberty requiring special approval, such as exchange of duty, absence during work hours, early liberty and extended liberty. Men having the duty are not normally allowed to go on liberty. They may do so only when they can furnish a substitute and when the request is approved by all officers and petty officers concerned. Exchange of duty is usually kept to an absolute minimum and may require the approval of the executive officer.

9. Evaluation and Promotion. One of the most serious obligations a battalion officer has to his men, his shipmates and the Navy is the semiannual evaluation of personnel under his command. These evaluations have important bearing upon promotions, advancement, assignments and other phases of a naval career. They must accurately reflect six months of effort on the part of the individual concerned.

In order to provide a meaningful comparison with all other enlisted personnel of his pay grade, the evaluations must be made in strict accordance with the guidance established by BUPERS and published in Chapter 7 of Part C of the BUPERS Manual. It must be remembered that perversion of this system will ruin the chance to recognize and reward truly outstanding men and will breed mediocrity in the SEABEE forces.

In addition to the regular evaluations that are due two times every year, there are special evaluations, some of which are mandatory, others optional. Marks must be assigned when a man is reduced in pay grade, is transferred more than 90 days from the last regular evaluation or will be on temporary additional duty when the next report is due. Marks may be assigned when his performance was outstanding in either the meriotorious or the derogatory sense such as, meritorious mast, recommendation for reenlistment, assignment of non-judicial punishment or conviction by civil authorities. Enlisted personnel are rated on the Enlisted Performance Evaluation (NAVPERS 792). This form, shown in Figure 29, permits evaluation in five distinct categories:

a. Professional Performance
b. Military Behavior
c. Leadership and Supervisory Ability
d. Military Appearance
e. Adaptability

The crew leader/fire team leader immediately responsible for a man makes the evaluation on an Enlisted Performance Evaluation Form. It is forwarded to the squad leader, platoon leader and then to the company commander for review. He prepares the form for the signature of the Commanding Officer.

Every time the company officer makes an evaluation he should ask himself "do I want to see this man promoted." Since evaluation and recommendation for participation in Navy-wide examinations are so closely allied, this question will help the company commander guard against inconsistency. Although there are many steps to promotion for enlisted personnel (See Figure 30), the key is recommendation by his company commander. Before a man is recommended he must be fully qualified in accordance with the Manual of Qualification for Advancement in Rating, NAVPERS 18068. This includes:

a. Demonstrating the ability to perform the required practical factors listed on the NAVPERS 760 form.
b. Completing the specified Navy training courses in technical and military subjects.
c. Passing leadership examination when required.
d. Meeting the required minimum time in grade requirements. Then he is recommended and the Commanding Officer's recommendation is entered in his service record.

FIGURE 29
EVALUATION REPORT

REPORT OF ENLISTED PERFORMANCE EVALUATION
NAVPERS 792 (Rev. 6-65)
0105-402-3001

PERIOD OF REPORT
_____ To _____

NAME (Last, First, Middle)		RATE ARR	PRESENT SHIP OR STATION

INSTRUCTIONS

1. For each trait, evaluate the man on his actual observed performance during this reporting period, evaluate him on what he did. Describe what he did in the Comments section.
   If performance was not observed, check the "Not Observed" box.
2. Compare him with others of the same rate.
3. Pick the phrase which best fits the man in each trait and check left or right box under it. (Left box is more favorable.)
4. If the major portion of his work has been outside his rate or pay grade

1. PROFESSIONAL PERFORMANCE His skill and efficiency in performing assigned duties (except SUPERVISORY)					
NOT OBSERVED ☐	Extremely effective and reliable. Works well on his own	Highly effective and reliable. Needs only limited supervision	Effective and reliable. Needs occasional supervision	Adequate but needs routine supervision	Inadequate. Needs constant supervision
	*				* | *

2. MILITARY BEHAVIOR How well he accepts authority and conforms to standards of military behavior					
NOT OBSERVED ☐	Always acts in the highest traditions of the Navy	Willingly follows commands and regulations	Conforms to Navy standards	Usually obeys commands and regulations. Occasionally lax	Dislikes and flouts authority. Unseamanlike
	*				* | *

3. LEADERSHIP AND SUPERVISORY ABILITY His ability to plan and assign work to others and effectively direct their activities					
NOT OBSERVED ☐	Gets the most out of his men	Handles men very effectively	Gets good results from his men	Usually gets adequate results	Poor supervisor
	*				* | *

4. MILITARY APPEARANCE His military appearance and neatness in person and dress					
NOT OBSERVED ☐	Impressive. Wears Naval uniform with great pride	Smart. Neat and correct in appearance	Conforms to Navy standards of appearance	Passable. Sometimes careless in appearance	No credit to the Naval Service
	*				* | *

5. ADAPTABILITY How well he gets along and works with others					
NOT OBSERVED ☐	Gets along exceptionally well. Promotes good morale	Gets along very well with others. Contributes to good morale	A good shipmate. Helps morale	Gets along adequately with others	Misfit
	*				* | *

6. DESCRIPTION OF ASSIGNED TASKS

7. EVALUATION OF PERFORMANCE (E-5 and above include comment on ability in self expression and command, orally and in writing, of the English language)

* 8. THESE ITEMS MUST BE JUSTIFIED BY COMMENTS IN ADDITION TO THOSE IN ITEM 7 ABOVE

9. REASON FOR REPORTING	10. DATE	11. SIGNATURE OF REPORTING SUPERIOR
☐ SEMIANNUAL ☐ TRANSFER ☐ OTHER _____		

```
┌─────────────────────────────────────────────┐
│                  FIGURE 30                    │
│    STEPS TO PROMOTION FOR ENLISTED PERSONNEL  │
│                                               │
│    1.   PRACTICAL FACTORS (NAVPERS 760)       │
│                                               │
│    2.   NAVAL TRAINING COURSES                │
│                                               │
│    3.   TIME IN GRADE                         │
│                                               │
│    4.   LEADERSHIP EXAMINATION (E-4 & E-5)    │
│                                               │
│    5.   RECOMMENDED BY COMMAND                │
│                                               │
│    6.   NAVY-WIDE EXAMINATION                 │
│                                               │
│    7.   HIGH MULTIPLE                         │
│                                               │
└─────────────────────────────────────────────┘
```

Examinations for advancement up to and including pay grade E-3 are conducted by the battalion. No quota exists for these rates. Written examinations are normally prepared by the S-1 using representative questions that demonstrate a man's knowledge of the rate to which he aspires. When successful the appropriate entries are made in the man's service record. Examinations for petty officers are normally service-wide. Several weeks before the scheduled date for the test, they are ordered for every individual who is qualified and recommended. The examination is monitored by battalion officers and the test and answer sheets are sent to the Examining Center for grading.

Passing the examination does not ensure promotion because there are quota limitations. Men are assigned a "multiple" based upon examination score, past performance, total active service, time in pay grade and medals received (See Figure 31). Those with the highest "multiple" are promoted first and so forth until all

available quotas are filled. Those men who were not promoted or who failed the examinations, receive an Individual Examination Profile Card which will assist them to prepare for the next examination. The card indicates those sections of the previous examination in which the man was strong and those in which he was weak.

10. Civil Matters. The reputation of the Navy in the civilian community is important, and it is established to a great degree by the actions of individual servicemen. BUPERS has policies for the guidance of the Commanding Officers in frequently occurring problem areas Chapter 11 of Part C of the BUPERS Manual and various BUPERS Instructions discuss civilian employment, marriage outside the Continental United States, payment of federal and state income taxes, support of dependents, indebtedness and financial responsibility, customs declarations, public liability insurance and political activity.

11. Incentive Programs. In order to keep highly qualified men at all levels of responsibility, the Navy must provide a path for the capable man to travel upward at a pace that will tax his ability. Therefore several incentive programs are offered, many leading to a commissioned status. These programs may apply only to a limited number of battalion personnel and therefore the personnel officer working through the Career Counselor and company commanders must select that particular program best suited for each individual. For example: the Selected Training and Retention (STAR) Program, BUPERS INST. 1133.13 series, allows men to receive a reenlistment bonus early, go to Class B School and possibly be promoted early; the Selected Conversion and Retention (SCORE) Program, BUPERS INST. 1440.27 series allows a man to receive a reenlistment bonus, to change rating without an examination (possibly receiving proficiency pay), to be assigned

```
┌────────────────────────────────────────────────────────────────────────────────┐
│                                   FIGURE 31                                      │
│                     COMPUTATION OF A MAN'S "MULTIPLE"                             │
│                                                                                  │
│   EXAMINATION SCORE..................X1 =..........│MAX = 80 (4.0 = 80)           │
│                                                                                  │
│   PERFORMANCE FACTOR.................X1 =..........│MAX = 50 (4.0 = 50)           │
│                                                                                  │
│   TOTAL ACTIVE SERVICE..............X1 =..........│MAX = 20 (NO CREDIT BEYOND 20 YEARS) │
│                                                                                  │
│   SERVICE IN PAY GRADE..............X2.............│MAX = 20 (NO CREDIT BEYOND 10 │
│                                                    │           YEARS IN GRADE)    │
│                                                                                  │
│   AWARDS...........................  ............│MAX = 15 (MEDAL OF HONOR = 10) │
│                                                    │         (GOOD CONDUCT = 2)   │
│   ─────────────────────────────────────────────── │                             │
│   FINAL MULTIPLE                     = ?           │MAX = 185                     │
└────────────────────────────────────────────────────────────────────────────────┘
```

to Class A and B Schools and have the opportunity to be promoted early; the Navy Enlisted Scientific Education Program allows a man to attend a four-year college course full time and upon successful completion attend Officer Candidate School (OCS) and receive a commission. Other officer programs include quotas for the Naval Academy and NROTC Colleges. Various Warrant Officer and Limited Duty Officer Programs are also available.

12. <u>Orders and Check Out</u>. When a man is due for shore duty, the battalion must plan on training a replacement for him. Normally the S-1 will notify the company commanders that a man is on SEAVEY and that he can expect orders within about six months. The next indication comes when PAMI picks him up in a SEAVEY status on the 1080 and sends him the Rotation Data Card to fill out stating his choice of duty. When the 1A Card arrives aboard the battalion,

the man knows his assignment by Naval District and the approximate transfer month. Then an assignment card is sent by EPDOCONUS which informs the battalion of the month for transfer and the new duty station and indicates the authority for the battalion to write the orders. The permanent change of Station Orders (PCS) are written and signed by the Commanding Officer. The men being transferred are given a check-out sheet that is to be initiated by representatives of the various staff officers. The principal purpose of this procedure is to ensure that all property of the battalion has been turned in. The library, armory, recreation issue room and supply office are the principal activities in this category. The list also shows that the medical, dental, disbursing and service records have been picked up, that a forwarding address is on file at the post office and that the liberty card and bedding has been turned in to the MAA.

Chapter IV
BATTALION TRAINING

SECTION A –
PLANS AND TRAINING OFFICER

1. **Command Responsibilities.** Articles 0704 of U.S. Navy Regulations charges the Commanding Officer to exert every effort to maintain his command in a state of maximum effectiveness for war service consistent with the degree of readiness prescribed by proper authority. Article 0710 further places the responsibility upon him to increase the specialized and general command by frequent conduct of drills, classes, instruction and use of appropriate fleet and service schools. The development of skilled and experienced petty officers can only be accomplished by a battalion training program that is correctly oriented and vigorously and continuously pursued by all levels of the command.

The success of an MCB or ACB operation is greatly dependent upon the efficient and organized operation of battalion personnel and equipment. The goal of MCB and ACB training and education programs is to promote the full and visible development of personal skill and knowledge in order that assigned missions may be competently accomplished. Continuous technical and military training of battalion men is essential to the teamwork and skills required of a construction battalion. Training and education programs assist men in assuming their responsibility to carry out their tasks efficiently. Through the effective use of training and education, the battalion functions as a powerful, coordinated unit. This chapter sets forth considerations and programs in the training of individuals and troop unit.

Many aspects of a construction battalion must be taken into consideration in formulating a successful training program. It is necessary to consider both long range and immediate goals. Training depends upon many factors, among them, the location and operational commitments of the battalion, directives from higher authority, the experience and previous training of personnel and the available training facilities. Usually, recently enlisted men lack construction experience as well as military training. Therefore most of their construction and all of the combat skill and experience must be acquired in the battalion itself. Not only must the training courses and schools be efficient, but day-by-day work tasks must be organized to provide effective training. Every day and every task should be utilized to contribute to the construction and combat capabilities of the battalion. During mobilization, the battalion personnel have usually had civilian construction experience, therefore, emphasis is placed upon rapid combat training.

Articles 0803 of the U.S. Navy Regulations states that the Executive Officer will supervise and coordinate the work, exercises, training and education of the personnel of the command. He supervises the training of battalion officers, coordinates the planning and execution of the training program and when necessary, acts to correct deficiencies in the program. He does this in his capacity as Chief Staff Officer. His principal assistant is the Plans and Training Officer.

Company Commanders are directly responsible for the training of their company personnel and for the fulfillment of training goals as established by the Commanding Officer. They help formulate training programs; supervise the training of subordinate officers; and direct the technical, military and general training of their companies. The battalion service department heads are responsible for the individual training of the personnel in the departments. They conduct training for Advancement in Rating and administer the OPNAV sponsored general training. Platoon leaders observe closely the training progress of the men in their platoons. They directly supervise on-the-job construction and military training. All petty officers assume the responsibility for training their men and must be able to conduct effective training courses utilizing lectures, discussions, project work and written methods.

2. **Introduction to the S-2.** The fourth senior CEC Officer in the battalion is usually designated as the Plans and Training Officer (S-2). He is head of the Training Department and serves as the S-2 on the battalion staff. He is also the battalion intelligence officer and is responsible for maintaining the war and contingency plans required by NCF units. He may be designated Top Secret Material Control Officer. In combat-support and disaster recovery operations, he may be assigned as Liaison Officer to the staff of the commander exercising direct operational control of the battalion because of his detailed knowledge of the capabilities include the preparation and execution of the battalion's technical, military and general OPNAV training programs as shown in Figure 32. In many cases he returns to the homeport ahead of the battalion in order to set up a training program with the many agencies involved.

The S-2 is generally assisted by a permanently assigned staff of 3 or 4 petty officers

FIGURE 32
BATTALION TRAINING

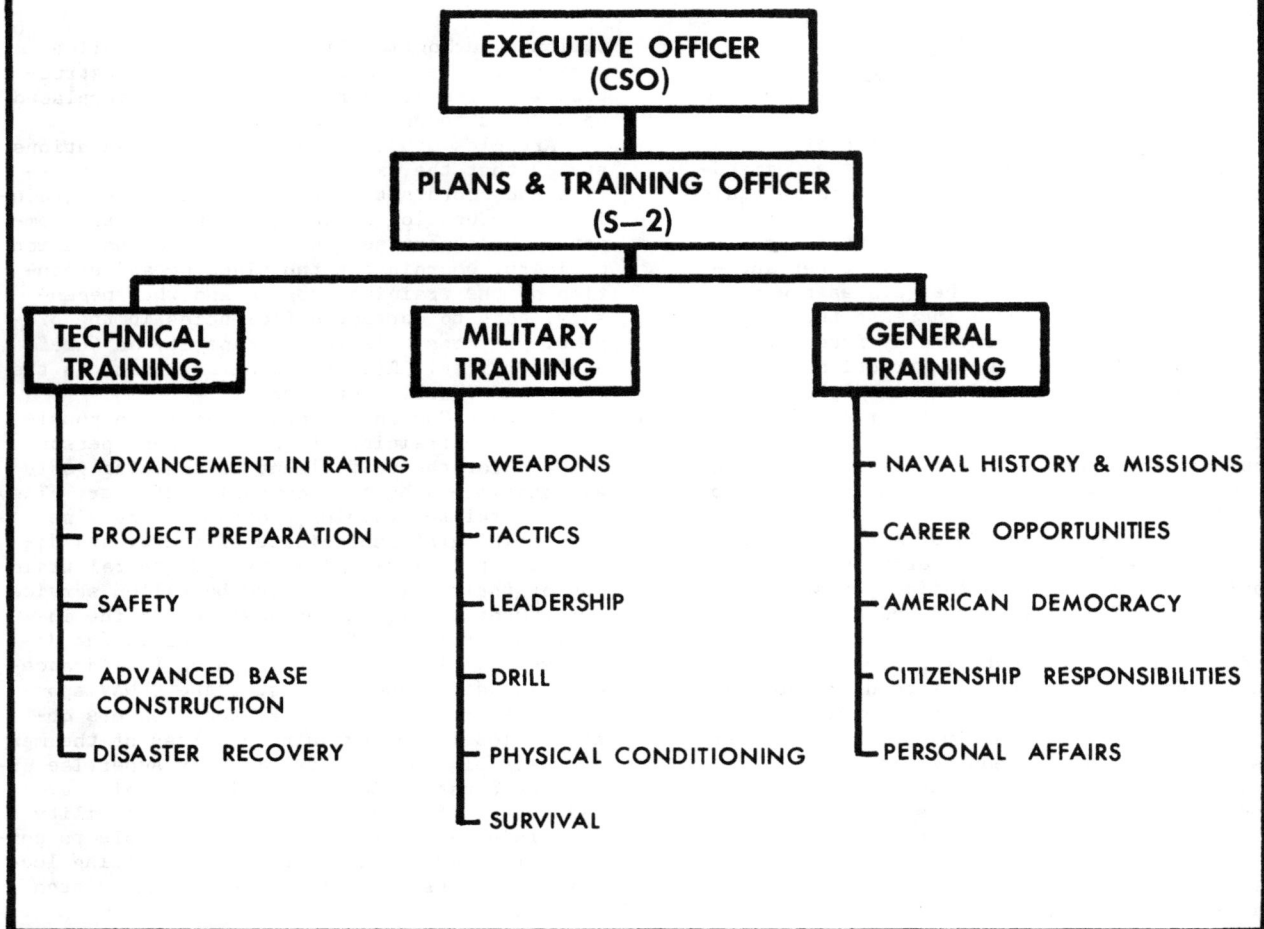

```
                    ┌──────────────────────┐
                    │  EXECUTIVE OFFICER   │
                    │       (CSO)          │
                    └──────────────────────┘
                                │
                    ┌──────────────────────┐
                    │ PLANS & TRAINING     │
                    │ OFFICER  (S—2)       │
                    └──────────────────────┘
            ┌───────────────┼───────────────┐
    ┌──────────────┐ ┌──────────────┐ ┌──────────────┐
    │  TECHNICAL   │ │  MILITARY    │ │  GENERAL     │
    │  TRAINING    │ │  TRAINING    │ │  TRAINING    │
    └──────────────┘ └──────────────┘ └──────────────┘
```

TECHNICAL TRAINING	MILITARY TRAINING	GENERAL TRAINING
— ADVANCEMENT IN RATING	— WEAPONS	— NAVAL HISTORY & MISSIONS
— PROJECT PREPARATION	— TACTICS	— CAREER OPPORTUNITIES
— SAFETY	— LEADERSHIP	— AMERICAN DEMOCRACY
— ADVANCED BASE CONSTRUCTION	— DRILL	— CITIZENSHIP RESPONSIBILITIES
— DISASTER RECOVERY	— PHYSICAL CONDITIONING	— PERSONAL AFFAIRS
	— SURVIVAL	

and by additional personnel on a parttime basis as directed by the battalion's formal training work load. This group is generally headed by a chief petty officer and its members often function as training instructors. Although responsible for the entire battalion training program, this group is primarily concerned with the formulation and administration of the formal military training program and that part of the technical training program which includes Advanced Base Construction and Disaster Recovery. The other aspects of technical training and the general training programs are formulated and administered within each company but should correspond to the general guidelines established by the S-2.

In the ACB, the Training Officer may serve as assistant to the Operations Officer. He arranges and schedules all formal training of officers and enlisted men, performing essentially the same duties as the S-2 of the MCB.

3. Duties of the S-2. The duties of the S-2 are to:

a. Maintain the battalion library of required contingency plans, intelligence information, maps and other publications related to possible wartime tasks.

b. Coordinate the preparation of supporting plans for combat support and disaster recovery operations.

c. Prepare OPORDS for field problems and exercises.

d. Gather information from training policies and directives of higher authority and make studies of the types of technical training most needed by battalion personnel for present and future projects and for wartime operations.

e. Prepare battalion-wide military and technical training programs, provide guidance to company commanders, observe training progress and recommend action to correct training deficiencies.

f. Keep informed on the availability of school quotas, training manuals, training aids, training films and other training facilities and request school quotas as directed.

g. Set up orientation programs for incoming officers and men.

h. Conduct training for battalion instructors and assist in the preparation of lesson plans and training aids.

i. Reserve classrooms, firing ranges and other space needed for scheduled training.

j. Disseminate pertinent training information.

k. Obtain and issue Navy Training Courses, examination blanks, training manuals, films, transparencies and other training aids.

l. Arrange for personnel to take enlisted correspondence courses.

m. Keep necessary training records and submit training reports.

n. Prepare Navy Training Course Certificates (NAVPERS Form 672) for necessary signatures, entry in service records and presentation by the Commanding Officer.

o. Prepare training programs for homeport deployment.

p. Serve on the Examination Board for Advancement in Rating and provide other testing services as requested.

q. Train and be prepared to direct the use of the Damage Survey Team (CD) and the Security Team (SA) in disaster control operations.

r. Collect, evaluate and display intelligence during combat operations or disaster control situations.

SECTION B- CONTINGENCY PLANNING

1. General. In the ACB, the duty of contingency planning falls under the Operations Officer. In the MCB, the Plans and Training Officer has this duty. The discussion presented in this section applies equally to the Plans and Training Officer in the MCB, and the Operations Officer in the ACB.

2. Planning Procedures. The object of planning is to save time and effort when responding to a crisis. It is important that at each level of an organization there is sufficient knowledge about duties, responsibilities and relationships with other parts of the organization to provide coordinated and rapid response to emergencies. While there are great differences in scope between the planning done by a fleet staff and a unit of the NCF, the need for definite planning is no less at the battalion level than at higher levels of command. Comments are frequently made that the preparation of definitive plans is a waste of time, since in virtually no operation do the actual operations carried out coincide precisely with the previously planned operations. However, performing detailed advanced planning provides a common means of briefing personnel of different organizations on the same operation, maintaining a continuity of thought despite change over in personnel, and by requiring supporting plans, ensuring that subordinate units have thought through their individual problems with respect to the proposed operation In addition, these detailed plans transpose operational concepts into actual material and personnel requirements when coupled with adequate intelligence information. Actual physical requirements may be spelled out so that action may be taken to provide the material and personnel to meet the planned requirement.

Operational planning is based on mission-type directives issued by higher authority followed by supporting operational plans prepared by the subordinate commanders. This planning goes through three phases; the estimate of the situation, the development of a plan and publishing a directive and the supervision and execution of the plan (See Figure 33).

Estimate of the Situation. An analysis of the mission includes the directive of the assigned mission, a statement of your own mission your superior's mission your relationship with other units in the operation and the objective of the enemy. Other factors must be considered for their effect on operations such as:

a. General situation and characteristics of the area of operations (which include political, economical, social and psychological factors along with terrain, topography, climate and weather).

b. Locations and distances.

c. Lines of transportation and supply.

d. Health and sanitation conditions.

e. Enemy capabilities, the alternates to your own course of action and the enemy's capability to disrupt any of your courses of action.

When all these things are considered, they are listed with the advantages and disadvantages and a final test for suitability, feasibility, and acceptability of each course of action is made. The selected course of action is then transformed into a formally stated decision.

Development of a Plan and Publishing a Directive. A directive is any communication that initiates or governs action, conduct or procedure. It must be clear, concise, complete, and it must be authoritative. That is, it must convey positively, the operational commander's intentions and will. The directive must be in the prescribed format, because complexities arising from unified and combined operations have led to the realization that some uniformity in directives is necessary. Common interpretation of instructions by all is a basic requirement for the success of any operation. Consequently, standard forms for directives have been agreed upon and have been promulgated. The standard five paragraph order is discussed in Annex D. The directive should state the assumptions upon which the plan is based whenever there is a lack of intelligence in any area. These assumptions should not be mere conjectures but should be based on the best information available. Forces are placed into a task organization with subordinate commanders designated and a chain of command established. The tasks are assigned to each unit and sufficient information is presented in the form of amplifying instructions for each unit to carry outs its assigned mission.

Supervision of the Operation. While the operation is in progress, the operational commander supervises the execution of his directive and modifies it as necessary to meet new developments in an ever-changing situation. He must not only recognize the necessity for a change in his plans, but he must recognize it in time to take effective action. Supervision during the operation may be necessary because of error or misunderstanding on the part of subordinates, or because of a change in the enemy situation or a change in the commanders own situation or mission.

3. Analysis of Plans from Higher Authority. Planning is a continuous process carried on at all levels of the Naval Construction Forces. In order to reduce to the minimum the time required to do planning at the various command echelons, the technique of concurrent planning should be utilized. This process allows direct liaison among subordinate units listed in the task organization of the OPLAN. Problems peculiar to these units are explained to all,

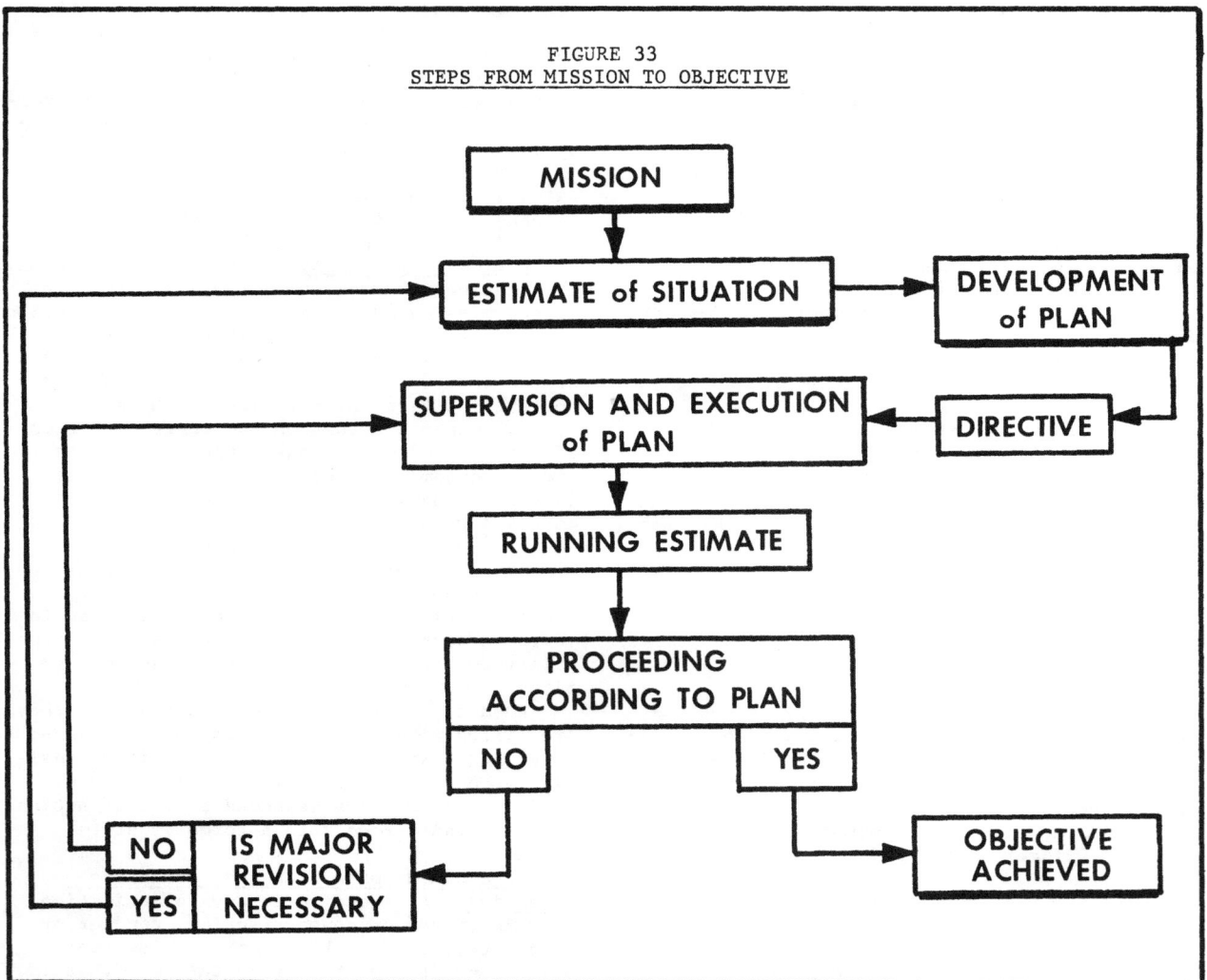

FIGURE 33
STEPS FROM MISSION TO OBJECTIVE

thus allowing mutual solutions to be reached when possible. When compromises must be made, all the concerned units are made aware of the effect such decisions have upon their operations and they can inform their superiors in the chain of command. Since many of the supporting plans may be executed by more than one NCF unit, the plans provide an invaluable source of guidance for all battalions.

The responsibility in planning as well as actual readiness (since the two are interdependent) must involve providing a unit fully capable of accomplishing the construction/military support mission summarized in Chapter I. Adherence to the following basic rules of readiness will assist the battalion to gain the flexibility to effectively deploy in support of any OPLAN within the required period. Planning normally is concerned with: (a) review and knowledge of the plans supported; (b) knowledge and effective utilization of the planning library and; (c) the preplanning necessary to enable the battalion to effectively and efficiently embark when directed and to accomplish its construction/military support mission upon arrival in the objective area.

In reviewing the OPLANS from higher authority, the S-2 must constantly try to extract engineer requirements and evaluate them with regard to the state of preparedness of his battalion. From these reviews the S-2 is able to realize what training is required to support wartime or contingency operations. In many cases the training for normal construction tasks does not prepare the battalion for all the tasks they can expect in wartime. It is here that the S-2 must ensure that the battalion has a training program to offset inexperience. For example; although the battalion may be required to harvest lumber for its own construction projects in wartime, lumber is supplied through normal supply channels in peacetime. Without maintaining the point of view of preparing for emergency or wartime deployments, the battalion might overlook the fact that some personnel should be trained in sawmill operations.

In many cases detailed planning for large operations does not reach the battalion level or the regimental level. This is true because the operating units cannot be burdened with vast amounts of planning. However, there is the inherent danger that essential items or functions will be ignored or inadequately provided for if some plans are not reviewed by lower echelons. In the case of the NCF units, it is not unusual for planners to overlook the need for consumables such as fuel, lumber and cement to support the SEABEE operation. In other cases unreasonable working conditions are encountered in the OPLAN because logistic planning factors, valid at division or corps level, were incorrectly applied at battalion or company level. Therefore the S-2 has the responsibility to inform the Commanding Officer not only of the training requirements generated by various OPLANS but also to point out any difficulties imposed upon the battalion by the plans of higher headquarters.

As the next step the S-2 should prepare written briefs of current operational planning and keep this information in his planning library along with facts concerning the organization and equipment of other units included in OPLANS that require SEABEE participation. This library should also contain Navy, Marine and Army publications on:

a. Construction in the combat zone, e.g., Engineer Field Data (FM 5-34) and Field Fortifications (FM 5-15)

b. Advanced Base Construction, e.g. Advanced Base Drawings (NAVFAC P-140) and Army Engineer Functional Component Table and Drawings (TM 5-301/TM 5-302/TM 5-303).

c. Embarkation Data, e.g., Ships Characteristics Loading Pamphlets for various class vessels and embarkation (FMFM 4-2).

A thorough documentation of the battalion allowance including how it is packaged, moved and loaded is also essential for a complete planning library. With these tools the S-2 may start his planning and layout the battalion training programs.

4. Readiness Requirements. The goal of planning and training in the battalion is full development of the teamwork and skills needed to maintain a state of operational readiness to provide rapid and effective construction support to Naval, Marine Corps and other forces as directed, to be prepared to conduct defense operations when required, and to be prepared to conduct disaster control operations. Therefore the S-2 is concerned with operational and material readiness. He must ensure that the battalion is trained in military and construction skills, and embarkation procedures. This would include the proper assignment of tasks to be performed by battalion personnel under each OPLAN supported and the knowledge by all hands of what these tasks entail. He must ensure that designated personnel have a working knowledge of all tools, equipment and supplies required for each OPLAN supported and that they understand the effort that is required to move the battalion equipage from one location to another. To gain this working knowledge of allowance items, each company commander and staff officer should maintain detailed cargo analysis of all equipment assigned to his custody. This should include a description of the item, its weight, cube, dimension, location state of maintenance, quantity and component parts. The embarkation officer can them summarize this information on the cargo and loading analysis sheets, tabulate weights, cubes, number of lifts and location by classes of supply. He can prepare embarkation plans for various classes of ships, aircraft and overland vehicles and determine time frames for assembly of allowance items needed in response to specific tasks. When this information is

available to the Commanding Officer, he can envision any emergency situation and direct his unit to respond. These exercises should be carried out round the clock to provide experience in 2 shift operations and should be made as realistic as possible in order that experience data can be compiled and can be used as a basis for refinement of the plans and for the preparation of future schedules. These exercises provide the optimum in training because they put classroom knowledge to work and test the theories and assumptions of the planners. By furnishing other battalions a copy of the "lessons learned" one NCF unit can make a significant contribution to overall SEABEE readiness.

The S-2 plans for air and surface mount-out involving the movement of battalion personnel, equipment and supplies to a forward area. Since actual embarkation depends upon many factors such as shipping availability, type of debarkation, scheduling and amount of overland movement, the battalion must be responsive to changes that occur on execution of the OPLAN. Firm shipping data cannot be provided in many cases. The required flexibility can only be accomplished through a detailed knowledge of the materials to be moved, effective mount-out procedures, discipline and proper pre-planning.

Upon receipt of the Operation Order, which normally contains the task assignment, the initial number of days of supplies to be carried and the shipping space allotted, the battalion officers can plan the movement in detail. The normal peacetime deployment schedule permits several weeks for accomplishment, but under emergency conditions only several days or perhaps hours are allowed. Therefore much of this work can be completed prior to the emergency, leaving only the details of a particular operation to complete the entire picture.

The battalions have therefore written standard embarkation plans for hypothetical conditions and have developed the concept of a Mount-Out Control Center (MOCC) to reduce the possibility of overlooking any details. The MOCC is designed to provide the battalion commander with up-to-date information on his unit's preparedness to mount-out and accomplish a contingency task. It is essentially a set of file folders which contain detailed check-off lists for staff and company officers. Further details of the MOCC are contained in Chapter VI.

SECTION C - TRAINING

1. _Program Formulation_. The training program of each battalion is formulated to provide the personnel on board with the skills needed for the accomplishment of the battalion's current and mobilization missions. The program is drafted by the S-2 in accordance with the pattern, priorities and tempo established by the Commanding Officer. The program covers many phases, from orientation courses to special technical courses. It depends upon such factors as operational commitments, policies and directives from higher authority, experience and previous training of the personnel and the training facilities available. Although much of the construction training has been provided by the Class A, B, and C construction schools, additional skill and experience must be acquired in the battalion. Therefore, it is essential that officers and petty officers be equipped to assume the responsibility of training the men on-the-job. The training program is built around the on-the-job construction effort supplemented by formal battalion courses of instruction and backed up by various special SEABEE training courses and BUPERS schools. However, it should be noted that although various activities are available to assist in the training of NCF units, the responsibility for training the command rests with the Commanding Officer and cannot be delegated. The training program emphasizes the supervised development of construction and military skills through orientation of newly assigned officers and men, operational readiness exercises, daily planned work assignments and special technical courses. The program also includes daily military formations and observances, periodic ground and disaster control exercises and regularly scheduled inspections. The battalion must provide competent instructors and adequate facilities support for the training program, but it should be arranged to take full advantage of local courses of instruction or specially-qualified instructors, available firing ranges, maneuver areas, confidence courses, etc., and military and commercial training films or other aids.

While on a normal peacetime deployment, the MCB is expected to spend approximately 50% of the available manpower on direct and indirect construction labor. This is on-the-job training for all Group VIII ratings. Another 25% of the available manpower should be directed toward formal training in all three major fields shown in Figure 32. Often when construction is impossible, for example during rainy weather, or when aboard ship, the slack time is used for training. These percentages must be adjusted when an MCB is employed on projects that have operational deadlines. Because of the nature of ACB operations, no percentages of on-the-job or off-the job training have been established.

On-the-job training plays a major role in the development of highly skilled SEABEES. Nothing is a suitable substitute for experience. Insofar as practical, the battalion should attempt to rotate personnel among the various common work items involved in normal construction operations so as to provide battalion personnel with the opportunity to broaden their individual capabilities. Since the overseas construction is subject to inspection and rigid scheduling, it makes sense to integrate scheduled technical training with on-the-job training. Preferably, the scheduled technical training should be given in field demonstrations of applied construction theory. Training of this type should be aimed at improving individual and crew proficiency and thereby obtain higher quality workmanship. Improved quality will have the net result of improved production since less corrective work will be required. When the desired level of proficiency has been reached, training should be directed at broadening individual and crew capabilities. Training for senior petty officers should include provisions for becoming familiar with the general aspects of the other construction ratings.

In the homeport the training programs become the primary mission. The MCB is expected to spend about 70% of the available mandays in formalized technical, military and general training. In addition the planning and estimating group may be considered to be involved with on-the-job training. Shortly before an MCB leaves a deployment site to return to the CBC, it sends an advance party to the Center to prepare the training schedule for the battalion's homeport stay. Although this advance party performs functions other than training that are necessary for the battalion's arrival, it is usually headed by the S-2. This officer must be fully cognizant of the individual leave schedule and training courses desired by the MCB with the names of personnel to attend each course. The advance party prepares the actual training schedule, arranges for instructors and material support when required and obtains quotas for schools. The S-2 is assisted by the regimental training officer at the homeport.

The ACB personnel are trained in essentially the same areas of technical, military and general topics. However, the program is tailored to meet the specialized mission of the ACB. It provides ACB personnel with the knowledge required by operational teams and crews in all phases of their primary missions, including seamanship, the installation and operation of causeway piers and fuel systems and beach salvage techniques.

2. General Orientation. Each battalion develops a comprehensive orientation program for new officers and men. Such orientation is usually part of or a continuation of the check-in procedure. Carefully planned orientation enables new officers and men to become full-fledged members of the battalion with the least delay. The orientation program may include:

    a. A word of welcome by the Commanding Officer or Executive Officer, and others.

    b. An orientation pamphlet.

    c. An explanation of battalion directives on leave, liberty, camp regulations, daily routine and other pertinent matters.

    d. Familiarization with base and battalion facilities.

    e. An explanation of the role of the battalion in naval operations, the principles of the construction/military organization and the battalion's present assignment.

    f. A review of the battalion's history. On-the-job orientation is usually planned and conducted by the newcomer's crew leader and project petty officer.

In some cases the battalion Commanding Officer may desire to run a more extensive orientation program for non-rated men. This might include the firing of the rifle for qualification, instruction in disaster recovery operations and basic military training in tactics. The program may last several weeks. When it does, the trainees are usually formed into an "X" division and are berthed together until graduation and assignment to one of the companies.

Another type of orientation is deployment orientation. This is given to the entire battalion just prior to leaving homeport. When possible, the personnel of the battalion are brought together for the last week before embarkation. Training has been completed, special details are secured and no one is on leave. During this week called, Battalion Week the men are briefed:

    a. By the Executive Officer on:
        (1) The tasks assigned in the OPORD.
        (2) Instructions and notices published for the deployment.
        (3) Daily routine and the watch structure.

    b. By the S-2 on:
        (1) History and customs of the area to which deployed.
        (2) Attitude of the people and social conventions.

    c. By the S-3 on:
        (1) Projects for the next deployment.
        (2) Construction organization and crew leaders.
        (3) Construction schedule.

    d. By the Chaplain on:
        (1) Behavior in foreign countries.
        (2) Religious facilities available.

    e. By the Medical Officer on:
        (1) Personal Hygiene.
        (2) First Aid.
        (3) Health hazards particular to deployment site.

    f. By the Legal Officer on:
        (1) Status of Forces Agreement.
        (2) Local Laws.
        (3) Local base instructions.

    g. By the Special Services Officer on:
        (1) Recreational facilities available at the overseas location.
        (2) New battalion gear or programs.

    h. By the Embarkation Officer on:
        (1) Schedule for Embarkation Day.
        (2) Routine while enroute.

3. Officer Training. The Navy recognizes that the wartime capability of the Naval Construction Forces hinges on the continuous, well-rounded development of SEABEE Officers.

Such development requires careful rotation of assignments plus intensive supplementary training in construction, combat support and related background subjects. Officers learn what is expected of the battalion in the readiness and mobilization concepts and how each part of the construction battalion functions. They become expert in the performance of their current construction and military assignments and staff duties, acquiring the general and technical knowledge needed for future assignments. Battalion officers learn how to mobilize civilian engineers and journeymen into organized effective construction and combat teams to meet any emergency. The individual and group training of the battalion officers is supervised by the Executive Officer.

Officers enroute or assigned to a battalion are frequently placed on TAD to attend regular or short courses at Navy Schools depending upon funds, current quotas and needs. The Civil Engineer Corps Officers School conducts a two-week course in Construction Battalion Operations designed specifically to acquaint company officers with SEABEE functions. Other courses in Naval Justice, Military Justice, NBC Warfare Defense, Embarkation, Basic Indoctrination in Chemical and Biological Warfare and Military Training are designed to aid officers in their assignments and prepare them for specialized staff duties. These programs range from several days to eight weeks in length.

While in the battalion, all officers should actively participate in a correspondence course program to broaden their professional background. Recommended executive, technical and operations courses are given in BUPERS Instructions of the 1500.49 series.

As a supplement to formal training and correspondence courses, the Commanding Officer may designate some time each week for officer seminars. Supervised by the Executive Officer and designed for group participation, these meetings are especially effective for the presentation of such subjects as:

    a. Construction-Combat Operations.
    b. Advanced Base Functional Components.
    c. Planning for Advanced Base Construction.
    d. Battalion Camp Administration.
    e. Engineering Reconnaissance.
    f. Construction Operations.
    g. Staff Functions and Relations with Companies.
    h. Base, Job and Camp Defense and Security.
    i. Disaster Recovery Operations.
    j. Battalion Policies.
    k. Enlisted Personnel Programs.
    l. Mount-Out Procedures.
    m. Instruction Techniques and Company Training Programs.
    n. Military Justice and Battalion Discipline.
    o. Safety.
    p. Duties of the OOD and the Battalion Watch.
    q. Effective Reading and Writing.
    r. Officer Responsibilities.
    s. Command Ability and the Art of Delegation.

It is important for each officer of the battalion to read all the selective material necessary to keep informed of development affecting his primary and collateral duties. He must keep up-to-date in order to prepare for projected work and training assignments. The Executive Officer may prescribe individual or group assignments on special subjects or selections from the Guide for Professional Reading for Officers of the Navy and Marine Corps. Journals such as the Engineering News-Record and Construction Equipment and Methods offer information of interest to construction battalion personnel. As each officer will have a heavy reading schedule, he should strive to develop efficient reading habits.

Each CEC officer is encouraged to maintain an "engineer's notebook" containing information such as data on materials being used, observations on methods and results, data on productivity of equipment and crews, design data, calculations, charts and tables, notes on difficulties encountered and solutions employed, job defense plans and check-off lists. This cumulative documentation becomes a personal notebook of specific information which will become increasingly valuable in deriving benefit from past experience.

4. Formal Schooling. During his career every SEABEE has ample opportunity to attend some formal classes of instruction. The requirements to keep trained men available for many different task assignments means that the battalion must rely heavily on formal schooling. Company officers nominate personnel to fill quotas obtained by the S-2. Nomination must be based upon the capability of the individual, his interest and his ability to use the knowledge gained for the good of the battalion and the Navy. School quotas should be granted to those who have worked hard to improve themselves and who are the most difficult to lose. Effective use of quotas can help the company commander increase the efficiency of his unit.

The U.S. Naval Schools, Construction (NAVSCOLCONST), Port Hueneme, California, is the principal school for technical training in construction. Such training also takes place at the Construction Training Unit, Davisville, Rhode Island. Most men sent to Class A Schools at NAVSCOLCONST come directly from a recruit training command (boot camp). However, the battalions are allotted a quota and can send qualified candidates to Class A and Class B schools whenever funds and quotas permit. An average NAVSCOLCONST Class A or B course lasts for 12 or 13 weeks. NAVSCOLCONST Class A courses are not intended to turn out finished journeymen. Their purpose is to teach the fundamentals of a rating so that practical aspects can be rapidly mastered on the job. They provide technical knowledge and skills needed to prepare for the lower petty officer rates.

Class B Schools are designed to cover technical and professional aspects at the supervisory level and prepare men for advancement to first class and chief petty officer. The Class C Schools are shorter, usually ranging from 4 to 8 weeks. They provide skills or techniques which are not peculiar to any one rating. They include blasting, automatic transmission, welding certification and planning and estimating. A graduate of Class C School is usually awarded an NEC. Quotas are allotted to the battalions upon request. Successful completion of a prescribed course at these schools usually satisfies the requirement for completion of the prescribed Navy Training Course, although it does not in itself assure advancement in rating. A list of schools for each rating appears in the Catalog of U.S. Naval Training Activities and Courses, NAVPERS 91769.

Special SEABEE Training Courses are usually available while the battalion is at its homeport or at a large base overseas. Subjects taught include damage control, railroad maintenance, welding and NBC warfare defense. Such courses usually last from a few days to two weeks. Quotas are established through the local training office. The S-2 arranges for personnel to attend. These courses are extremely important to the battalion training program because of the number of personnel who may attend. In addition the battalion instructors are trained by acting as assistant instructors.

Other schools available to Navy men are those of another service or those of industry. The Army School, located at the Army Signal Corps Laboratory, Fort Monmouth, New Jersey, is an example of a service school where Navy construction electricians and other personnel who work with telephone equipment may receive valuable training. The Infantry Training Regiment at Camp Lejuene, North Carolina and at Camp Pendleton, California provide advanced infantry training and weapons training for all battalion personnel. In addition several manufacturers run specialized training courses on the operation and maintenance of their equipment. These schools are intended primarily for dealers and company service personnel, but upon request, are often made available to the government.

5. Technical Training. Technical Training consists of advancement in rating, specialized skill training for a particular project, construction safety, contingency or advanced base construction and disaster recovery.

Advancement Training. Every enlisted man in the battalion is in training for advancement in rating. This training is supervised by battalion officers and petty officers in accordance with the BUPERS Manual and directives. A man is eligible only after he has served the prescribed time in pay grade, has enough total naval service and has fulfilled any sea duty requirements. He must also have completed satisfactorily Navy Training Courses and be fully qualified in practical factors. and examination subjects as prescribed in the Manual of Qualifications for Advancement in Rating, NAVPERS 18068. In addition, he must be considered capable of performing the duties of the higher rate by senior petty officers and have been recommended by the Commanding Officer. The promotion steps are described in Chapter III.

Navy Training Courses (NTCs) present information needed by enlisted men to acquire all-around military proficiency and to prepare for advancement in rating. These courses are issued by the Bureau of Naval Personnel with the technical assistance of interested commands. NTCs may be used with a minimum of supervision and help. However, they should be interrelated as closely as feasible with work assignments and other training. Officers and petty officers must be thoroughly familiar with the scope of the NTCs used by their men in order to assist them. There are three types of courses:

a. The Basic Courses serve as the introduction to a program of self-study in a number of ratings since they cover fundamental skills and knowledge common to these ratings. Among the basic courses are Basic Hand Tools, NAVPERS 10085, Basic Machines, NAVPERS 10624, Basic Electricity, NAVPERS 10086, Mathematics, NAVPERS 10069, 10071 and 10073, and Blueprint Reading and sketching, NAVPERS 10077.

b. The Rating Courses help enlisted men acquire the knowledge needed to advance to the next higher rate and a series of rating courses have been prepared for each rating. For example, the Builder series includes Constructionman, NAVPERS 10630, Builder 3 and 2, NAVPERS 10648, and Builder 1 and C, NAVPERS 10649.

c. The Military Courses must also be completed to qualify a man for promotion. Basic Military Requirements, NAVPERS 10054, is one such course.

Because of the tempo of battalion operations and the requirements for other training, these courses are usually completed by battalion personnel as correspondence courses. Applications for these courses are processed through the company commander and the S-2 to the Naval Correspondence Course Center, Scotia, New York. However, the battalion may decide to conduct the courses in groups. In this case, Navy Training Courses are kept by the S-2 until issued to the company commander or department head. Each time a man has successfully completed a Navy Training Course, the company commander notifies the S-2 who prepares a Navy Training Course Certificate (NAVPERS Form 672) for the signature of the Commanding Officer. An entry is then made in the man's service record by the S-1. Normally, prescribed Navy Training Courses must be satisfactorily completed before a man is eligible for advancement in rating. However, this requirement may be satisfied by the successful completion of an equivalent course at a Navy

School.

Practical factors are a series of manual tests where a man physically demonstrates that he can perform a certain operation. These factors are listed on a NAVPERS 760 for each rating. When a man demonstrates his ability to complete a factor, the company commander places his initial and the date beside that factor. Normally a senior petty officer designated by the company commander performs the actual checkout.

Project Training. An analysis of labor requirements for a forthcoming deployment might reveal that the battalion is short on some needed skills. For example, a large airfield may involve much earthmoving. Thus the S-2 may set up a program to train extra men as equipment and vehicle operators. In such cases, it may be necessary to train builders or others for temporary service as dump truck drivers.

Other jobs may call for intensive training in certain skills within rating. For example, if a battalion must develop its own water supply on its next deployment, a program may be needed to train certain equipment operators in well-drilling techniques, certain utilitiesmen in water purification and distribution, and so on. The next development may involve unusual climatic conditions, such as an arctic operation. In this case all hands are indoctrinated in the special hazards of the area and techniques for coping with them. Special

training in all ratings may be needed to learn the methods of construction peculiar to that region.

Whenever an uncommon work item of sufficient magnitude is encountered during a deployment it should be accomplished in a manner which will allow the majority of career personnel of the appropriate rate and rating to gain experience in the operation. For example, if a battalion is engaged in pontoon assembly type work, all applicable rate/ratings should be given the opportunity to participate in the project for sufficient time, if possible, to gain a basic knowledge of and minimum proficiency in the operations. If for valid reasons it is not possible to rotate personnel through an unusual operation, it is a good idea to make the operation a subject for formal technical training.

Safety Training. The technical training program should vigorously stress safety and provide personnel with a solid background in practical safety techniques plus an awareness of the inherent dangers of construction work and the proper procedures to combat these dangers. Battalions should attempt to enroll the maximum number of crew leader petty officers as possible in Industrial and Construction Safety Courses conducted by Navy Schools, CBC Industrial Relations Division or industrial firms. A Safe Driving Training Program along with many other aspects of safety training is conducted at homeport and on deployment.

FIGURE 34
EXPEDIENT CONSTRUCTION

When the training deployment does not include the type of construction that would be required in a wartime situation, instruction and practice on temporary structures should be included in the technical training.

Further details can be found under the Safety Officer in Chapter II.

Advanced Base Construction. The construction projects assigned to the battalions in peacetime are usually of permanent and semi-permanent type construction. This is true because many of the projects are facilities required as part of the Navy's overseas bases and temporary construction poses too much of a maintenance problem. While this provides excellent on-the-job training for the battalion personnel, certain tasks that are related only to the construction of temporary facilities are not included. Thus a program to supplement the on-the-job training has to be conducted in the battalion schools to bridge the gap between peacetime construction tasks and emergency or wartime construction tasks. Each battalion must train to maintain the capability to:

    a. Tactically load amphibious ships and cargo aircraft.

    b. Install and maintain the SATS complex.

    c. Install and repair railroad track.

    d. Construct timber bridges.

    e. Repair and expand waterfront structures.

    f. Perform reconnaissance missions.

Additionally men must be trained to assemble, operate and maintain:

    a. Aircomatic welding machines.

    b. Portable well drilling rigs.

    c. Portable rock crushing machinery.

    d. Portable sawmills.

    e. Camp component items.

    f. Functional components.

This training is directed toward developing a hard core of career petty officers knowledgeable in advanced base construction, materials and techniques.

Therefore, each battalion will send some officers and chief petty officers to Embarkation Courses conducted by the U.S. Naval Amphibious Training Schools or the Marine Corps. Other courses such as Railroad Maintenance and Repair, Quarry Operations, Logging and Sawmill Operations, Functional Components, etc., have been developed by NAVSCOLCONST as fleet training courses. In addition to this training the battalion conducts classroom instruction and field application in such areas as: erection and camouflage of tent camps; construction of functional components such as aviation tank farms, air base magazines, drum filling plants, etc, engineering data collection and the construction of road blocks, tank traps and improved defensive positions. Many of these tasks are included in battalion-wide field problems and in exercises conducted with other fleet units in order to provide practical experience to battalion petty officers.

Disaster Recovery. Included within the mission of the construction battalions is the task of assisting in recovery from the effects of natural disasters and from nuclear, biological and chemical warfare actions. To provide personnel with the training necessary to per-

mit the battalions to perform this mission, a Disaster Recovery Training Program has been organized at the Construction Battalion Centers. Its contents are balanced between individuals, classroom and team training and culminates in the full scale, integrated recovery of an advanced base from the effects of a realistically simulated nuclear, biological and chemical attack. Maximum use of this program is urged in order to upgrade and maintain battalion base recovery potential. However, due to the length of the Disaster Recovery Training Course and the limited number of qualified instructors available, it is impossible to provide all battalion personnel with the required training during a single homeport deployment. It is good practice, therefore, that the limited number of personnel who are selected to attend during each CONUS deployment be carefully selected in order that they may serve as instructors and provide training to other personnel while the battalion is deployed away from the homeport.

In order to ensure the maximum recovery of the battalion personnel and facilities and their subsequent availability to higher authority for further assignment, the entire battalion should be trained in:

    a. Characteristics and effects of NBC Warfare.

    b. General survival information.

    c. The use of protective masks, protective clothing, vessicant protective ointment, atropine syrettes, and radiation dosimeters.

    d. First aid.

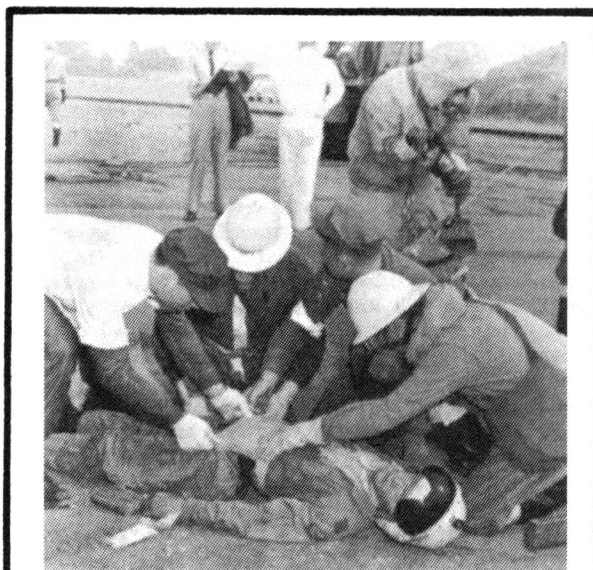

FIGURE 35
DISASTER RECOVERY TRAINING

Disaster recovery training includes realistic field training to acquaint battalion personnel with some of the problems they must overcome in emergency situations.

e. Fire fighting.

In addition, specialized training is required by:

a. The entire NBC element or company.

b. The entire control element.

c. Construction company personnel in principles and procedures involved in decontamination.

d. Rescue teams.

e. UT and CE supervisors of construction company personnel.

f. Staff officers in special problems created by the disaster, e.g., casualty treatment.

Classroom instruction of the various teams is tied together by conducting periodic battalion exercises in emergency recovery. The exercise is as realistic as possible and can be tailored to include the peculiarities of the deployment area. The disaster assumptions picked for the exercises can vary to include a "limited war" situation where the enemy is using tactical nuclear, biological, and chemical weapons to an accidental release of any one of the three in friendly territory.

6. **Military Training.** The Naval Construction Forces must provide a military unit which is capable of defending itself while carrying out the primary construction support mission either as a unit or in combination with other forces. It must also be capable of integrating with other forces for the defense of advanced Naval bases. The military training programs for the Naval Construction Force is aimed at maintaining the necessary ability to perform this mission.

Military Training will be conducted at the homeport and at Marine Corps facilities (when possible) where emphasis will be placed on acquiring basic military skills by all battalion personnel and at the deployment site where practical application designed to retain and to re-enforce the fundamentals learned should be emphasized. All battalion personnel must participate in training since all personnel will be performing some function in support of the battalion during combat. For example, the battalion cooks must learn to prepare meals on field ranges and act as reserve riflemen. In order that medical personnel (including dental personnel) may effectively support their unit when engaged in a military operation, it is vital that they receive certain minimum training with that unit. This training is not intended to circumvent their non-combatant status nor is it contrary to the principles of the Geneva Convention for the amelioration of the wounded and sick in armed forces in the field, 12 August 1949.

Battalion programs are backed by military training conducted at the homeport by the military training group of COMCBLANT/COMCBPAC and by periodic deployments to the Infantry Training Regiments at Camp LeJeune, North Carolina, and Camp Pendleton, California. The Marines have provided training periods from two to six

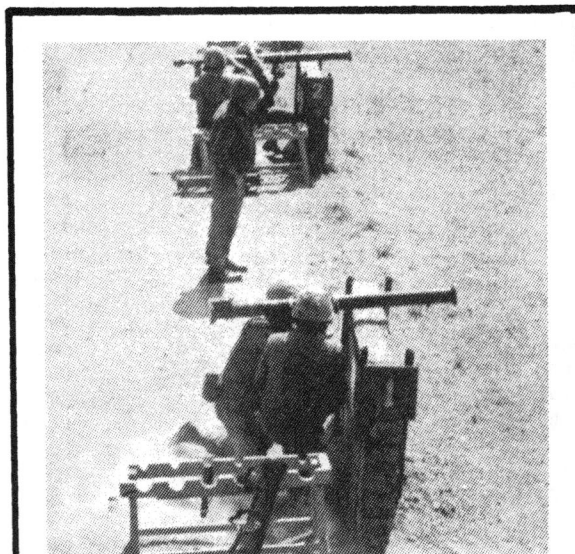

FIGURE 36
WEAPONS TRAINING

The military training program includes instruction on individual and crew served weapons. The weapons are live fired during deployments to Camp LeJeune and Camp Pendleton and when suitable ranges are available overseas.

weeks for the battalion in such topics as:

Individual combat training
Advanced combat training
Qualification firing
Field Medical School
Small arms repair
Field messing and cooking school
Demolitions
Communications
Instruction orientation

Since increased military training at the sacrifice of construction effort and technical training is not possible, personnel with technical skills developed for the construction mission are used in the tactical organization where their skills are required. For example, the CEs handle field communications and EAs are assigned fire control duties.

The military training of each battalion encompasses weapons training, tactics training, drill, physical conditioning and survival training.

Weapons Training. Weapons training should be given to all battalion personnel to enable the individual to be completely confident in his ability to use his assigned weapon. This training should include:

a. Care and cleaning of the weapon.

b. Field stripping.

c. Safety precautions

d. Combat firing positions

e. Sighting and aiming procedures

f. Immediate action procedure

g. Nomenclature of the parts of the weapon.

h. Characteristics of the weapon including weight, range, etc.

i. Fire command response

The crews of crew-served weapons should be able to conduct gun drill. Unit leaders must be able to issue fire commands and effectively employ individual and crew-served weapons.

Tactics Training. Instruction in basic tactics and survival in combat should be given to all, stressing defensive tactics to enable small units up to and including battalion size to defend specific areas of operation. Tactical training should include:

a. The principles and general information concerning camouflage of individual vehicles and field fortifications.

b. A basic knowledge of map reading including map orientation, grid coordinates, military signs and symbols, map contours and marginal data.

c. The proper field hygiene and sanitation practices necessary to keep healthy in the field while camped or on the march.

d. The use of patrols, how to issue a patrol order and make a patrol report.

e. Forming and controlling a tactical column and the duties of each element.

f. The principles of defensive combat including:

(1) How to move into a hasty defense and dig a proper fighting hole.

(2) How to use a password.

(3) How to establish sectors of fire and principle directions of fire.

(4) Difference between primary, supplementary and alternate positions.

(5) Reason and definition of fire discipline.

(6) Types of obstacles and mine fields.

g. Some principles of offensive combat which include the control and coordination of fire team and squad movements, and an understanding of fire superiority.

Leadership Training. The training program should stress small unit tactics (company and below) as well as military staff functions. Squad leaders and even fire team leaders must exercise command and control on the battlefield that is normally not delegated to them in other less chaotic situations. Therefore they must be prepared through careful training in the lower rates. Provisions for the training of officers and chief petty officers should also be included because lack of understanding at the level exercising command over these larger units of the battalion can have disastrous results in a combat or emergency situation. Instruction should include:

a. Defensive and offensive tactics at the platoon, company and battalion level.

b. The six troop leading steps.

c. The five paragraph order.

d. Weapons employment and coordinated fire planning.

e. Terrain analysis and organization of the ground.

f. Defensive fortification and barrier planning.

Drill. Close order drill should be administered regularly and for short periods whenever possible. This type of training is performed not only to ensure that the men drill properly but as a valuable disciplinary training aid. It will instill individual habits of precision and response to a leader's orders, provide practice for petty officers and officers in commanding troops and enhance unit morale by developing pride in its ability and appearance. The Landing Party Manual contains the procedures to be followed when drilling or participating in ceremonies.

Physical Conditioning. Physical conditioning programs should be conducted in conjunction with regular training by having specified periods of calisthenics, obstacle course training and organized athletics for all personnel. These periods need not be lengthy but will be most effective if they are of a short duration and regularly scheduled.

Survival Evasion Resistance and Escape Training. Normally only SEABEE Teams and special detachments are given intensive training in Survival, Evasion, Resistance to Interrogation and Escape (SERE). However, all battalion personnel are given orientation training in SERE to acquaint them with the problems and to assist them to understand the principles contained in the Code of Conduct. This training may include lectures, demonstrations and field problems in survival techniques, first aid, survival psychology, improvised clothing, shelters and equipment, procurement of water, procurement and preparation of food, interrogation techniques, resistance methods, POW compound organizations, medical aspects of POW survival, escape techniques and evasion procedures.

These courses are conducted at the Marine Corps Mountain Welfare Training Center, Bridgeport, California, and by the Landing Force Training Units of the Amphibious Training Commands at Little Creek, Virginia, and San Diego, California.

7. General Training. There are in effect within the Navy numerous programs which have as a common objective the preparing of Naval personnel to fulfill the objectives of their oath of Naval service and the providing of guidance and information in matters which affect Naval personnel as individual citizens and as members of a military organization in the service of their country. These programs have been gathered under the title, "General Military Training" and are listed in OPNAV INST. 1500.22. In order to keep these OPNAV

programs separate from military training as commonly understood among SEABEES, these programs will be called General Training in this manual.

The information phase of the General Training program assists Naval personnel in furthering their understanding of the principles of American democracy and of the importance and responsibility of the individual citizen in American life. It also aims to increase the men's comprehension of the individual and collective missions of the Armed Forces and to familiarize them with current issues in national and international affairs. The individual's responsibility in promoting international goodwill is emphasized. The information program may include lectures, discussion of news highlights, public address system broadcasts of news and special transcriptions, displays of posters and photographs indicating the location of important current events and news bulletins. Written material used may include publications issued by Naval activities, the Department of Defense, and other government agencies, as well as locally prepared news summaries, short articles and condensations of longer articles.

The educational phase of the General Training programs is designed to help all hands prepare for career advancement, continue education begun before entering the Navy and perform their duties more efficiently. The core of the voluntary education program for military personnel is the United States Armed Forces Institute (USAFI), a correspondence school and educational supply agency. USAFI furnishes materials for counseling and class programs. These materials, issued on a loan basis, include foreign language material (including records and texts), self-teaching texts with study guides, instructor's course outlines and comprehensive examinations for the measurement of general educational maturity. In addition, it furnishes reports to educational institutions and prospective employers on the scores made on courses and tests. USAFI offers:

a. Correspondence courses in high school, technical and college subjects.

b. Group study classes with instructors.

c. General educational development tests (GEDs) which are designed to enable men to receive credit for education acquired informally. Men who take the tests designated for high school level and make a specified score are considered for all military purposes to have the equivalent of a high school education. The Navy accepts this certificate of completion as equivalent to the corresponding high school course.

The general training for SEABEE units can be divided into the five phases described below:

Naval History, Customs, Traditions, Roles and Missions. It is the policy of the U.S. Navy that all Naval personnel and their dependents will be presented the information needed to gain an understanding of the historical role of our Navy in support of national policy and security, an appreciation of Naval customs and traditions and a knowledge of the current roles and missions of the Navy. Therefore, this phase is designed to develop within Naval personnel an understanding of:

a. The purpose and worth of the Navy and its way of life.

b. The importance of the individual and his unit to the Navy.

c. The unique role of the Navy in support of foreign policy.

d. Naval customs, courtesies, and traditions, their role in the modern Navy.

Naval Career Opportunities. The Navy tries to ensure that its officers and men be aware of the opportunities, responsibilities, intangible satisfactions, and tangible benefits to be accrued from a career as a member of the honored Naval profession. This phase is concerned with developing within each individual, an identification with, and an understanding of, "The Naval Profession." It should deal with the relationship of one's oath of office, code of ethics and the standards of personal and technical qualifications as related to the Naval profession and a Naval career. Thus this phase should include:

a. Meaning and qualifications of a Naval profession and its ethical, mental, physical, and technical standards.

b. Qualifications and requirements for promotion and advancement in one's individual career.

c. Educational opportunities available to the Naval careerist.

d. The benefits to be derived from honorable service.

American Democracy and Hostile Forces. This subject area is designed to broaden an understanding of the foundations of our democratic form of government and the American way of life and provide personnel with the necessary information to understand and recognize those nations or forces which threaten the security of the United States of America. This can be done by training Naval personnel to:

a. Comprehend the historical and constitutional foundations of the U.S. Government.

b. Understand current issues in national and international affairs.

c. Realize the importance of the individual in the American way of life and the responsibilities and obligations of a military citizen.

d. Recognize the individual's responsibility in promoting international goodwill and understanding.

e. Understand the nature and policies of international communism and other hostile forces.

f. Implement policies in the conduct of the Cold War.

Naval Citizenship Responsibilities. This phase is concerned with informing Naval personnel of their individual duties and responsibilities as United States citizens and members of the Naval service, both at home and overseas. It is directed toward instilling an individual awareness of the importance of personal conduct in affecting public attitudes towards the United States and the U.S. Navy. It is also directed toward motivating individuals to participate in federal, state, and local elections, authorized fund drives, people-to-people projects and community activities. This indoctrination consists of:

a. Instruction in a program designed to develop respect for the customs, culture and attitudes of the inhabitants of foreign countries in which personnel are serving through formal indoctrination and encouragement of direct person-to-person contacts with local citizens.

b. Information concerning the nation's policy that there shall be equality of treatment and opportunity of all persons in the Navy and Marine Corps without regard to race, color, religion or national origin.

c. Guidance regarding responsible operation of private vehicles and observance of civilian laws and ordinances.

Personnel Affairs. This phase of the program is designed to instruct personnel and their dependents on Navy policies to promote habits of thrift and to encourage all members of the Naval service to conduct their financial affairs in such a manner as to reflect credit upon the service, and to keep Naval personnel and their dependents informed of their rights and benefits. This training includes explaining:

a. U.S. Savings Bond Program.

b. Life and Liability Insurance.

c. Navy policy on indebtedness.

d. Navy policy regarding support of dependents.

e. Medicare.

f. Retired Serviceman's Family Protection Plan.

g. Service from the Navy Relief Society.

h. Veteran's Rights and Benefits.

i. Survivors, Retirement and Disability Benefits.

Chapter V
BATTALION OPERATIONS

SECTION A - OPERATIONS OFFICER

1. **Introduction to the S-3.** The Operations Officer is usually the third-ranking CEC Officer of the battalion and serves as the S-3 on the battalion staff. He is responsible for coordinating the activities of elements of the battalion when they are engaged in performing task assignments under any type of mission be it amphibious, construction, defense or disaster control. These responsibilities include recommending the task organization, establishing priorities for allocation of resources, adjusting assignments to battalion units and keeping the Commanding Officer informed of the situation as it develops.

Military action, changes in the combat situation, natural disasters, delays in shipping, or loss of men, equipment, or materials may drastically affect battalion operations, and in wartime, such changes may prove to be the rule rather than the exception. Thus, the battalion must be ready to regroup its resources and improvise plans in response to any emergency. In such situations commanders rely heavily on SEABEE ingenuity and resourcefulness at every level of command.

In the MCB the construction tasks vary widely in type and scope. Often projects are not defined until after the battalion has arrived on site and priority of effort may be changed by the commander exercising operational control. In the ACB the problem of scope definition is not as acute because of the specialized nature of the construction, but the high degree of timing and proper sequence of events are critical operational problems.

The functions of the S-3 involve engineering, inspection, testing, planning, estimating and safety. The department is staffed with engineering aids, photographers and similar personnel, in addition to experienced field construction men. To meet the demands of variable workloads, individual assignments in the operations department depend upon qualifications rather than rating, allowing interchangeability of personnel among sections.

2. **Duties of the S-3.** In furthering the mission of the battalion, the S-3:

a. Performs liaison on task assignments with the command having operational control over the battalion and may assist higher commands in planning the work to be performed by the battalion.

b. Studies each Operation Plan or Order directed to the battalion for execution and analyzes the arrangements for logistic support requesting changes where necessary.

c. Determines status of drawings, specifications, area intelligence, etc. influencing each assignment and directs design and site adaptation for projects assigned to the battalion.

d. Determines specific camp facilities needed by the battalion for the deployment and either arranges for accommodation of the battalion at existing bases or plans for the construction or improvement of the battalion camp.

e. Instructs battalion reconnaissance parties and advance parties, setting priorities for information to be gathered or indicating the work to be done before the main body arrives.

f. Supervises surveys, site adaptations, quantity takeoffs, bills of materials, labor and equipment estimates, shop drawings and other engineering work assigned to the battalion.

g. Estimates workload on each rating and class of equipment recommending the distribution of personnel throughout the battalion and requesting extra manpower and equipment as needed.

h. Selects construction methods and issues job orders for all construction work to be performed by the battalion, keeping close check on the execution of these job orders from the standpoint of deadlines, manhours and machine-hour expenditures, safety and quality of workmanship.

i. Reassigns work, reschedules work, or otherwise revises battalion job orders, and advises the Commanding Officer on the effect of changes in any job orders.

j. Coordinates the work of the battalion and arranges for the employment of local labor and use of local materials when approved.

k. Supervises compilation of Manpower and Equipment Utilization Records and drafts Monthly Operations Reports, Project Completion Reports and other reports on operations for approval of the Commanding Officer.

l. Coordinates the use of the teams in the engineer and transportation elements in disaster control operations.

m. Assists the S-2 to devise the overall battalion training plan by providing guidance as to operational requirements and deficiencies.

While the general duties of the S-3 in the ACB and the MCB are similar, they differ in detail and in approach because the ACB has surface, underwater and beach assignments in contrast to the shore construction assignments of the MCB, and therefore will be treated separately in the following sections of this chapter.

A typical breakdown of the functions of the MCB Operations Department is shown in Figure 37.

3. The Engineering Officer. Directly under the S-3 comes the Engineering Officer who, in his capacity as assistant to the S-3, supervises all work of the battalion engineering office. This arrangement frees the S-3 from the office during a deployment and allows him to oversee the activities in the field. In carrying out his duties the Engineering Officer supervises the work of the planning and estimating, resource control, technical support, records, photography, and safety sections. He supervises the preparation, modification and any required reproduction of drawings, sketches, specifications, quantity take-offs, bills of material and manday estimates. He assists the Operations Officer in planning the camp, inspecting the construction progress and measuring the work-in-place. He directs topographic surveys, soil and material tests and supervises the compilation of progress and performance records and other reports as required.

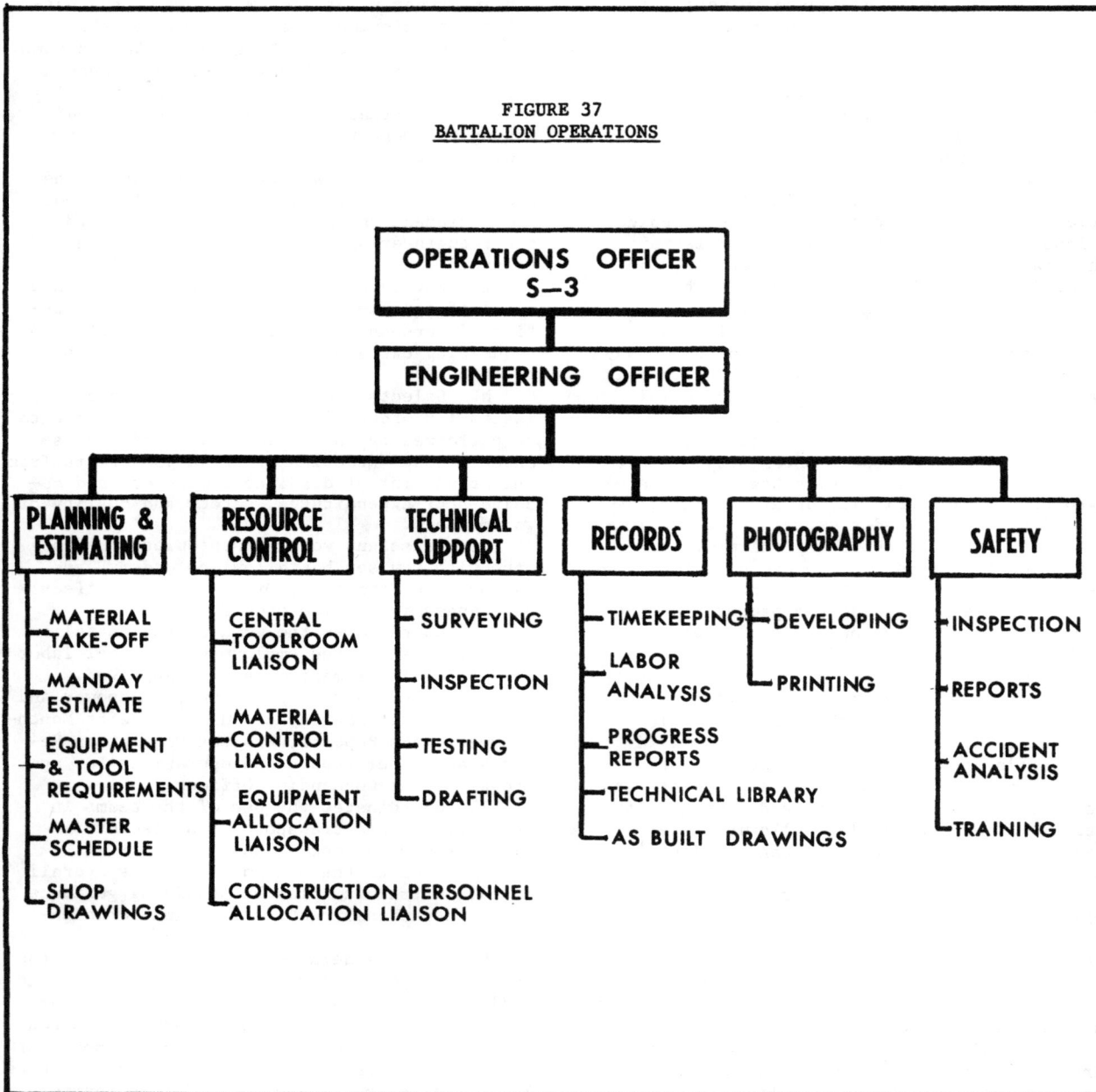

FIGURE 37
BATTALION OPERATIONS

SECTION B - MCB DEPLOYMENT PLANNING

1. <u>Workload Section</u>. In peacetime the workloads are generated by overseas bases as part of new construction, repair, alteration, replacement or maintenance. Project justifications are drawn up by the overseas base or station. From these descriptions COMCBLANT/PAC selects a balanced workload of projects for SEABEE accomplishment.

This selection is based upon the size of the battalion and the number of mandays available for direct labor in each rating. An attempt is made to pick projects that will provide work for all ratings and a proper mix of classroom training and on-the-job training. Consideration is also given to operational rotation of the unit and its possible emergency utilization, the length of deployment and the construction season at the site. Preference is given to projects that most directly contribute to preparation for mobilization tasks. Therefore, jobs of a highly specialized nature requiring special skills or equipment are not usually picked nor are tasks of a repetitive nature with low training value. As soon as the projects are approved for SEABEE accomplishment and funded, the preparation of plans and specifications is started. These plans and specifications are usually drawn up by an architect and engineering firm under contract to the Navy or by the Public Works Department of the sponsor station. When received by COMCBLANT/PAC, copies of the plans and specifications are sent to the battalion assigned to do the work for a detailed workload analysis and to the material take-off team for material procurement.

In wartime the projects are usually assigned by higher authority based upon operational requirements. Little or no consideration can be given to training, balanced workload, weapon or length of deployment. Plans and specifications as such do not exist but the battalion will be required to do the engineering to adapt standard functional components to the site, utilizing the NAVFAC P-140, Advanced Base Drawings and intelligence information available about the area.

Projects assigned to a battalion are detailed in an Operation Order (OPORD) prepared and issued by COMCBLANT/PAC for each battalion for every deployment. This OPORD tells the battalion when and where it will deploy, what is needed and what equipment, tools and materials are available. The how of project accomplishment is entirely in the hands of the battalion Commanding Officer. The OPORD is in the format of the standard five-paragraph order discussed in Annex D of this manual. The OPORD will usually contain annexes on Command Relationships, Intelligence, Training, Logistics, Project Descriptions, etc. The OPORD

is the culmination of all the planning work done by the COMCBLANT/PAC staffs and should contain all the information needed by the battalion to carry out all assigned tasks.

Since the battalion reports to the base or area commander for operational control when deployed, the priority of the projects or the jobs themselves may be changed by the commander exercising operational control. While this is usually avoided in peacetime because of the difficulties of getting project materials on short notice, it may prove to be the rule rather than the exception during emergencies as the operational commander adjusts his plans to take advantage of the changing situation. In some cases during periods of war or emergency, work will be assigned to the battalion directly by the area, task force or advanced base commander.

2. <u>Engineering Reconnaissance</u>. A thorough, accurate engineering reconnaissance of a proposed construction site is essential for efficient construction. When such reconnaissance is possible, it provides the data needed for:

    a. A topographic map of the proposed site.
    b. A feasibility study.
    c. A determination of climate, weather, soil and other geographic factors.
    d. A study of the availability and suitability of local labor, existing structures, water sources and local building materials.
    e. A study of other factors affecting construction, military operations and logistic support and selection of construction methods.

Each of the above factors has a major bearing on design, predeployment planning, logistic support and selection of construction methods.

In peacetime the battalion normally sends the S-3 and a few senior petty officers for a predeployment site visit. They will inspect the proposed project sites to supplement the information available in plans and specifications, learn what equipment is needed, what living facilities are available and what emphasis should be placed on the technical training in preparation for the deployment. This group may remain on site and constitute the battalion's advance party or they may return to homeport. If they do not return to homeport, a letter or message is sent to the battalion Commanding Officer giving him pertinent information to assist in planning and scheduling the projects.

In wartime, advance reconnaissance of proposed sites may be non-existent or limited to fragmentary and dubious intelligence data. In some instances, advance site reconnaissance may be limited to uncontrolled aerial photographs. In other cases topographic maps, area studies, trip reports or intelligence reports may be available to provide some specific answers to questions asked by the planning and estimating team. In many cases actual sites cannot be selected until the military situation permits ground reconnaissance by invasion troops or construction troops. In such cases advance parties are extremely useful, even when they land only a few hours or days before the main body of the battalion.

Reconnaissance missions must be thoroughly planned and coordinated. Reconnaissance parties must be carefully selected, equipped and trained. Reconnaissance orders list the specific information desired (by priorities) and the time limit for the job. Reconnaissance reports are usually written. Figure 38 contains a typical outline. When opportunity permits, maps, overlays, simple sketches, photographs and check lists are used to present a picture of actual conditions. When reconnaissance reveals unexpected conditions, immediate action must be taken to resolve any major problems thus presented.

FIGURE 38
SAMPLE OUTLINE FOR CONSTRUCTION RESOURCE COLLECTION

A. GENERAL AREA

    - TOWN NAME; MAP COORDINATES; PROVINCE.

B. UTILITIES

    - SOURCE OF WATER SUPPLY; STORAGE CAPACITY; PRODUCTION OF SYSTEM; CONDITION OF MACHINERY; PRESENT CONSUMPTION RATE; PURIFICATION SYSTEM.

    - ELECTRICAL GENERATION CAPACITY; TYPE, MODEL AND CONDITION OF GENERATORS AND ENGINES; PRIMARY AND SECONDARY DISTRIBUTION VOLTAGE; CYCLES.

    - TELEPHONE AVAILABILITY AND RELIABILITY.

    - TELEGRAPH AVAILABILITY AND RELIABILITY.

    - RADIO AVAILABILITY AND TYPE OF EQUIPMENT.

    - SEWAGE SYSTEM.

C. SUPPLIES

    - REPAIR PARTS AVAILABILITY FOR U.S. VEHICLES.

    - LOCATION AND CAPACITY OF MACHINE SHOPS.

    - TYPES OF HARDWARE AND TOOLS AVAILABLE.

    - TYPES OF HOUSEKEEPING SUPPLIES AVAILABLE.

    - OTHER CONSTRUCTION CONSUMABLES AVAILABLE SUCH AS BRICKS, SHEET METAL, CEMENT, REBAR, ETC.

    - TYPE AND QUALITY OF LUBRICANTS AND FUELS.

D. BULK CONSTRUCTION MATERIALS

    - LOCATION, CONDITION AND PRODUCTIVE CAPACITY OF SAWMILLS, CONCRETE BATCH PLANTS; ROCK CRUSHERS; ASPHALT HOT-MIX PLANTS.

    - LOCATION, DEPTH, EXTENT AND QUALITY OF QUARRIES; GRAVEL SOURCES; SAND SOURCES; SELECT FILL.

Reports from intelligence agencies, combat reconnaissance parties, advance parties or other sources may call for changes in the Operation Plan. Changes may be ordered even after actual construction has begun. Such changes may be caused by the tactical situation or other factors. Thus, certain structures may be added, deleted, altered, relocated, delayed or otherwise changed. The battalion must expect such changes and be ready to adjust to them as a matter of course. This is true whether the project consists of functional components or specially designed structures. Pavement and other work such as foundations must be individually planned for each component. When assigned responsibility for site adaptation, the construction battalion must provide for all design changes made after quantity take-offs, bills of material, requisitions or shop drawings have been prepared. Immediate action is necessary when these design changes effect work in progress particularly because materials sometimes are being supplied from sources which are thousands of miles away.

3. Workload Analysis. The MCB is capable of any kind of construction in a variety of situations and each employment poses a unique set of problems. However, the basic elements of the construction cycle remain fairly constant. Rapid, low-cost construction is the result of effective management by the command. The most important factor is the relationship between planning and control.

The Operations Officer assembles all the information that is available about the tasks assigned and defines the scope of the construction effort. This workload analysis is initially done by the planning and estimating section. They study the plans and specifications and all information gathered about site conditions, prepare manday estimates for each work element and do a material take off. While assistance from personnel outside the command may be available and utilized, it is the responsibility of the S-3 to prepare a plan of action for approval by the Commanding Officer of the battalion. While evaluating a future deployment, the planning and estimating section is augmented with senior petty officers of each rating. Some of these petty officers should be graduates of the Class C, Planning and Estimating School. It is helpful if the petty officers who will supervise the actual construction are part of the planning group.

Most drawings and specifications are furnished to the battalion. However, it is often necessary to adapt structures to a site, prepare plans of existing structures, design and draw alterations, adopt local or non-standard material to standard plans, design new structures and perform other design tasks. So deployment planning will include: review and possible revision of original drawings and specifications; preparation of working draw-

ings, shop drawings, topographic maps and other necessary sketches; assembly of bills of material from quantity take offs; estimates of labor and equipment requirements; schedules for each work element and for the entire deployment and the assignment of work to subordinate units. To assist the battalion to do this estimating, NAVFAC has published the NAVFAC P-405, SEABEE Planner's and Estimator's Handbook. This book contains basic man-hour and material tables and a brief description of the Critical Path Method. Normally, CPM will be employed on all projects involving large expenditures of man-hours and material. Conventional methods, as set forth in the first part of P-405, will be employed for small jobs which can be estimated quickly and accurately by using the Handbook. Also helpful are books in the NAVFAC P-700 series, Engineering Performance Standards.

Major savings in time, shipping space and expense are often possible if suitable raw materials or building materials are available close to an overseas construction site. The availability of suitable local materials may even warrant design changes. If local labor is employed, the workers may be more familiar with such materials and may be able to erect structures more suitable to the local weather and climate.

In a combat area, raw, semi-finished or finished building materials may be acquired by capture of local purchase. In a friendly area arrangements are made for local purchase.

Whenever possible, a study of the capability and availability of local labor is made before the construction methods are made final. Local labor is usually engaged on a contract basis, and in all instances, its use must be approved and funded by the cognizant authority. Therefore, alternate methods of construction for a given project should be considered at the earliest possible moment. Each phase of the project must be analyzed in terms of established deadlines and the manpower, material and funds available. Whenever possible, arrangement for extra manpower, special training, augmented equipment, supplemental functional components and other special needs should be made well in advance of the battalion deployment. For example, it is important to make early decision on: the location and productivity of quarries, sand pits, rock crushers, batch plant and other special facilities; the cycle time for earthmoving operations; the project tasks that can be prefabricated; the number of shifts to be used and the staffing and phasing of jobs.

This all starts when each project is broken down into its basic work elements such as erect forms, place reinforcing, place and cure concrete, etc. An estimate is made of the number of builder hours, steelworker hours and equipment hours that will be required to do this work. The workload is then expressed in manhour estimates for each rating and machine hour estimates for each type of machine. The

FIGURE 39
PEACETIME TRAINING PROJECT

Most peacetime construction projects needed by overseas stations such as this enlisted mans club. These projects provide excellent on-the-job construction experience for all SEABEES.

FIGURE 40
ADVANCED BASE CONSTRUCTION PROJECT

Wartime construction projects usually involve the construction of functional components. These two 1,000 barrel water tanks were built for the Marines in Vietnam.

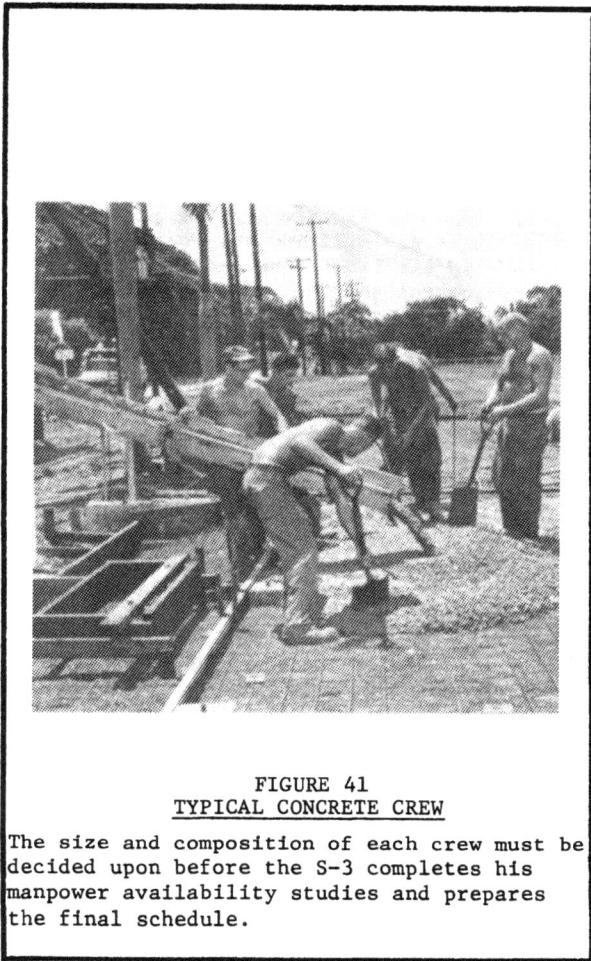

FIGURE 41
TYPICAL CONCRETE CREW

The size and composition of each crew must be decided upon before the S-3 completes his manpower availability studies and prepares the final schedule.

original Labor and Equipment Estimates are made in pencil on tracing paper to facilitate reproduction and revision. The form provides for entry of whatever work element codes are used locally, and quantities are based on the best available data at the time the information is compiled. Manday (M/D) estimates are made by leading petty officers of each rating after the method of construction has been decided by the S-3. These estimates are summarized by rating for each project and then for all projects scheduled for the deployment.

The estimated man hours (or machine hours) are based on handbook standards, which are adjusted for job and management factors. Job factors may include weather, specification tolerance, size of work area and skill level of the men. Management factors to be considered might be the morale of the men, the amount of care and repair of equipment and the sequence and timing of the schedule. Careful attention must be given to these factors because when they are less than perfect, efficiency is reduced and manday estimates must be adjusted upward to compensate. For example, if the estimated total workload was 20,000

M/D, but efficiency was calculated to be about 83% because of the factors listed above, the true workload would be slightly over 24,000 M/D.

Next the S-3 must look at the manpower he has available for direct labor on the projects. He usually knows the length of the deployment or time allotted to do a project, and from past experience or records he knows the percentage of men available to perform direct construction labor. For example, assume a 6 months peacetime deployment of approximately 182 days. Subtracting Saturdays, Sundays and Holidays and 5 days on either end of the deployment for mobilization and demobilization one would be left with 125 working days. By dividing 24,000 by 125 the S-3 would find that he needed 190 men on direct labor every day if the projects were to be completed in six months utilizing a 40-hour week. A similar calculation must be made for equipment. The analysis is not complete, however, until it is broken down by rating, type of machine and job sequence.

These adjusted estimates are projected against the man and machine hours available for each period. When a long-range imbalance in the workload occurs, uncommitted work crews are trained for critical specialties. When the imbalance is temporary, uncommitted crews are scheduled for prefabricating, service work, fill-in assignments or training. Every effort is made to maintain the integrity and individuality of each construction platoon. Therefore, the S-3 must balance his estimated workload against the battalion resources of skill and equipment and prepare a construction plan for approval by his Commanding Officer and others in the chain of command.

4. Manpower Availability. The MCB represents an enormous investment in human and material resources. Full responsibility for the management of this investment rests with the officers and petty officers of the battalion. While the officers and petty officers of the MCB have little direct control over the assignment of funds from higher command, each of them has it in his power to double the output and halve the cost of his operation. The ineffective man requires as much housing, as much food, as much shipping space, and more administrative and logistic support than the effective SEABEE. By making profitable use of every hour, of every step, of every piece of equipment, of every communication, of all supplies, the individual SEABEE can multiply the real strength and cut the real costs of battalion operations. What each man in the battalion actually accomplishes each working hour is the key factor in effectiveness. Reliable data on the ratios of manhour input to construction output is an essential tool in selecting peacetime projects, estimating manpower, training and material requirements, and scheduling prime and subcontract work.

81

The ratio of actual strength to effective construction strength is one of the most significant indexes of MCB operations. Not all of the actual strength of a construction battalion is available for construction work because the battalion must be capable of:

a. Operating as an independent unit.

b. Feeding all of its own personnel.

c. Providing off-the-job training.

d. Participating in combat-support and disaster control exercises.

e. Operating its own camp.

f. Providing its own administration.

g. Providing religious services, medical and dental treatment and other personal services.

h. Engaging in other military activities not related to construction.

Thus, many of the men assigned to construction battalions specialize in work not directly related to construction, such as, postal clerks, gunnersmates, etc. Even men with Group VIII (construction) ratings must perform many non-construction duties, such as camp maintenance, shore patrol and compartment cleaning. For peacetime deployments, the distribution of labor charges has been found to vary with the size of the battalion and the extent to which the battalion is supporting itself.

Personnel on active duty are in a 24-hour duty status, and therefore their military duties at all times take precedence over their time, talents and attention (See Article C11101 of the BUPERS Manual). However, for reporting and estimating purposes, the unit of work is a manday of 8 hours. The usual work week in peacetime including formal military and technical training time is 44 hours (5½ days) however, in combat the time spent on the projects is increased. Thus the actual work day and work week are varied to suit local conditions or operational needs. All accountable time is charged to manhour accounts. Manhour accounts falls into the following general categories and account for approximately these percentages of the total available manpower of the battalion.

	Training Deployments	Operational Deployments
Direct Labor	30%	45%
Indirect Labor	20%	20%
Military Readiness and Training	25%	10%
Overhead	25%	25%

The performance figures are based on sustained effort extending over several months. Individual battalions have exceed those performance rates over short periods under pressure or emergency conditions, but unfavorable weather, bad terrain, and health conditions have resulted in performance rates equally below average.

Direct labor is defined as that labor which is expended directly on a project and that contributes to the placement of the end product while indirect labor is that labor required to support a project but which does not produce an end product in itself, such as the labor expended on project supervision, equipment maintenance, locally processing materials and engineering. Military readiness and training labor is all labor expended on military, disaster control and technical training. All those functions necessary to the battalion's organization, such as medical, dental, administration, disbursing, security and special services are accounted for under overhead as well as leave, confinements, sickcall, TAD, etc. The details for defining the various labor codes are set forth in COMCBLANT/PAC Instructions of the 5213.1 series, Progress and Performance Reporting.

From the figures above, the S-3 makes his initial evaluation of M/D availability. For a training deployment he would apply the 30% for direct labor to the battalion onboard strength. (The standard complement of 563 will be assumed for this example.) He would find that only about 169 men are available for direct labor. This is less than the 190 men he calculated would be needed to do 24,000 M/D of work in a six-month deployment. Therefore, he knows early in the planning cycle that he must make some modifications to his construction plan. This analysis must also be done by rating. Calculating the percentage of available BU, EO, UT, CE and SW mandays, he must compare these figures with the estimated workload for each rating.

Since the S-3 is the officer who is most familiar with the coming deployment and the construction workload on each part of the battalion, he may recommend a distribution of Group VIII personnel among the companies. When approved by the Commanding Officer, this would create an unofficial company personnel allowance. The battalion personnel officer would ensure that new men are assigned so that each company has its proportionate share of personnel as defined by company allowances. An example of a possible distribution of skills to the companies is given in Figure 42.

5. Material Take-Off. Material Take-Off (MTO) for the MCBs may be done by the P&E section of the battalion, or it may be done by a team outside the battalion such as an A&E firm, the Public Works Department of a station requesting the work, a NAVFAC Field Division, the CBC, the regiment or the COMCBLANT/PAC staff. However since the MCB is the construction agent, the S-3 has the responsibility of checking the material take-off to be certain that all the material needed is on order. He may do this by requiring his P&E section to perform an independent MTO or when appropriate, he may assign members of the P&E section to assist the team doing the MTO. To perform a detailed MTO, a great deal of liaison is required with the constructing agent because work methods

FIGURE 42
EXAMPLE OF COMPANY PERSONNEL ALLOWANCE FOR ONE DEPLOYMENT OF A 563-MAN MCB

RATING	ON BOARD	HQ	A	B	C	D	RATING	ON BOARD	HQ	A	B	C	D
CUCM/BUCS	4	2	0	0	1	1	BM	4	4	0	0	0	0
BUC	7	1	0	0	3	3	GM	3	3	0	0	0	0
BU1	17	1	0	0	8	8	YN	8	8	0	0	0	0
BU2	39	1	0	2	18	18	PN	3	3	0	0	0	0
BU3	45	3	0	2	20	20	PC	2	2	0	0	0	0
TOTAL BU	112	8	0	4	50	50	SH	5	5	0	0	0	0
SWCS	1	0	0	1	0	0	SK	12	12	0	0	0	0
SWC	3	1	0	0	1	1	DK	3	3	0	0	0	0
SW1	7	1	0	2	2	2	CS	14	14	0	0	0	0
SW2	7	0	0	1	3	3	SD	6	6	0	0	0	0
SW3	13	1	0	2	5	5	HM	8	8	0	0	0	0
TOTAL SW	31	3	0	6	11	11	DT	2	2	0	0	0	0
EACS/EAC	2	2	0	0	0	0	PH	1	1	0	0	0	0
EA1	1	1	0	0	0	0	SF	2	0	0	2	0	0
EA2	4	4	0	0	0	0	MR	2	0	2	0	0	0
EA3	4	4	0	0	0	0	ET	2	2	0	0	0	0
TOTAL EA	11	11	0	0	0	0	JO	1	1	0	0	0	0
EQCM/EOCS	2	0	2	0	0	0	SN	14	14	0	0	0	0
EOC	5	0	5	0	0	0	NON GROUP VIII	92	88	2	2	0	0
EO1	16	1	15	0	0	0	TOTAL PER CO.	563	151	141	101	85	85
EO2	22	0	22	0	0	0							
EO3	30	2	28	0	0	0							
TOTAL EO	75	3	72	0	0	0							
CMCS	1	0	1	0	0	0							
CMC	2	0	2	0	0	0							
CM1	7	0	7	0	0	0							
CM2	11	0	11	0	0	0							
CM3	14	0	14	0	0	0							
TOTAL CM	35	0	35	0	0	0							
CECS	1	0	0	1	0	0							
CEC	2	0	0	2	0	0							
CE1	6	0	0	6	0	0							
CE2	12	0	0	12	0	0							
CE3	15	1	0	14	0	0							
TOTAL CE	36	1	0	35	0	0							
UTCM/UTCS	1	0	0	1	0	0							
UTC	2	0	0	2	0	0							
UT1	6	0	0	6	0	0							
UT2	11	0	0	11	0	0							
UT3	14	1	0	13	0	0							
TOTAL UT	34	1	0	33	0	0							
CN/CA	137	36	32	21	24	24							
TOTAL SEABEES	471	63	139	99	85	85							

NOTE: The 63 SEABEES in the HQ Company may include:

1	CUCM	Battalion Leading Chief
1	CUCM	Operations Chief
1	BUC	Material Expeditor
1	BU1	Inspector
1	BU2	Inspector
1	BU3	"C" Company MAA
1	BU3	"D" Company MAA
1	BU3	Special Services
1	SWC	Safety Chief
1	SW1	Assistant Safety Chief
1	SW3	Messdeck MAA
11	EA	Operations Section
3	CN	EA Strikers
1	EO1	P.O. in Charge Special Services
1	EO3	"A" Company MAA
1	EO3	Messdeck MAA
1	CE3	Special Services
1	UT3	"B" Company MAA
2	CN	Special Services
28	CN	Messmen
1	CN	Headquarters Company Yeoman
2	CN	MAA Office

will affect the amount and type of material required. Thus the most satisfactory arrangements are those where the MTO is done at the homeport by representatives of either the CBC of the regiment working with the MCB. This method not only reduces duplication of effort but provides battalion personnel with the knowledge of the manner in which the take-off was made and the materials ordered. This knowledge will be helpful during construction. Under emergency conditions the battalion personnel will usually do the entire take-off.

As soon as plans and specifications are obtained by COMCBLANT/PAC, the material take-off is started. The items that have long lead time either because of cost or complexity are worked on first, for example, heating units, generators, electrical panels, etc. Next are the items that would be needed early in the deployment, for example, form lumber, reinforcing steel, cement, etc., and then the remainder of the needed materials are calculated. These materials are listed on bills of material or material summary sheets and identified by federal stock number or a set of specifications when one is not a standard stock item. In general, standard construction materials and construction methods are utilized for all projects. However, deviations must be readily accepted when necessity demands. These material lists serve as a basis for the procurement, identification and issue of construction materials. Although standard bills of material are available for each functional component, even these lists must be tailored to meet actual job requirements.

Major savings in time, shipping space and expense are possible when construction materials can be procured locally, therefore, arrangements for the use of local materials should be made at an early stage of the planning process, preferably before any CONUS materials are ordered for the project. When possible, local U.S. sources and foreign sources are screened against the bills of material. In the case of foreign sources, reliable tests of materials must precede any decisions regarding their use. When local materials are used, provisions must also be made for any extra men or equipment needed to process the material.

When funds are provided, requisitions are prepared from the bills of material and forwarded to the cognizant supply agency, for issue, procurement or processing. Most materials are requisitioned by the Supply Department of the CBC but may be requisitioned by the battalion S-4 or by regimental, brigade, base or other designated supply officers. In some cases supply action may precede the battalion's assignment to the project. If it does, copies of all bills of material and procurement documents are made available to the battalion at the earliest possible stage of the assignment.

After the battalion has deployed, the material take-off team may also serve as an expe-

FIGURE 43
POSSIBLE AUGMENT TO THE 563-MAN MCB

The Basic 563-man MCB may be augmented at some point during contingency execution. Operations in Vietnam pointed up the need for a horizontal augment while operating in underdeveloped areas. The additional support personnel required for an augmented MCB are small in number. Below is the rate/rating breakdown of a standard 175-man Heavy Construction Company (HCC) augment including required support personnel:

NON-GROUP VIII		GROUP VIII	
GM -	1	EOC -	3
YN -	1	EO1 -	8
PN -	1	EO2 -	26
SK -	1	EO3 -	26
DK -	1	EOCN -	15
CS -	1	CM1 -	5
HM -	3	CM2 -	12
RM -	1	CM3 -	12
SN -	17	CMCN -	8
		CN (under) -	33

diting agency to see that the project material is being processed and sent to the battalion within the established time frame. Periodic material status reports are sent to the MCB indicating those items on order, those received by the Supply Department and those actually shipped. The battalion checks this report, marks it to show the materials that they have received and returns it to the MTO team. This procedure provides a positive material control from take-off to receipt at the overseas site. This procedure has been drawn on Figure 44.

In wartime, the MCB may have to rely upon the force they are supporting for all supplies including project materials. This is usually the case when MCBs are providing engineering support to an expeditionary force. The battalion must draw necessary construction materials from supply dumps hastily established. Unless a great deal of preplanning with the logistics groups of the expeditionary force has been done, the MCB will find that many supplies needed for construction have not been loaded aboard ships of the task force and therefore are not available.

6. Equipment, Tool and Training Requirements. During the review of the plans and specifications, the P&E section of the MCB should be alert to spot any unusual design or method of construction that would require the battalion to use special equipment or tools. When minor design changes can be instituted to permit the MCB to use a standard procedure, they should be requested. However, there are many situations where this cannot be done or where better machines or methods have been developed. In these cases requests for special equipment or tools should be made by the S-3 as early as possible. The required equipment may be available in the deployment area, from other units assigned to the project, from the COMCBLANT/PAC equipment pools or from Pre-Positioned War Reserve Stock. These sources are usually inves-

tigated before special procurement action is taken. In all cases requests for special equipment must be approved by higher authority When special equipment is authorized, action must be taken to arrange for the spare parts support, supplies, extra manpower and special training needed.

Special training may also be necessary because of new materials or unique work methods in addition to the procurement of new tools or equipment. In many cases battalion personnel can be sent to Navy schools or to fleet training courses to learn these techniques. However, manufacturer representatives may be requested to conduct short classes or demonstrations for those crews who are going to use the new equipment, material or methods.

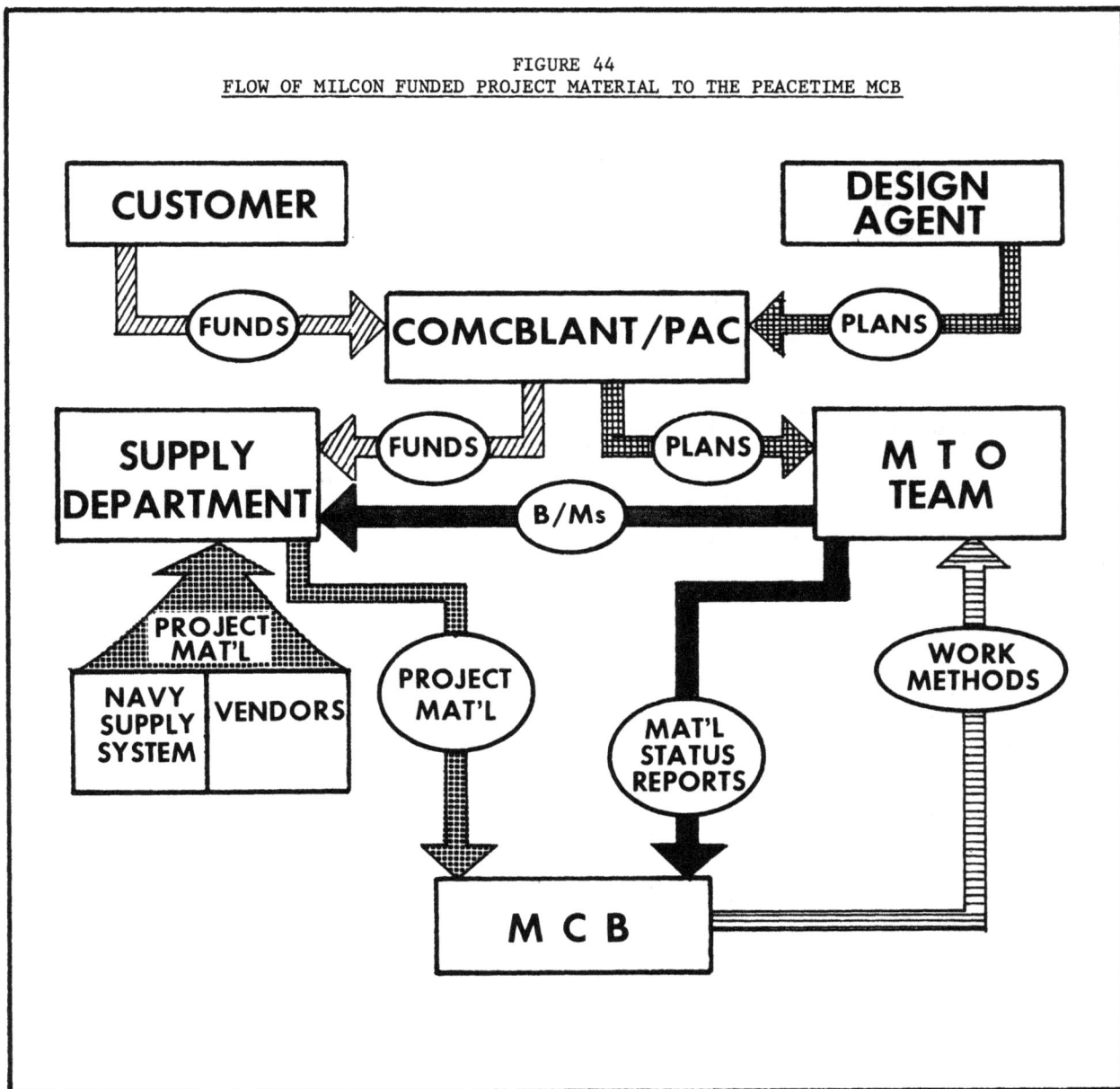

FIGURE 44
FLOW OF MILCON FUNDED PROJECT MATERIAL TO THE PEACETIME MCB

7. <u>Assignment of Projects to Company Commanders</u>. A, B, C, and D companies are trained, organized and equipped for construction, military operations and disaster recovery, and it is essential to preserve the organizational integrity of a battalion throughout the complete cycle of construction and combat-support arrangements. All manpower allocations, from work details to an involved project, should be made on the basis of construction/combat-support teams, squads and platoons.

The S-3 tentatively assigns each task to a company, indicating when the work must be started and completed. When work assignments are approved by the Commanding Officer, the company commander distributes tasks among his platoons, designating petty officers responsible for each portion of the work. In the case of C and D companies which perform most of the prime contract work, the project petty officers are selected, and they prepare tentative workload schedules. In the case of A and B companies, the platoon leaders select crews to be sent to various projects and coordinate a schedule for each crew with the project petty officers concerned. These schedules are then submitted to the company commanders. The tentative workload schedules for each company are recapitulated by the Planning and Estimating Section for the purpose of projecting workloads against available man and machine hours and construction deadlines. This data enables the S-3 to redistribute or reschedule work assignments, arrange for extra personnel, equipment or special training as needed.

8. <u>Scheduling</u>. After all the information from the plans, specifications and reconnaissance have been analyzed and the workload defined, the S-3 must begin the process of scheduling the work. Although standard industrial procedures used by contractors are the basis for SEABEE schedules, modifications must be made for those characteristics that are unique to battalions. First, the MCB has a fixed number of men that must be productively employed at all times. A contractor can hire men for peak periods and lay them off when the work reaches a bottleneck. Second, the MCB has essentially a fixed amount of equipment and has little chance to vary the numbers or types after they are deployed. Finally, for peacetime projects, the deployment is of a fixed length. So the major problem of the S-3 is to schedule for a balanced workload, ensuring that enough men and equipment are available for each work element and that alternate proposals are available if material shortages develop.

There are certain obvious rules for scheduling that are basic to scheduling any project:
a. Scheduled tasks must follow the sequence of work required for the particular job in question.
b. Scheduled operations cannot exceed the capability of the unit to accomplish the work.
c. The schedule must provide continuity

of effort for each project.

They are listed here because they are often overlooked, especially during emergencies, in an effort to get the project started. A schedule is a management tool of the command and may take many different forms from a simple bar chart to a sophisticated computer program, as long as it serves the purpose of providing the means to organize and direct the work.

Part 1 of Chapter 8, of NAVFAC P-405, contains instructions for the preparation of schedules which will result in the most efficient use of manpower, equipment and materials on simple construction projects. A method for scheduling called Critical Path Method (CPM) is used when a project is more complex and the usual type of scheduling does not provide the control or clarity required. For example, in many cases jobs or deliveries which might otherwise be overlooked are found to be, in fact, critical to the overall time to completion, while others which might be thought critical are not and can be scheduled at alternate periods. This will provide flexibility to a schedule. This method, which can effectively be used for SEABEE construction, is described in Part 2 of Chapter 8 of NAVFAC P-405. The Critical Path Method (CPM) can be solved either manually or by the use of computers. When available, computer-run solutions are advisable because of the savings in time and the ability to handle complex relationships. Usually when the deployment schedule consists of 200 activities or more, it is economical to employ a computer.

It is convenient to discuss battalion schedules in the form of four stages: Stage I is a general overall picture of the entire deployment. It lists all the projects assigned, shows a schedule for each of them and also indicates the planned rate of accomplishment for the entire deployment. It is normally used for reporting progress to higher authority. Stage II is a general schedule for each project It contains all the major work elements and a schedule for each element. Stage III is a detailed schedule for each project used by the company commander to manage his project. It shows such details as crew sizes, material delivery dates and the periods where special tools or equipment are needed on the project. Stage IV is an informal, perhaps even unwritten, schedule presented to the company commander by his project petty officers which provides for day-to-day adjustments in the Stage III schedule.

Schedules are usually developed along these lines:
a. List all the projects assigned by the OPORD for the entire deployment.
b. Taking each project individually, break it down into the major work elements.
c. Estimate the mandays (by rating), tools and equipment needed to do each work element.
d. Determine optimum crew sizes.
e. After a schedule has been developed for each project, begin to fit these schedules

together to form a schedule for the entire deployment.

f. Total mandays by rating, equipment and tools by type, etc., for each day and compare these tools to the men and machines available.

g. Employing some crew leveling techniques, adjust the work elements of the various projects so that peak demands are reduced to availability levels and idle time is minimized.

h. Redraw the schedules for the individual projects making sure that the logical sequence of work has not been destroyed.

This process may have to be repeated several times to approach the best workable solution. It is essential that the company commander and his project petty officer do most of this detailed planning. The functions of the S-3 is to make certain that the schedules of the individual projects are compatible with the overall battalion construction program. When the schedule is completed, it is submitted to the Commanding Officer for approval.

The schedule must be a working document that is easily adjusted to accommodate changes Some factors which might upset the schedule are:

a. Assumptions do not hold true.

b. On-site conditions vary from those estimated, e.g., weather, soil, availability of utilities.

c. Incorrect estimates made of mandays required.

d. Project material does not arrive on schedule.

e. Skill, experience or morale of men is not as anticipated.

f. Equipment is not available because of breakdowns.

g. Delays were underestimated.

h. Plans or specifications have been changed.

i. Additional work has been assigned.

FIGURE 45
CONSTRUCTION METHODS INFLUENCE MATERIAL TAKE-OFF

The S-3 must determine the construction methods early in the planning and estimating cycle so that project material and special equipment and tools may be procured prior to embarkation to the deployment site.

SECTION C –
CONDUCT OF MCB CONSTRUCTION OPERATIONS

1. __Advance Party Functions__. An advance party is a group of battalion personnel that precedes the main body of the battalion to the deployment site. The size, composition and equipment of an advance party depends entirely on the particular need. An advance party is organized and equipped on the same principles which apply to a detachment.

In wartime their mission is primarily reconnaissance. The party checks priorities and determines what arrangements must be made to make the battalion living area habitable. If the battalion must erect its own structures for a camp, the advance party determines the locations where they can best be erected or how any existing facilities can be modified to suit battalion needs. They may start on camp construction, stake out construction sites, establish supply dumps, set up beaching facilities, provide a water supply, test local materials or site-adapt construction drawings.

A battalion normally sends an advance party consisting of some MAA personnel to set up the berthing facilities, some SKs to receive any project material or allowance gear shipped ahead and some Group VIII personnel to perform any required maintenance on the camp facilities. If the battalion is relieving another, the advance party checks the equipment to be left at the job site for serviceability and learns about any special problems associated with equipment maintenance at that site. The advance party checks the status of any unfinished projects, ascertaining command relationships, supply procedures, location of project material, work remaining, etc., and receives the plans, specifications, field notes and other data about the project. Unusual design or problems encountered are discussed with the advance party. The advance party usually establishes liaison with the various departments of the base and transmits pertinent instructions and notices to the battalion.

2. __Quality Control__. It is incumbent upon each construction battalion to ensure that quality control consistent with good engineering practices is exercised on all projects assigned whether for operational or training purposes. The Engineering Officer, working for the S-3, is the officer who is concerned with overall coordination of the quality control program from the initial project layout to the inspection and testing of materials, equipment and workmanship.

When construction of a project is started, a field engineering team usually performs the preliminary, topographic and location surveys, and then the construction stake-out and calculations for establishing line and grade.

They can also be used to:

a. Make regular measurement of the quantity of the work-in-place and earth work quantities.

b. Establish the actual location of structures, pipe lines, etc., for the preparation of "as built" drawings.

c. Provide solutions to other survey problems such as property, triangulation, hydrographic or determination of azimuth.

The construction battalion usually has complete responsibility for inspection of construction. To maintain control of the quality of work, the Engineering Officer acts in effect as a resident officer in charge of construction He is assisted by a staff of inspectors and is equipped with materials-testing facilities. For inspection of specialized work, he may be assisted by electricians and utilitiesmen of the Planning and Estimating Section, the Safety Inspector or various field parties. Inspectors make soil tests as needed, and test and control all locally-produced materials, such as concrete, lumber, concrete block, crushed stone, sand and gravel. For example, the inspectors may take soil samples and make laboratory tests to assist in determination and control of optimum moisture content and compaction for earthmoving operations. They measure the physical characteristics of aggregates used for concrete, such as size and gradation, moisture content and hardness; calculate and control the proportioning of concrete mixes; and test concrete cylinders and beams. A knowledge of NAVFAC TP-AD-5, Inspection of Construction, will be most helpful.

Another function of quality control is to report to higher authority products or methods which do not produce satisfactory results for one reason or another so that they can be corrected or avoided. Things that should be reported include:

a. Material that differs in quantity or dimensions from that required by specifications.

b. Damaged material.

c. Material conforming to specifications but which is found unsatisfactory for construction.

d. Plans or specifications requiring materials or methods that are obsolete, inefficient, expose personnel to hazardous working conditions or that cause excessive maintenance costs.

It should be noted that while the Engineering Officer has overall coordination of the quality control efforts, the company commander and project petty officer must conduct inspections to determine that the various elements of the project are being properly executed in accordance with plans and specifications and in an efficient and orderly fashion. These inspections should be a continuous function of any supervisor as he carries out his routine duties of directing the work.

3. Material and Tool Control. The S-3 is responsible for the proper utilization of project materials and project funds given to the battalion to purchase local material and construction consumables. The S-4 is responsible for procurement, stowage and issue of these project materials and consumables. These officers must work together closely if proper and efficient control is to be maintained. The Company Commanders are each responsible for:

a. Using Materials properly to eliminate waste or unauthorized use.

b. Maintaining complete cognizance of all materials required for the job, whether on-hand or ordered.

c. Drawing materials in accordance with established procedures.

d. Initiating requests for materials not ordered by the material take-off team.

Unless the battalion relies directly on a base, regimental, brigade or other supply department, all building materials except those produced by the battalion are received and issued by the Materials Liaison Section. Upon receipt of MCB job orders, prime and subcontractors check bills of material for adequacy. They next check the status of required materials with the Materials Liaison Section which uses the master bills of material or material summary sheets to note the requisition number, date material required (DMR), receipts, location and quantities of each item allocated and issued against each job. The Materials Liaison Section advises the Operations Officer and the Company Commander of anticipated delays or other foreseeable supply problems.

When requesting material listed on the bills of material and held in the battalion warehouse for that job, the company commander, project petty officer or designated representative will fill out a DOD Single Line Item Requisition System Document, DD Form 1348 in several copies. One copy goes to the Operations Department where it is used to update the material status sheets. The others go to the Supply Department where the original is marked to show the quantity and date of issue and returned to the Company Commander. This serves as his receipt. Requests for substitutions, extra materials or materials currently in short supply must be approved by the S-3.

To request material not ordered by the material take-off team, the Company Commander must completely fill out the DD 1348 identifying the material required and the date by which it is needed. All copies of the DD 1348 go to the Operations Department where the S-3 must approve the request before further action is taken. If approved, the project funds are cited, the material status sheets are marked and the request is sent to the S-4 for procurement.

On-the-job supply crises are inevitable unless each officer concerned gives immediate and close attention to all pertinent details of the operation order (including construction priorities), the drawings, specifications and shop drawings, the quantity take-offs and bills of materials.

The Allowance List provides for hand tools in kit form backed-up by Central Tool Room (CTR) stocks. The need to maintain a rapid mount-out capability makes it mandatory that these tool kits be kept intact and complete at all times. Financial and special limitations make it necessary to use these kits on training deployments; furthermore, the peacetime use of these kits has the inherent advantage of fixing responsibility for all battalion tools whether in the CTR or in the field. The tool kits essentially contain all the craft hand tools needed by one 4 man-fire team of a given rating to pursue their trade. These kits may be augmented with additional items as dictated by the crew mission, however, the kits will not be reduced in scope and should be maintained at 100% kit assembly allowance.

The crew leader is authorized to draw the tools needed by his crew. He is responsible for:

a. Maintaining complete tool kits at all times.

b. Assignment of tools within his crew.

c. Proper use and care of assigned tools by his crew.

d. Preservation of tools not in use.

e. Security of assigned tools.

To ensure that tools are maintained in the proper state of readiness all tools are inspected periodically, by someone other than the crew leader. Further, all power tools are given a preventive maintenance inspection by qualified shop or CTR personnel.

The Central Tool Room, as the cognizant battalion subdivision, shall be responsible for the general custody and upkeep of all battalion tools and the maintenance of proper inventory levels. Damaged and worn out tools are turned into the CTR for replacement in kind. CTR personnel determine whether these tools should be repaired or replaced. Tools requiring routine maintenance such as saws are turned in to the CTR and when repaired are reissued. When the balance of tools on-hand reaches the low limit established by the Commanding Officer for a project, the Central Tool Room petty officer fills out a request that goes to the Supply Officer for approval. Requests for special tools are submitted by the Company Commanders via the Operations Officer to the Supply Officer for authorization. The CTR is a subdivision of and responsible to the battalion supply and logistics officer.

4. Evaluation of Progress. As soon as the deployment starts, every crew leader on through to the Commanding Officer wants to know how the work is progressing with respect to the schedule. It is not surprising that this same information is needed by COMCBLANT/COMCBPAC and others in the chain of command. Therefore, the performance of the battalion and its

FIGURE 46
REPAIR TO WATERFRONT STRUCTURES

Battalion divers are often used to assist in construction and repair or waterfront facilities.

subordinate units is evaluated by many methods. This evaluation helps to determine the effectiveness of the battalion with respect to its past performance or capability. It will also provide facts on whether available manpower, equipment and other resources are being used effectively along with generating data for analyzing the effectiveness of equipment, training, personnel allowances, construction techniques and other variables.

When battalions are working on similar types of construction, the evaluation is useful for establishing performance standards and will provide a degree of competition to determine "best of type", thereby increasing the overall effectiveness of the NCF. There are many yardsticks that can be used to measure effectiveness. Some of them are:

a. Making a comparison between actual work-in-place and the scheduled work-in-place at any point in time.

b. Comparing the labor expended versus the estimated mandays.

c. Measuring the labor expended per unit of work-in-place.

d. Plotting the percentage of productive work versus the percentage of overhead work.

Many of these are used by battalion officers in trying to determine the effectiveness of their programs.

The platoon leader who serves as prime contractor is responsible for keeping the job on schedule and keeping the subcontractors informed on job progress. During slack periods each platoon performs advance, preparatory of other authorized fill-in work. Man and machine hour expenditures are charged against the proper job order and, as soon as an obvious discrepancy is noted between the estimate and the performance, it is brought to the attention of the Operations Officer. If a major change in the scheduling of a job is required, the job order is completely revised and reissued as Issue No. 2, etc.

One of the most complete means of evaluating the construction program is to prepare

90

a report to higher authority. This requires a thorough examination of work patterns, progress rate, safety records, etc. Usually the reports of construction operations fall into two classes, the Monthly Progress Report and the Completion Report. When specified in the OPORD, the unit may be required to submit situation reports (SITREPS). All these reports are described in COMCBLANT/COMCBPAC Instructions of the 5213 series. Reports will require the battalion to devise some internal reporting system to produce information. These internal reports may include time cards for every man in each crew, shop or office or a project petty officers report of work performed daily or weekly and a projection of the work schedule for the next week or a special report from Company Commanders on something that would improve the capability of the battalion, such as, recommendations to improve allowance tool kits, or supply support to a project.

In the final analysis the performance of a unit is measured by the amount of field work successfully completed per unit of time. But because projects and conditions vary so greatly, it is not always practical to gage performance by a mere comparison of complete projects. Any evaluation figure is a relative one rather than an absolute one, and thus, many factors are considered in assessing performance. Although any one method serves only as a guide, these figures and reports give battalion officers an opportunity to determine whether the manpower utilization is out of balance and perhaps indicate what corrective action is possible.

5. Timekeeping and Labor Analysis. In order to record and measure the number of man hours the battalion spends on various functions, a labor accounting system is necessary. This system must permit the day-by-day accumulation of labor utilization data in sufficient detail and in a manner that allows ready compilation of information required by the unit in the management of its labor forces and in the preparation of the required reports to higher authority. Each battalion must account for:

a. Forty (40) man-hours per week (normal work week) for all enlisted men assigned to the battalion for the full work week; plus,

b. all man-hours actually expended above the normal work week; plus,

c. All indigenous civilian man-hours actually charged to the battalion.

While the system may vary slightly between battalions, they are so similar that the one described in this manual can be considered typical.

During the planning and scheduling phases each work element of each project is given an identifying number. For example, "clear and grub site" may be the first task of project R-15 and therefore would be identified as work element R-15a. All man or machine hours used in clearing and grubbing that project site would simply be listed under R-15a. In like

manner indirect labor would have a series of codes as would training and overhead. An example of some indirect codes are listed below:

X01 Construction Equipment Maintenance Repair and Records.
X02 Operations and Engineering. Includes planning and estimating, material take-off, drafting, surveying, testing, photography, safety chief, inspectors and all other phases of project support.
X03 Project Supervision. Labor spent by field and shop personnel in supervising project work. When over 50% of a man's time is spent in supervision, all his time shall be charged to supervision.
X04 Project Material Expediting (Shop Planner). Labor spent in arranging for equipment and tools, scheduling for utility outages, coordinating with other crews, etc.
X05 Location Moving. Moving equipment, tools, field offices, etc., to and from project sites as well as time spent on mobilizing resources at the job site.
X06 Project Material Support. Includes material liaison functions, receipt, storage, issue and on-site delivery of project materials.
X07 Tool and Spare Parts Issue. Includes issue and tool room repairs and maintenance.
X08 Other. Used for reportable items not covered above and should be explained fully when used.

These codes are used by the crew leaders when reporting the hours spent by each member of their crew. Reports are usually made as simple as possible and take the form of a card such as the one shown in Figure 47. Crew leaders submit these cards to the Operations Department daily. The man hours expended by all crews in the battalion are summarized by the operations clerk. All man-hours expended on each work element are correlated by the S-3 with measurements of work-in-place to determine progress with respect to the schedule. These figures can also be used to determine the daily percentages of direct, indirect and overhead labor. At the end of each month the mandays in every category are summarized and used to fill out the Labor Distribution Report which is usually an enclosure to the monthly report. An example is included as Figure 48. Further guidance on how the timekeeping records are to be kept and reported is contained in COMCBLANT/COMCBPAC Instructions of the 5213 series.

6. Reports and Records. Each MCB submits a monthly report of operations, in letter form, to COMCBLANT/COMCBPAC. Copies are sent to the Commander, Naval Facilities Engineering Command

FIGURE 47
TYPICAL TIMEKEEPING CARD

CREW LEADER SMITH UTI | CREW | CREW SIZE 6 | TRANSFERS THIS DATE NONE

PROJECT POL SYSTEM | DATE 15 NOV 1965

NAME	PRODUCTIVE						OVERHEAD					
	DIRECT		INDIRECT		MILITARY		ADMINISTRATIVE		MISCELLANEOUS			
	LABOR CODE		LABOR CODE		LABOR CODE		LABOR CODE		LABOR CODE			
	R9C	R9D		X04						Y01	Z04	
SMITH	4	4										8
AARON		6		2								8
FRITZ		8										8
GONDA		7								1		8
SANTZ	8											8
MANUEL	7										1	8
TOTAL	19	25		2						1	1	48

DAILY LABOR DISTRIBUTION

and to administrative, military and operational commanders concerned. This report is a concise review of the battalion's activities during the month, regarding accomplishments, problems and capabilities. It includes such information as planning, construction, welfare, morale, discipline, safety, training and equipment. The numbers of officers and enlisted men assigned to the battalion and to all detachments are shown. Any movement of the battalion is reported specifying the method of movement. The following topics are usually included as separate enclosures:

    a. Administrative Summary
    b. Equipment Status Report
    c. Training Report
    d. Labor Distribution Report
    e. Construction Summary
    f. Financial Summary

Additional enclosures usually include a series of photographs of the projects and a graphical presentation of overall project status usually in the form of a copy of the Stage I schedule.

A comprehensive, narrative report is usually submitted following the completion of each authorized construction project. This report should include all information and data of general interest concerning the project including, a description of all work accomplished, methods of constuction, problems encountered together with recommended solution, etc. Since this will constitute a final record report of the project or program, it should be comprehensive in scope and adequate in pertinent details. Usually the sponsors estimate of the cost for the project if done by contract is compared to the actual cost. This actual cost is compiled by calculating the cost of the labor expended on the project, direct, indirect and overhead and adding it to the cost of the project material and the cost of equipment operation. This comparison serves as a cross check on the unit's efficiency.

FIGURE 48
SAMPLE LABOR DISTRIBUTION REPORT FORM

LABOR DISTRIBUTION REPORT
MONTH OF   19

	DESCRIPTION	MANDAYS			
		MILITARY	OTHER	TOTAL	%
	CONSTRUCTION				
DIRECT	1. Project Description				
	2. Approved Fill-in Projects				
	TOTAL DIRECT LABOR				
INDIRECT	X01 Const Equip.Maint, Repair & Records				
	X02 Operations and Engineering				
	X03 Project Supervision				
	X04 Project Material Expediting				
	X05 Location Moving				
	X06 Project Material Support				
	X07 Tool & Spare Parts Issue				
	X08 Other				
	TOTAL DIRECT LABOR				
TRAINING	M01 Military Operations				
	M02 Military Security				
	M03 Embarkation				
	M04 Unit Movement				
	M05 Mobility Preparation				
	M06 Contingency Readiness				
	M07 Administration Functions				
	M08 Mobility & Defense Exercises				
	D01 Disaster Recovery Ops.				
	D02 Exercises				
	T01 Technical Training				
	T02 Military Training				
	T03 Disaster Recovery Training				
	T04 Leadership Training				
	T05 Safety Training				
	T06 Training Admin				
	TOTAL TRAINING LABOR				
	TOTAL PRODUCTIVE LABOR				
OVERHEAD	Y01 Personnel and Administration				
	Y02 Medical & Dental Departments				
	Y03 Ships Service & Special Services				
	Y04 Disbursing and Supply				
	Y05 Commissary				
	Y06 Camp Upkeep and Repair				
	Y07 Security				
	Y08 Leave and Liberty				
	Y09 Sickcall				
	Y10 Personnel Affairs				
	Y11 Lost Time				
	Y12 TAD				
	Y13 Other				
	TOTAL OVERHEAD LABOR				
	TOTAL ALL LABOR AVAILABLE				100%

ACTUAL NUMBER WORK DAYS IN MONTH ____   LENGTH OF NORMAL WORK DAY (HOURS) ____
TOTAL OVERTIME INCLUDED ABOVE(MANDAYS) ____

This completion report may have as enclosures:

    a. The financial statement.

    b. The letter of acceptance by the sponsor.

    c. Completion photographs.

    d. A final set of drawings showing the "as built" conditions.

Interim completion reports are submitted when one battalion relieves another before the project is completed, at the close of each fiscal year or upon termination of a deployment when work has been done on a project but it is not yet completed. The format for the interim completion report is identical to the completion report.

Situation Reports (SITREPS) are usually required only when the project is so important that daily or weekly progress reports are desired or because normal mail service is not available or not dependable. They are usually sent by message and have no special format. However, they must be numbered in sequence.

Significant experience is gained in every construction project. Problems are solved, performance is recorded and methods are tested. To afford others who may be faced with similar problems the benefit of these successes and failures, project folders along with progress and completion reports should be maintained. When the work is completed, the project folder is reviewed by the operations officer. Supplementary data may be added. Certain information summarized and material of no permanent value removed. A complete project history is thus compiled in as condensed a form as is consistent with a clear understanding of its governing features. The history can be typed and bound together with the sketches and photographs that illustrate it and placed in the technical library. The primary concern is to summarize all factors which affected construction progress. Examples of various subjects mentioned in the project history are as follows:

    a. Description of the project.

    b. Description and dates of authorized changes.

    c. Geographic conditions.

    d. Sources and characteristics of local materials.

    e. Mixes used in concrete and other materials.

    f. Results of all tests of soil, concrete and other materials.

    g. Supply problems and delays.

    h. Number and classification of men assigned to the project.

7. **Job Site Safety**. The S-3 is usually charged with the responsibility to ensure that an active safety program is part of the operational effort of the battalion whether engaged in construction, combat support or disaster recovery operations. Company Commanders are in turn responsible to the S-3 for carrying out a vigorous and effective safety program on their job sites. Safety training will be conducted in accordance with command policies.

A chief petty officer whose sole duty is to make inspections of the job site and ensure that proper safety precautions are being observed is assigned to the Operations Department. In addition to assisting the S-2 in the preparation of training programs in job safety, he makes full reports concerning accidents which occur on the job site and completely analyzes all accidents so that measures may be taken to prevent their reoccurrence.

SECTION D -
 CONDUCT OF AMPHIBIOUS OPERATIONS

1. **Mission and Control**. The construction mission of an ACB consists of the assembly, maintenance and operation of various ship-to-shore equipment. The actual construction involves the assembly of pontoon units or other components as a means of achieving the ACB mission namely, the transfer of men, equipment and supplies to and over the beach. This transfer includes causeway operations, lighterage and transfer barge operations, warping tug operations, beach salvage, buoyant and bottom laid ship-to-shore Amphibious Assault Bulk Fuel Systems (AABFS) and limited construction. The limited construction includes minor construction tasks incidental to an amphibious assault, such as road grading, preparing beach exits. camp site improvements, NBC recovery operations and temporary repairs to harbor or pier facilities. The scope of the mission depends upon the size and equipage of the task force, the climate, terrain and weather conditions, the existing facilities at the landing site and, during opposed landings, the outcome of the initial enemy contact by the attacking forces. The battalion task organization must be tailored prior to each mission to suit several alternate sets of circumstances and the personnel must be trained to cope with any immediate emergency measures.

Because a series of teams are task-organized for every operation and are embarked aboard various ships in the task force, control of the teams becomes a serious concern. Normally because of the diverse nature of tasks assigned to an ACB and the interdependency between the individual teams and other units in the mission, the operational control passes from the battalion to the Task Group Commander for reassignment to subordinate commands for transportation. Upon arrival at the objective area, the control is passed to various commanders, such as, the Beach Party Commander in the case of causeway teams and the fuel system teams or the Cargo Control Officer in the case of transfer barges. This is done at predetermined stages when a conflict of effort among individual teams could impede or otherwise adversely affect the combined effort. The Task Force Commander's operation order contains specific instructions for assault launching assignments.

2. _Deployment Planning_. The operation begins with pre-deployment planning, initiated upon the receipt of an operation order or letter of instruction from the Type Commander. The Operations Officer establishes liaison with other participating units via the Naval Beach Group to determine the extent of the ACB contribution to the mission. The accuracy and effectiveness of the detailed planning are predicated on a compilation of the following information:

a. Specific tasks to be performed; number and length of causeways and piers; number, length and type of bulk fuel systems.

b. Number of beaches to be supported and the size of the landing or attacking force.

c. Type of surface transportation, (LST, LSD).

d. Approximate date of embarkation and the travel time to the objective.

e. Approximate duration of operation.

f. Accurate maps, topography and reconnaissance reports.

With this information the Operations Officer can determine the size and strength of the detachment required. Working with the company commanders concerned, he compiles a list of all billets and equipment necessary. When the list is approved by the Commanding Officer, the detachment is designated and the Officer-in-Charge proceeds with embarkation planning. Any changes in the operation order must be proposed to the Beach Group Commander who solicits approval from the Task Force Commander.

Prior to an operation, it is the Operations Officer's duty to see that a briefing meeting is held for all officers and leading petty officers involved. If the mission is for training, the briefing will serve to arouse interest of the participants and establish a purpose in the minds of those concerned. If an actual mission is involved, a briefing is important to give as much of the overall picture as possible as an aid to each man in doing a better job. The briefing should include:

a. Task organization.

b. Concept and scheme of maneuver.

c. Intelligence information including beach gradient, sand bars, obstructions, mines or other opposition.

d. Embarkation and debarkation procedures.

e. Ship-to-shore movement of causeways, barges and bulk fuel systems.

f. Protection measures for NBC attack, heavy weather or fire.

g. Personnel preparation and training and provision for ready reserves.

h. Amount of support required and the division of support for rations, special clothing, ammunition and back-up supply.

i. Communication equipment and personnel required, frequencies assigned and the codes to be used.

j. Inspection procedures for clothing and equipment.

k. Reports required by the Commanding Officer.

l. Estimated time for travel and completion of the mission.

3. _Embarkation Movement and Debarkation_. Prior to embarkation, the Officer-in-Charge conducts liaison with the ships executive officer during which operational problems are resolved and the OIC learns what will be expected of the detachment personnel. Arrangement can be made at this meeting for the welfare of the personnel and the cargo handling facilities can be reviewed. A check-off list of pre-embarkation tasks, compiled by the Officer-in-Charge, should contain the following items to be discussed with, and clarified by, the ship's officers: berthing assignments, messing arrangements, custody and upkeep of service and pay records, use of ships facilities, such as laundry, etc., storage space assigned and security watches. If causeways or barges are to be side loaded, a check must be made of the ship prior to loading time for the proper hardware. The inspection should include the hoisting pads, pad eyes, clover leafs, lashing sockets, shelf angles and the proper winch.

Prior to embarkation the Operations Officer prepares the embarkation tables for submittal to the Naval Beach Group Operations Officer. After consolidation of all Beach Group units' requirements, the Beach Group Operations Officer forwards the embarkation tables to the Embarkation Officer involved, who then makes out the ship's loading plans. Careful consideration must be given to the order of requirement during the mission. Before the equipment is loaded, it is mandatory that the Embarkation Officer know the exact order in which every item of equipment is needed so that it may be stowed aboard in the reverse order.

Prior to loading aboard ship, the CPO in charge reports to the OIC the following items:

a. All equipment checks out and operates to the satisfaction of the individual operators and mechanics.

b. Adequate supplies of spare parts, grease, crankcase and hydraulic oils, tool kits, special tools and batteries are in readiness.

c. All tractor blades are equipped with bumper timbers and tires for protection when moving heavy objects.

d. All wire is of proper length and size and is in good condition.

e. If vehicles are to be operated through surf, each unit is equipped with a properly installed water fording kit.

While underway, ACB personnel stand security watches around the clock. Enough men shall be used so that, in the opinion of the Officer in Charge, adequate protection is afforded all equipment. Special details of competent men shall be assigned to check periodically the sideloaded causeways and barges and their securing lines and fittings. Engines in equipment must be started and run daily. Mechanisms such as the steering clutches on bulldozers should be sufficiently operated as to keep them free.

The priority order of debarkation during an assault is predicted on necessities which may vary with mission and even change during a mission. The initial responsibility for the debarkation pattern belongs to the Task Force Commander and is specified in the operation plan. When necessitated by early events of the assault phase, the Beach Party Commander may request changes in priority.

4. <u>Causeway Operations</u>. Causeway sections are constructed by assembling the four standard pontoon units to meet the requirements of the standardized structures described in NAVFAC P-401, and the instruction book for the P-series pontoons. The four standard pontoons of the T-series, which were in use from the time of World War II until after 1960, are being replaced by the P-series which were first issued for general use in 1964. The two series are not interchangeable. The size of the basic pontoon, 5x5x7 feet, is the same for both series, but the method of attachment differs The new series consists of P-1, the basic pontoon, P-2, pontoon with straight line sloping bow; P-3, slope deck; P-4 ramp end; and P-5M and P-5F, which are the P-2 with end to end hinge-type connectors (See Figure 49)

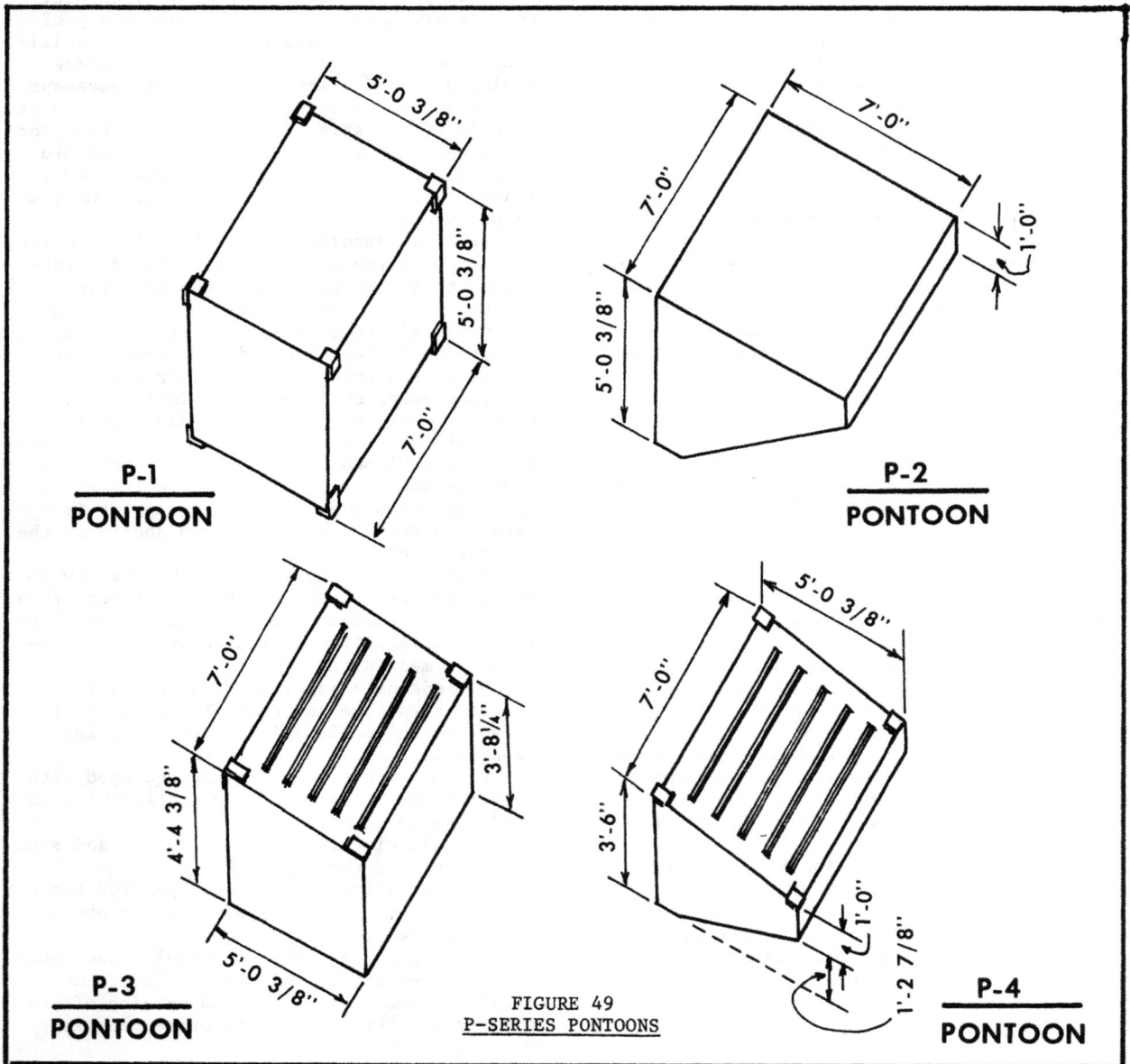

FIGURE 49
P-SERIES PONTOONS

FIGURE 50
TANK LANDING SHIP

A LST loaded with heavy equipment and sections of a pontoon causeway heads for an assault area.
Once in the staging area, the pontoon causeway sections will be dropped into the water, put
together and beached, forming a mobile pier over which heavy equipment from the LST can be
driven ashore to support land action.

FIGURE 51
AMMI PONTOON
Two Sections of Ammi Pontoon.

FIGURE 52
AMMI PONTOON

A Ammi Pontoon Section in use as a Floating Artillery Base

The new series is simpler to fabricate, uses no proprietary items, has fewer parts and reduces assembly time. Strength factors are equal to, or greater than, those of the old series. Deflections and stresses are less. The main body of the causeway consists of P-1 pontoons and the ends are varied depending upon the use of the structure. General information on construction techniques, along with drawings is contained in NAVFAC P-401.

The crews for handling the causeway elements are tailored by the operations officer prior to detachment to meet the special circumstances of the mission. A detachment of men would be based on the number of ships involved in carrying a complete causeway pier. A complete causeway may vary in length from four to twelve sections. A standard causeway detachment usually consists of one officer and from twenty to twenty-eight enlisted men and their supporting equipment. Figure 53 shows a 100-ton floating crane loading a causeway section on an LST.

Launching procedures differ for the various classes of LST. A method of controlled launching has been developed for the 1171 class LST in which the causeways are lowered to the water. On earlier classes of LST a chopping block method is used, resulting in the causeway being dropped violently from the side of

the ship. Specific details for both methods are described in the 5400 series of NAVFAC instructions. Causeway sections, after launching, are connected end to end. Standard operational procedures for connection are described and illustrated in operational handbooks of the 5400 series of instructions for each ACB. The times required for connection varies with the length of the causeway, the method of connection, weather conditions and other operational factors. In a calm sea, in good weather, a four section 3 x 15 causeway can be connected end-to-end in approximately 60 minutes.

The connected causeways are established on the beach either by momentum beaching or by being towed into position by warping tugs. The choice of method depends upon beach and surf conditions, weather and length of the causeway. "Momentum beaching is used for causeway up to four sections in length." (Four sections are maximum that can be side loaded on one LST). In this method, the causeway is rigged alongside the LST, which proceeds to ground itself on the beach, casting off the causeway just before grounding. The causeway being of shallow draft continues on to beach by itself. Longer causeways are towed into position by warping tugs.

Upon beaching the causeway is held in position by tender craft until the assigned dozers are in position on the beach to take a strain on the anti-broaching wires. These wires are attached to the causeway and extended at approximately 45° to the causeway and are attached to two dozers. On causeways over six sections in length, 3000-pound anchors are used to replace the dozers after the causeway is secured thus releasing the dozers for other tasks. As soon as the antibroaching wires are secured, the sea anchors are laid. These 3000-pound anchors are attached to the seaward section, one on either side, and tend seaward at approximately 60° to the causeway. After both sea anchors are in position, the intermediate anchors may be laid as necessary. The intermediate anchors may be from 500 pounds up to 3000 pounds depending on length of the causeway, surf conditions and duration of installation. Since the prime consideration in establishing a causeway is to permit unloading of ships, anchorage provisions are limited to those absolutely necessary to keep the causeway intact and prevents broaching.

5. Ship-To-Shore Bulk Fuel Operations. Ship-to-shore bulk fuel handling systems have been developed to meet the need for large quantities of fuel required by a division in landing operations. The two types of ship-to-shore bulk fuel delivery are the positive buoyancy system, using rubber hose, and the bottom laid system, using metal pipe. Both systems are designed to work with ships of the AOG class, and both have been used to transfer fuels over distances of more than a mile. Beach terrain, expected duration of the operation and related operational considerations will determine the type of system used. Storage facilities for either system are the responsibility of the receiving command.

The positive buoyancy system employs 4 or 6-inch lightweight rubber hose, made in 50-foot sections, with quick couplings at both ends. "D" rings are incorporated at the ends of the hose to attach the suspension component which consists of 1/2-inch diameter wire. This wire supports the hose during pumping operations,

FIGURE 53
FLOATING CRANE LOADING A CAUSEWAY SECTION

An LST can side-carry four causeway sections. These sections are loaded either by the Ship's winch or by using the 100 ton pontoon floating crane that is built and operated by ACB Personnel.

FIGURE 54
BUOYANT FUEL SYSTEM

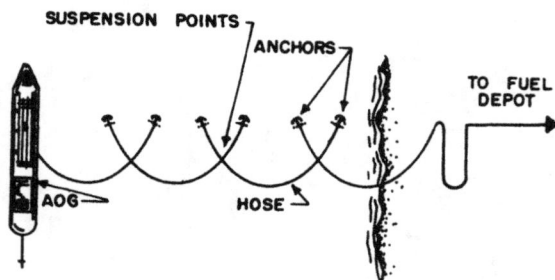

SUSPENSION POINTS

ANCHORS

TO FUEL
DEPOT

AOG

HOSE

SIZE OF HOSE.........4 IN...(INSIDE DIA)

LENGTH.....................5,000 FT. MAX.

CAPACITY........180 TO 600 GPM, at 100 PSI
TANKER DISCHARGE PRESSURE
(Depending on length of
hose and type of fuel
pumped)

and takes lateral stresses imposed by currents. Suspension points are provided at intervals of 250 feet on the four (4) inch system and 200 foot intervals on the six (6) inch system for the attachment of the mooring components. These consist of a 250-pound anchor secured to the mooring shackle by 210 feet of wire and a buoy attached by a wire line.

In general either LCMs or LCUs (modified for the task), or barges, warping tugs or causeway sections can be used for hose laying (See Figure 54. The system can also be installed with a diesel powered hose reel mounted on the deck of an LCU, warping tug, barge or causeway section for the installation. Specific instructions for installation are contained in the 5400 series of instructions of the battalion.

Estimated shore delivery rates in gallons per minute, assuming a tanker discharge pressure of 100 psi and a residual pressure at the shore end of 15 psi, are as follows:

AABFS 4" SYSTEM

Length	Gasoline	Jet Fuel/Diesel
1000 FT.	600 GPM	450 GPM
2000 FT.	460 GPM	325 GPM
3000 FT.	375 GPM	250 GPM
4000 FT.	315 GPM	210 GPM
5000 FT.	280 GPM	180 GPM

AABFS 6" SYSTEM

Length	Gasoline	Jet Fuel/Diesel
3000 FT.		800 GPM
4000 FT.	800 GPM	700 GPM
5000 FT.	700 GPM	600 GPM

Temperature and other factors may alter these rates, particularly those for diesel fuel. System recovery is accomplished by reversing the installation procedures. Gener-

100

ally, recovery time will be twice the time required for installation under comparable moderate sea conditions.

The bottom-laid fuel system consists of 5000 feet of 4 or 6 inch seamless well casing in 30 foot lengths, 300 feet of 4 or 6 inch refueling hose in 6 sections and installation equipment. The complete system is loaded on two 4-wheel pipe trailers. Other equipment required for installation, which must be landed at the same time as the pipe, are two TD tractors with blade and single drum winches, one 4x4 utility truck, one 2½ ton 6x6 cargo truck with camping equipment, one 315 cfpm air compressor, one 400-gallon water trailer, one power tong, one 4x4¼-ton jeep and one 5 KW lighting unit. A warping tug or LCU is also required in order to pull the line seaward. After landing, tractors are used to prepare an assembly and launching area. The pipe is assembled by connecting five sections of pipe together on an assembly way and rolling it on to the main launching way as the LCU pulls the pipe seaward (See Figure 55). After the desired anchorage is reached, the pipe is tested for leaks using 100 psi from the air compressor.

Operational capacities are approximately the same as for the positive buoyancy system. In recovery the pipe is pulled to the beach by the tractor, uncoupled into 30-foot lengths and loaded on the trailers. The LCU retrieves moorings, cables and other sea gear used in the operation. Under comparable sea conditions, the recovery time will be 1½ times the installation time. Further details will be found in NCEL TN-408, Techniques of Installation for the 4-inch Bottom Laid System.

FIGURE 55
BOTTOM LAID FUEL SYSTEM

In this picture of the bottom laid fuel system, the assembled 100-foot lengths of pipe are on the right. The warping tug which is pulling the line seaward from the main launching way is in the middle of the picture.

6. Beach Salvage Operations. ACBs provide personnel and equipment to the Beach Party for salvaging and assisting small craft during an operation. Beach salvage teams from the ACB are ordered to the beachmaster unit before activation of the beach party and become portions of the beach party teams when they are activated. Beach salvage cannot be classified into definite rules and procedures; the nature of the work requires men of skill and sense who can adapt themselves to variable and unpredictable circumstances. Beach salvage covers a broad range of assistance to landing craft, for example: assisting broached boats in retracting from the beach; helping to raise inoperative ramps on boats; lifting boats out of the water to affect simple repairs such as clearing fouled screws and hoisting boats and transporting them inland where repairs can be made. In addition when it is possible, assistance is given to vehicles stalled in the surf or bogged down in soft sand. These tasks, and any others that may be assigned, are directed by the Beachmaster. In a landing a salvage officer from the ACB will be assigned to the Beach Party Group Commander. Each beach salvage team is under the control of the Beachmaster and receives logistic support from the Marine shore party. Each salvage team consists of one officer and 12 enlisted men. Each team has one size 5 tractor with timberfendered dozer blade and rear-mounted single drum winch and one 20-ton crane, powered by another tractor. Other equipment provided to this team includes 2 2½ ton cargo (6x6) truck, a weapons carrier, a 400-gallon water trailer and a 5-KW floodlight trailer. During salvage operations additional equipment such as DUKSs, may be assigned to salvage teams by the Beachmaster.

7. Lighterage and Transfer Barges. The barge most commonly used for lighterage and transfer work is the "3x15" pontoon barge equipped with a single propulsion unit. This size barge meets as effectively as possible the demands of speed, maneuverability and transportability in a single unit. Other barge units used in landing operations are described in NAVFAC 401. These units vary in size from the 2x7 barge with a single propulsion unit, to the 6x30 (Rhino) barge with 2 propulsion units which can carry 500 tons or a maximum of 80 vehicles. Normally, four "3x15" barges are sidecarried by an LST to the scene of operations. Side-carried barges are ready for use within ½ to 2 hours after arrival, depending upon launching conditions. Barges may also be deck-loaded on an LST, carried in the well of an LSD or towed.

Normally in lighterage operations, the barges will be sent to an AKA or an APA to receive bulk cargo or vehicles for the beach. This method of transporting should be used only where light surf conditions exist and a suitable landing site has been established.

The lighterage element commander is assigned responsibility for the barge's operation and maintenance and is capable of effecting minor repairs

The primary function of the transfer barge is to transfer cargo from landing craft or amphibian vehicles to other landing craft. A 3x15 pontoon barge with a crawler crane lashed on deck is used as a transfer barge. This is known as a transfer line operation and is used where an offshore reef, a shallow gradient or other condition prohibits landing craft from reaching the beach. When the 3x15 barges are no longer required as transfer barges, the cranes may be removed and the barges reactivated as a self-propelled barge unit.

8. Warping Tugs, Tender Craft and Floating Cranes. The warping tug is primarily used for causeway tender and beach salvage duties, but its maximum line pull of from 40,000 pounds to 142,000 pounds, depending on its winch, and better maneuverability than an LCU have made it adaptable to other tasks. The basic warping tug structure is a 3x14 pontoon assembly using two outboard propulsion units. However, LCM-6s have been modified for use as warping tugs as well as causeway tender craft. A double-drum winch, driven by a 225 hp diesel engine is employed to handle the stern anchor line and the "A" frame itself stands twenty feet above the deck of the barge, and extends 7½ feet over the bow. This basic structure has been modified somewhat by each ACB to meet existing operating conditions.

The warping tug is used for the installation of causeways and, during unfavorable weather, a warping tug is necessary for tending long causeways to keep them from broaching. Using the "A" frame, warping tugs can salvage anchors weighing up to 8 tons. Using the stern anchor, the winch and the engines, the warping tug is able to free stranded craft from beaches. A warping tug can also be used for installing the bottom-laid fuel system or for salvaging vehicles lost overboard during the amphibious operation.

The normal method of transportation to the AOA is in the well of an LSD. This method has the advantages of easy loading without disassembly, easy launching and readiness for use when the tug reaches the landing area. LCM-type warping tugs can be deck loaded but are also normally carried in the well of an LSD, however, the tug may be side loaded on an LST by removing all deck gear from the tug.

Two types of floating cranes, both mounted on pontoon structures, are in use. These are a 75-ton crane mounted on a 6x18 pontoon barge and a 100-ton crane mounted on a 10x30 pontoon barge. Both are used for sideloading causeways and pontoon barges along with other lifting tasks which require their capacity. For descriptions of their other characteristics, uses and limitations, see the applicable NAVFAC instructions in the 5400 series.

9. Beach Improvement and Limited Construction Operations. Each Amphibious Construction Battalion is equipped for beach improvement and limited construction operations near the beach area. Sometimes beach improvement must begin before other operations can proceed. It is often necessary to land dozers from LCUs to prepare the beach area prior to the establishment of causeway connections. Later improvements, after beachhead establishment, include earth-moving tasks such as construction of sand ramps or slots for landing ships and craft, improvement of beach exits (in cooperation with the shore party) and improvement and maintenance of camp sites, sanitation and messing facilities. Other construction tasks arising from the nature of the mission may be assigned by the beachmaster. Conditions of climate, sanitation, local improvements and the overall strategy of the operation will dictate the kind, type and duration of improvements and construction required. In situations which involve so many variables, it is not feasible to present standard operational procedures.

Chapter VI
BATTALION LOGISTICS

SECTION A - THE SUPPLY OFFICER

1. **Introduction to the S-4.** The senior officer of the Supply Corps assigned to the battalion is the S-4 of the executive staff and is
head of the battalion supply department. (See
Articles 0901-0909 and 0982-0984 and Chapter
19, U.S. Navy Regulations.) He is detailed to
his billet by the Chief of the Bureau of Naval
Personnel. His responsibilities are to procure, receipt, store, issue, ship, transfer and
account for all equipage, repair parts and construction material. Operation of the general
mess and disbursement of and accounting for
government funds for battalion purchases and
military pay are also functions of the S-4.
When the battalion is operating at full
strength there are two Supply Corps Officers
assigned. In this case the junior officer usually serves as Disbursing and Commissary Officer. In the NCF units the Supply Officer is
expected to perform many but not all of the
duties usually associated with the broader concept of logistics. For example, the Supply Officer usually is appointed as Embarkation Officer and he is the major custodian of the allowance. However, he is not expected to be the
Transportation Officer, since this is left to
the Equipment Company Commander. Thus a SEABEE
S-4 will find that his tasks lie somewhere between the concept of a shipboard Supply Officer
and a Marine logistics officer. Figure 56 indicates his major functions. The Supply Officer is the key to the Navy Supply System. With
his technical advice, the battalion can properly requisition required allowance items or
project material.

In combat and disaster control operations,
the functions of the S-4 remain essentially
the same except that, because of the increased
tempo of operations and the urgency of supplies
such as water and ammunition, the company commanders designate petty officers at company
and platoon levels to assist the S-4 in the delivery of needed supplies to the units in the
field.

The function of the S-4 and thus the organization of his department vary considerably
depending upon where the battalion is operating. When deployed to the homeport or to an
established base in peacetime, the battalion
is not permitted to operate a clothing and
small stores or a ships store. The battalion
may utilize the disbursing facilities and the
general mess of the base to which deployed.
When this is done, men of the CS and DK ratings are ordered to the base on a TAD basis.
However, when operating a separate camp the

battalion runs a SEABEE mess and handles its
own disbursing. When deployed to an isolated
location the S-4 functions in all the capacities listed in Figure 56. The supply responsibilities and other logistic arrangements for
each deployment are published in the operation
order. On a joint operation, a major share of
logistic support may be furnished by the force
supported, be it Marine, Army, Air Force, or
other supply systems.

In preparing for a mission the S-4 must
avail himself of such information as the amount
of back-up supply support expected of the battalion supply department and the amount which
shall be forthcoming from the units of the task
organization. In back-up supply, consideration
must be given to all combinations of loss by
enemy action, heavy seas or inept handling.
Although some differences exist between the
supply systems of the ACB and that of the MCB,
the organizational pattern, general responsibilities, and functions are basically the same.
The variations in supply procedure for the most
part, involve types and quantities of materials
arising from the differences in the objectives
of the two battalions. Handling, stowing and
shipping problems likewise vary in complexity
and scope.

2. **Duties of the S-4.** The S-4 is expected to:
    a. Advise the Commanding Officer regard-
ing supply and related logistic matters.
    b. Maintain liaison with the COMCBLANT/
COMCBPAC Logistics Officer and Supply Officers
on regimental, brigade, division and advanced
base staffs along with other supply activities.
    c. Obtain, issue and account for allow-
ance equipment, supplies, and stores (Class
II) keeping appropriate records and inventories and maintaining stock levels of certain
supplies.
    d. Coordinate the annual Supply Overhaul
Assistance Program (SOAP) to determine the
condition of the allowance.
    e. Arrange for the supply of consumables
such as fuel and lubricants (Class III), rations (Class I) and project material (Class
IV).
    f. Assist with budget preparations and
administer O&M funds and project funds, keeping memorandum accounts and submitting records
and reports as required by COMCBLANT/COMCBPAC.
    g. Supervise the movement of battalion
material, the collection of prepositioned
supplies and the preparation of the required
embarkation forms.
    h. Operate warehouses and material com-
pounds for project material and shop stores
items.

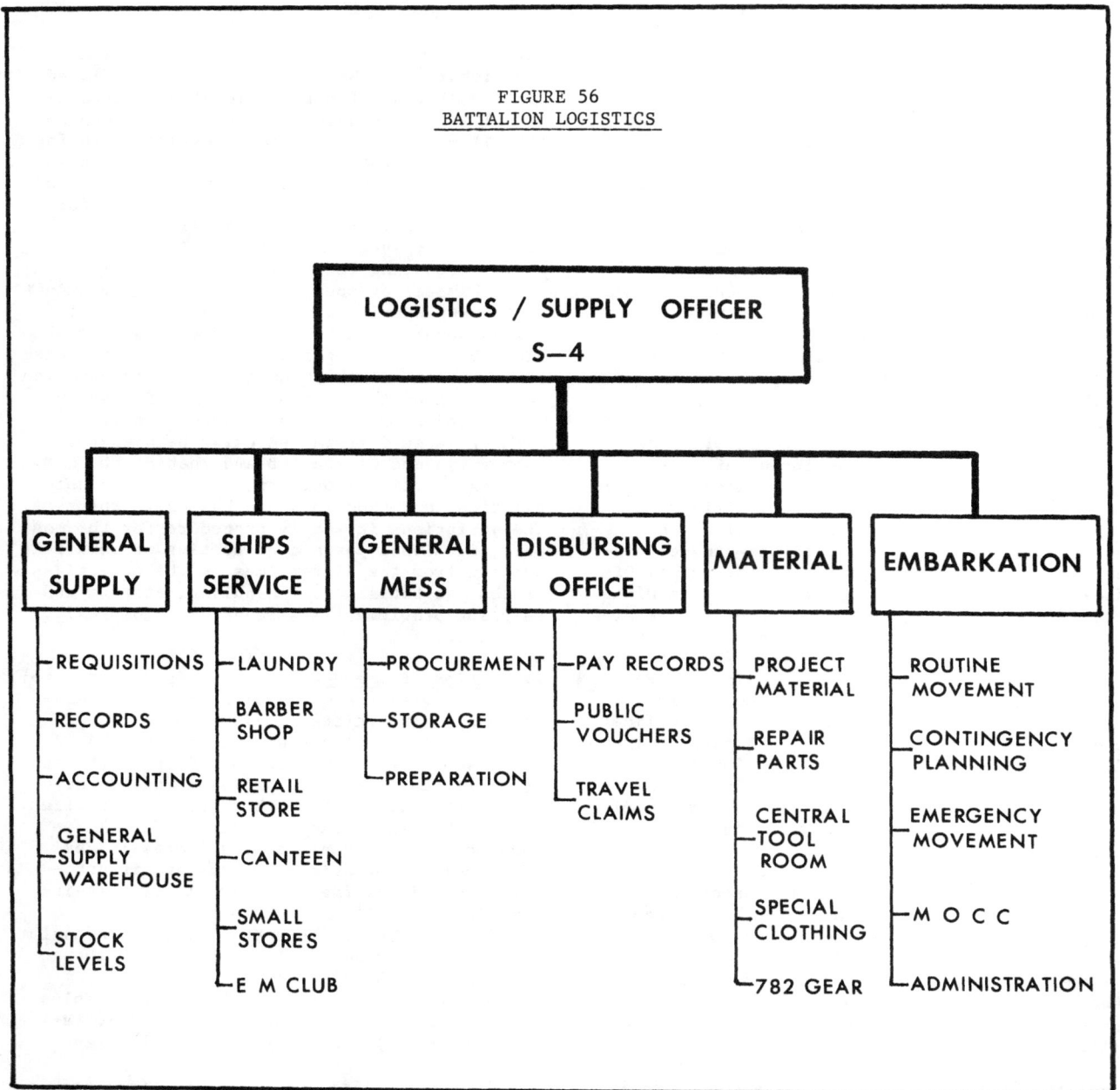

FIGURE 56
BATTALION LOGISTICS

LOGISTICS / SUPPLY OFFICER
S—4

GENERAL SUPPLY
- REQUISITIONS
- RECORDS
- ACCOUNTING
- GENERAL SUPPLY WAREHOUSE
- STOCK LEVELS

SHIPS SERVICE
- LAUNDRY
- BARBER SHOP
- RETAIL STORE
- CANTEEN
- SMALL STORES
- E M CLUB

GENERAL MESS
- PROCUREMENT
- STORAGE
- PREPARATION

DISBURSING OFFICE
- PAY RECORDS
- PUBLIC VOUCHERS
- TRAVEL CLAIMS

MATERIAL
- PROJECT MATERIAL
- REPAIR PARTS
- CENTRAL TOOL ROOM
- SPECIAL CLOTHING
- 782 GEAR

EMBARKATION
- ROUTINE MOVEMENT
- CONTINGENCY PLANNING
- EMERGENCY MOVEMENT
- M O C C
- ADMINISTRATION

i. Disburse government funds for military pay, travel claims, local purchase and local hire.

j. Operate a general mess (when required).

k. Operate a Ships Store (when authorized).

l. Operate a Clothing and Small Stores (when authorized).

m. Prepare the supply department to assist the battalion to meet all combat or disaster control commitments.

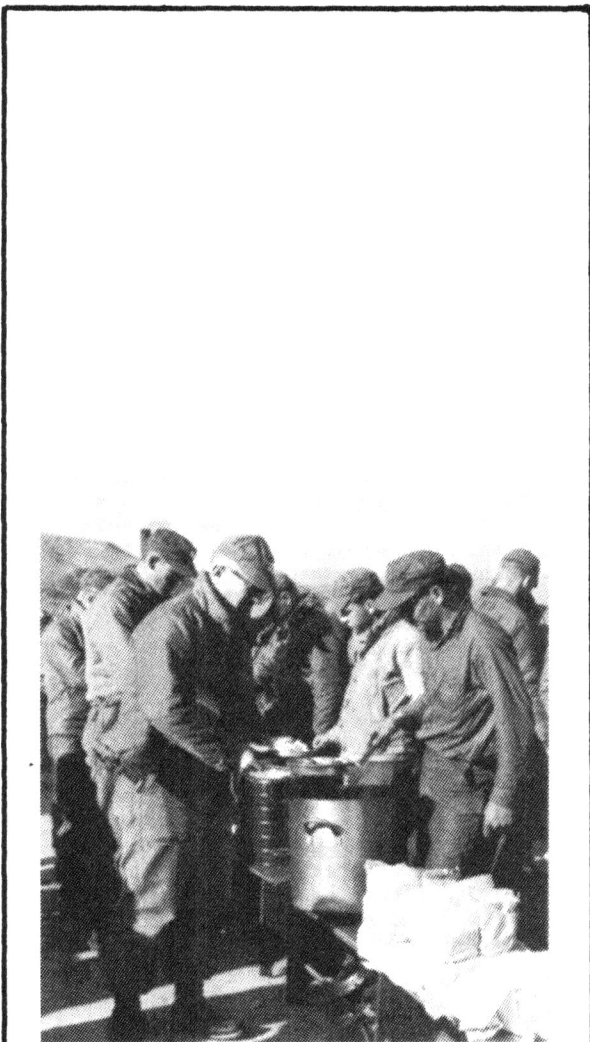

FIGURE 57
SEABEE FIELD MESS

As soon as he can set up the field cooking equipment ashore, the S-4 will provide hot food for battalion personnel. When operations permit, a galley and mess hall are constructed.

SECTION B - GENERAL SUPPLY

1. <u>Records</u>. The supply office may be supervised by a senior storekeeper (SK) who reports directly to the S-4. This office maintains the supply publications including those sections of the Federal Stock Catalog frequently utilized by the battalion and prepares letters, reports, requests for equipment and other communications to supply agencies. They also prepare, log, process and follow-up requisitions and post receiving documents against these requisitions. The accounting for battalion and project funds is done by members of this office. They also maintain a check on all stock levels, particularly of consumables under their jurisdiction and keep proper custody and inventory control of battalion equipage. Excess material, broken tools, etc., are disposed of by the supply office as directed by survey.

2. <u>Requisitioning</u>. Battalion operations require a complex variety of material. Equipment and automotive parts are usually the most critical shortages. Long-lead times and long-distance shipment require careful advance planning and action for procurement as soon as needs are established. In time of emergency mobilization, all items become comparatively scarce; therefore waste, hoarding or oversupply in one area can act to the detriment of others. Shipping space, port capacity and materials handling capacity are determining factors in military supply operations and only essential cargo can be carried. Under such conditions, many items become critical and their improper requisition or use can have severe adverse effects. Delivery of serviceable material to the location where it is needed requires co-operation among planning staffs, supply sources, and using activities. Because of the long-lead time, needs must be anticipated and acted upon as quickly as possible under the operational circumstances. The initiative for this logistic support lies with staff planners who have at their disposal information which cannot easily be made available to field activities. However, failures may happen between the source and the user through human oversight, unforeseen wear, improper maintenance, accidents, careless use of materials, storms or enemy action. To the extent that his situation permits, each MCB and ACB officer must know what arrangements have been made to support his operation, and take action to provide for any development which was not anticipated. He must have sufficient familiarity with the working of the supply system to obtain the material he needs, to maintain and use it, to repair or replace it if necessary, and to get along without it in an emergency.

To assist the Supply Officer to obtain material needed by battalion officers they must submit their request on DD Form 1348, DOD Single Line Item Requisition System Document (Manual), indicating the date the material is needed and completely identifying the items required. Since many tools, etc., are called by various names, the originator of the request must describe the item as it is listed in the Federal Stock Catalog. For example, to describe a tool as a "jackhammer" would cause delay in proper identification and processing by the supply system because, although "jackhammer" is a commonly used construction term, this tool is listed as a "pavement breaker" in the catalog. Identification must also include the specific stock number. For example, suppose the Supply Officer received a requisition for 2,000 board feet of 2 x 4s. He would find in the catalog, numerous stock numbers for 2 x 4s depending upon the type and grade of the wood and the number of finished sides. Since he would not know how the lumber is to be used, he must go back and ask the originator for more information.

When the Supply Officer must go to another Navy Supply Activity for the material, he prepares the requisition on the DD Form 1348, citing the accounting data and placing a priority number on the requisition. The number is a combination of a Force/Activity Designator (FAD) assigned by higher authority which reflects the battalion's employment (combat, positioned, ready or reserve and support) and the Urgency of Need Designator which runs from emergency requirements for material without which the activity is "unable to perform assigned operational missions" to material required "in preparation for a scheduled deployment." These designators, and their use are described in OPNAV INST. 4615.1 series.

Requests for material available at local government salvage yards or for surplus supplies of other activities on a "no funds involved" basis, will also be prepared on DD Form 1348, and submitted to the Supply Officer. Procurement of any surplus or salvage material or equipment of any nature whatsoever from any source must follow standard procedures. The requirements for custody and records on equipment procured from salvage yards are the same as for items procured new through regular channels.

3. <u>Accounting</u>. The Commanding Officer of the battalion is responsible for administering funds allocated by NAVFAC in accordance with the regulations governing the receipt, expenditure, and accounting for public money and stores (See Article 0717, U.S. Navy Regulations). Memorandum records of funds available to the battalion are kept by the S-4. Normally the O&M funds are given to the battalion by operating target (OPTAR) quarterly. Weekly, the obligation documents are sent to the CBC where the official allotment records are kept. The obligation documents are adjusted if necessary to conform to actual vice estimated costs and a monthly statement of the OPTAR is provided to the battalion. See Figure 95 for the battalion funding cycle. Project funds are handled in a similar manner. Instructions for the proper accounting for funds, limitations as to "flow of gold" restrictions etc., can be found in the NAVCOMPT Manual, the BUSANDA Manual and the directives of seniors in the battalion's chain of command.

4. <u>Custody Control</u>. Under current regulations and policy, all government property not of a consumable nature is controlled by a system of custody records. Items of equipage issued by the supply department are placed in the custody of a company commander or staff officer whose signature is required upon such issue. Company commanders in turn, maintain sub-custody records carrying the signature of personnel in their company, indicating the direct custodian of the equipage. Other accountable items of issue such as special clothing and ordnance gear are controlled by custody records signed by the individuals using them. Periodic inventories of all accountable items are held upon direction of the Commanding Officer, Supply Officer or company commanders.

All regular inventories and those special inventories occurring upon the relief or detachment of custodians will be accurate and physical. Survey proceedings will be promptly instituted for missing or worn out items. No custodian, officer or enlisted, may leave a command prior to the clearing of all records indicating his responsibility for accountable items. The importance of rigid control of accountable items of government property cannot be overemphasized.

5. <u>Battalion Allowance Stock Levels</u>. Effectiveness of deployed battalions is, in part, directly related to the proper provisioning, completeness and technical adequacy of the P-25 allowance. In the past the responsibility for maintenance and overhaul of Part 1 of the P-25 (See Annex H) has been a battalion S-4 responsibility carried out under the Supply Overhaul Assistance Program (SOAP) in accordance with CBPAC/CBLANT instructions. A complete inventory and inspection of Part I is made by battalion personnel under type commander guidance. This procedure commits a substantial number of battalion personnel to the SOAP process and thereby interrupts essential homeport training. In addition, the lead time involved in the maintenance/overhaul process frequently results in battalions deploying with less than a full pack up.

Experience gained as a result of RVN operations pointed up the fact that full outfitting capability, maintained in a ready-for-issue condition, is an absolute essentiality. Such availability is necessary to properly outfit redeploying battalions as well as outfitting new or additional battalions. Without this capability, the ability of the NCF to deploy at any time to any place to meet specific tasks and construction requirements is compromised.

Accordingly, the P-25 is now the subject of a central management program which includes all actions necessary to insure timely outfitting, maintenance and repair of the total component. To this end Part 1 of the P-25 will be maintained, overhauled and issued on a unit exchange basis. This process will exchange a completely refurbished/replenished Part 1 for a retrograde Part 1. Exceptions are confined to those elements of the Part 1 which are retained by a battalion to meet operational requirements and subject to the specific approval of the Type Commander.

The central management program in no respect is intended to abrogate the need for constant and continued "Fleet" supervision over the composition of the Part 1 Allowance. Central management of the Part 1 allowance does not change the present concepts of mobility and a full allowance of Part 1 equipment and supplies. The full allowance should be maintained on hand or on order. COMCBPAC/CBBLANT is responsible for detailed guidance on inventory and replenishment procedures to preclude excessive drawdown on the allowance while deployed. Compliance with these procedures will be examined during staff administrative visits to deployed battalions.

6. Disposition. A survey is required by articles 1947-1953, U.S. Navy Regulations when any Navy property, classified as equipage, must be condemned as a result of damage, obsolescence, or deterioration or be acknowledged as nonexistent as a result of loss or theft. These necessitate expenditure of the accountable material from the records of the holding activity. A survey of material involves the following steps:

a. Company commanders or department heads requesting a survey of material, set forth in writing to the Commanding Officer, via the S-4 the description of the material, its value, and a detailed statement of condition, cause, responsibility and recommended disposition. If an article is lost or stolen, this request will be made immediately (within 24 hours) upon discovery of the loss.

b. The Commanding Officer decides whether or not a survey shall be made and if so, whether a formal or informal survey is appropriate. He then appoints the necessary survey board.

c. The originator then transfer the items to be surveyed to the S-4 who will assume custody of the material and order it to be physically moved into a secure area. If the item is one which cannot be conveniently moved, such as a quonset hut or a pontoon assembly, it shall be tagged with the words "this (building or structure) being held for survey, unauthorized persons are prohibited from altering it in any manner."

d. The appointed Survey Board shall meet and conduct a survey in accordance with instructions set forth in the BUSANDA Manual and applicable instructions and directives. It will in all cases make specific detailed recommendations for the disposition of material and include a statement concerning the stripping of usable parts for return to battalion stock. It will issue an order to expend lost or stolen items from the record.

e. The Commanding Officer shall either approve or disapprove the actions of the Survey Board. If he disapproves, a new Survey Board may be appointed and a new survey made. If the second survey is still unsatisfactory to the Commanding Officer, the matter will be referred to the cognizant technical systems command for action.

After the survey, the Supply Officer will prepare all expenditure invoices. The Supply Officer or an assigned storekeeper will inspect material immediately before it is moved and compare the invoice with the material. A receipted copy of the expenditure invoice will be returned to the supply office.

7. Administrative Supplies. The responsibility of the S-4 for the general stores of the battalion includes maintaining allowance levels and controlling the issue of administrative supplies and special clothing along with maintaining custody and issue control of repair parts and project material. Administrative equipment and supplies consist of typewriters, adding machines, duplicating machines and the standard office supplies such as paper, ink, tape and staples. Since most of the office supplies are consumables, the S-4 must reorder sufficient quantities to maintain stock levels, taking into account the anticipated lead time.

SECTION C - SHIPS SERVICE

1. Ships Store. At most locations, the facilities of the base Navy Exchange or Ship's Store are available to battalion personnel. A Navy Exchange is operated with non-appropriated funds and is staffed mainly by civilians while a Ship's Store is operated with appropriated funds and is staffed by enlisted personnel, (Ship's Servicemen). Navy Exchanges are located at most permanent naval bases in the United States and overseas. BUSANDA regulations prohibit a Navy Exchange and Ship's Store from being operated at the same base. Thus, battalions operate a Ship's Store only when no other Navy facility exists for selling general merchandise. Both Navy Exchanges and Ship's Stores sell such items as candy, toilet articles, stationery, and clothing. They may also operate a canteen, barber shop, vending machines, laundry, tailor shop, cobbler shop and photography shop. Each battalion usually operates a barber shop in order to reduce the time that a man is absent from the job. The shop is normally located in the battalion camp. On a deployment where facilities are limited, the canteen is an important part of the ship's store, usually it includes a snack bar, soda fountain, and an enlisted men's club.

FIGURE 58
SUPPLY OVERHAUL ASSISTANCE PROGRAM

LAYOUT OF BATTALION TOOLS

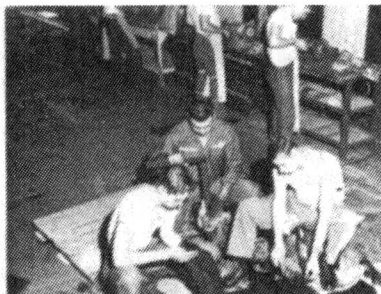

REPAIR and PRESERVATION

INVENTORY OF EACH BOX

REORDER MISSING OR
BROKEN TOOLS

TOOL KITS PREPARED
FOR INSPECTION

REPACK IN MOUNT-OUT KITS

Instructions for setting up and operating a ship's store are contained in the BUSANDA Manual.

2. Clothing and Small Stores. Clothing and Small Stores commonly referred to simply as "Small Stores" are established for the purpose of keeping standard uniform items and certain other articles such as towels readily available for sale to naval personnel. The battalion rarely operates a small stores because one is available at most standard deployment sites. But even when detachments or battalions are deploying to remote areas for short periods of time, the unit commanders usually ensure that each man has sufficient uniform items to last until he returns to an established base in lieu of setting up a small stores issue room.

SECTION D - GENERAL MESS

1. Commissary Section. Numerically, the largest section under the S-4 is the Commissary Section which operates all phases of the general mess or field mess including food storage, food preparation, food services, and the administration of commissary records, and returns. This section includes a bulk provisions storeroom, a commissary issue room, refrigerated spaces, butcher shop, bake shop, galley, vegetable preparation room, mess room and scullery. Assignments among the commissary men (CS) are based on rate, experience, training and ability. In addition to the commissary men, mess hall Masters-at-Arms and messmen are temporarily detailed to the general mess.

2. Master-At-Arms and Messmen. Mess hall Masters-at-Arms are assigned to the general mess under the direct supervision of, and responsible to the Commissary Officer for the performance of their duties. Personnel assigned to duty as mess hall MAAs become temporary members of the supply department during the period of assignment to the commissary section. Mess hall MAAs are responsible for the direction and supervision of the messmen in the performance of their duties, and maintenance of good order and discipline in mess spaces.

Messmen are the enlisted men temporarily detailed to the supply department by company commanders for food service duties in the general mess; they maintain cleanliness and sanitation of mess hall spaces and equipment, operate the scullery and dispose of garbage. In addition, messmen may be detailed to work in the vegetable preparation room, to wash cooking and baking utensils in the galley and bakery, or to perform any other duties required for the operation of the general mess. They will not be detailed to duties requiring the keeping of records. It is Navy policy to assign only non-rated men to duties as messmen. They are assigned on the basis of one messman

for each 20 men in the battalion. Duty in the mess is rotated so that non-rated men spend only one to three months as messmen.

3. Inspection by the OOD. Since food and food service have an important effect on morale, officers at all levels of command are interested in the general mess. The battalion Commanding Officer usually directs his officer of the day to sample the meals in the battalion mess. The OOD then inspects the galley and scullery and makes appropriate comments concerning quantity and quality of the food, the quality of the service, cleanliness of cooks and messmen, utensils and the general area. To properly make this inspection the OOD must be aware of the limitations imposed on the Supply Officer in trying to satisfy the desires of a large group of men. For example, the Supply Officer must provide a well-balanced diet for all three meals each day for about $1.15 per man per day. This daily ration may vary slightly from year to year but usually it will remain near this figure. It should be noted that provisions are made in the BUSANDA Manual for requesting additional funds when a unit is operating in a remote area or working hours other than normal. Additionally, the Commissary Officer must serve specified percentages of beef meals, chicken meals, etc., and is required to purchase an entire side of beef ranging from steak to hamburger. Therefore, battalion officers must be able to distinguish complaints about the food that stem from actions that can be corrected from those that come from a sense of monotony.

SECTION E - THE DISBURSING OFFICER

1. Disbursing Functions. When the battalion is operating at full complement, the junior Supply Corps officers serves as Disbursing Officer, otherwise the disbursing functions are done by the S-4. When deployed to an established base the disbursing facilities of that base may be used and the battalion disbursing clerks (DK) would then be assigned to the base. Normally, when deployed, the battalion operates a disbursing office. The disbursing clerks assigned assist the S-4 in maintaining pay records, paying bills chargeable to the battalion for locally purchased material, tools or labor, paying travel claims and per diem, and accounting for the disbursement of Government funds. The Disbursing Officer will not be charged with the handling, custody or accountability for other than public funds.

2. Military Pay. Before each pay day the Disbursing Officer determines how much is due each member of the battalion and posts the list in a conspicuous place. Regular paydays are prescribed by the Commanding Officer but special pay may be granted in emergency cases.

Payments may be made in cash, military payment certificates (MPC), foreign currency or by check. Area commander regulations usually govern method of payment and the responsibility of the Disbursing Officer to convert payment from one form to the other. The pay line forms at a time designated by the Commanding Officer, usually just before or after working hours. Payment is made at an easily accessible location such as the mess hall or at a window in the disbursing office.

SECTION F - MATERIAL OFFICE

1. Receipt Control. The Supply Officer is responsible for the receipt, checking, identification, and distribution of all incoming stores (except medical supplies and ammunition) and for processing receipt papers. Articles received from sources other than supply officers are opened and the contents verified before a receipt is signed. Company commanders will submit to the Supply Officer names of personnel who are authorized to sign for receipt of material for their company. The receiving storekeeper identifies and checks material using:

    a. Copies of procurement papers (requisitions, purchase orders, letter requests).

    b. Advance copies of incoming material papers (invoices, inspection reports, bills of lading) mailed by supplying activities.

    c. Packing copies of invoices accompanying material.

    d. Delivery slips furnished by the carrier with the material.

2. Project Material. While the Supply Officer is responsible for project material receipt, stowage and issue, the Operations Officer also has a vital interest in the flow of project material to the job site. Therefore, a good working relationship is essential. Most units assign a CEC officer to work for the Supply Officer as Material Liaison Officer.

The material control section is responsible for:

    a. Preparation, recording, and follow-up of requisitions for materials based on approved bills of material.

    b. Receipt of construction materials and record keeping against requisitions and bills of material.

    c. Operation and security of supply dumps, warehouses, compounds, and scrap and salvage yards for building materials.

    d. Direct phased delivery (whenever feasible) of building materials from ships to locations at the job sites approved by the Operations Officer.

    e. Controlled issue of materials to authorized persons in authorized amounts.

    f. Prompt notification to the Operations Officer of anticipated shortages or delays.

Members of the section may be designated to assist the planning and estimating section when the battalion is responsible for preparing material take-offs prior to deployment. If the battalion is drawing its building materials directly from a base, regimental, brigade, or other supply departments, members of this section may be placed on temporary additional duty (TAD) with the supplying activity.

Requests for supplies and materials for battalion use are made to the supply department on the basis of properly authorized DOD Single Line Item Requisition System Document (Manual) (DD Form 1348) from the companies. Company commanders have the authority to request stores for use in their companies within the monetary limitations assigned by the Commanding Officer. Routine requests for project materials are presented to the supply department except requests for transit-mixed concrete and local materials processed by the battalion (e.g., aggregate) may be made to the Equipment Company Commander. When the material is on board and set aside for that project, requests are usually filled from stock. The requisition also serves as a receipt. In some cases, the Supply Officer arranges for certain battalion personnel to draw certain supplies directly from base, regimental, brigade, or other supply departments.

3. Central Tool Room. The Central Toom Room (CTR) performs the issue and maintenance of all construction tools and equipment with the exception of those assigned to "A" Company. The Central Tool Room provides a central place where all construction tools can be gathered, inspected, maintained, repaired, packed for shipment, unpacked, and issued under an approved custody control system. Shop personnel perform tool repairs within the capability of the tool shop and maintain custody control of all battalion tools. The tool clerk requests assistance of other battalion shops for repair work when necessary. For example, major repairs on the motors of power-driven tools are performed on work requests by the electrical shop. The tool shop keeps a stock of tool handles, saw blades, and other consumables as well as special tools for emergency recovery operations. Shop machinery is issued on equipage custody directly to shop superintendents. Kits and tools needed continuously are issued to individuals or crews on custody receipts, which are filed alphabetically by the user's name. Other tools are issued on tool chits for the time needed.

4. Special Clothing. Special clothing is designed to provide protection and comfort under various climatic conditions, when the standard uniform is inadequate. Allowances are established for tropical and winter climates. When a new man reports for duty, he is given an initial issue of special clothing authorized for that geographical location. However, this clothing is the property of the Government, and is issued on temporary custody and is returned during the checking out process.

Special clothing includes the green working uniform, combat boots, safety shoes, utility caps, sun glasses, rain jackets, foul weather jackets, and the like. The Commanding Officer may authorize the standard allowance to be supplemented for men who work under extraordinary conditions. For example, a mechanic might be issued extra trousers and shirts or special items, such as coveralls may be procured to prevent excessive wear on the regular uniform or special clothing.

The Supply Officer has charge of the stores of special clothing and maintains a special room for its issue. He stocks special clothing for about six months normal usage. The storekeeper in charge of this room makes issues and exchanges as authorized. Each man signs a custody receipt for each article. All clothing is marked with an indelible number prior to issue. The safeguarding and care of special clothing is a responsibility of the man to which it is assigned.

5. Repair Parts. The Repair Parts Sections provides ready issue of repair parts to the mechanics of the Equipment Repair Platoon, and is operated by Storekeepers (SKs) and Construction Mechanics (CMs). The repair parts issue room is located in or near the equipment maintenance shops, but is controlled by the S-4. The repair parts section carries assem-

blies of parts for automotive, construction, weight-handling, materials handling and service equipment based upon Mobilization Allowance Lists (MALs) drawn up by the Ships Part Control Center (SPCC), at Mechanicsburg, Pennsylvania. These MALs are predicated upon 90-day initial supply and another 90 days of resupply. Manufacturers recommendation, equipment population and ACB/MCB usage data are used to develop and modify the MALs. There are two types of assemblies of repair parts: parts common and parts peculiar. The assembly of parts common contains parts of consumables that can be used on many equipments such as bolts, copper tubing and gasket material, etc. The content of this assembly is listed on MAL-1. The other assembly contains a supply of repair parts peculiar based on MALs for each make and model of equipment that requires repair parts support. The content of the assemblies may be adjusted on the basis of realistic estimates furnished by the Equipment Company Commander taking into account the age, condition, and known pecularities of the equipment to be supported.

All repair parts are arranged in "pack-up kits", see Figure 59. These kits consist of a series of standard sized boxes that are readily stacked on pallets. The insides are configured so they become the shelves of an issue room when the boxes are set up and the lids removed.

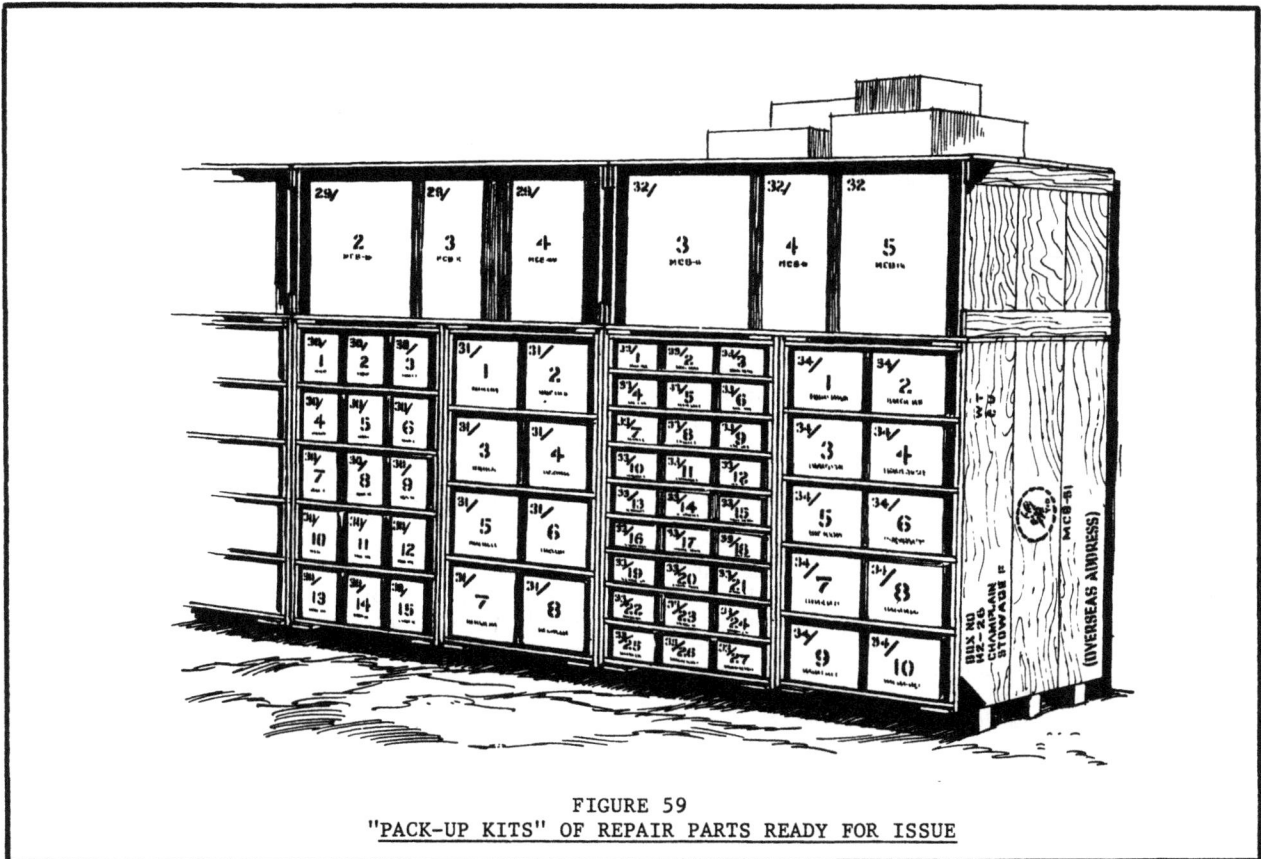

FIGURE 59
"PACK-UP KITS" OF REPAIR PARTS READY FOR ISSUE

Box number one is the control and contains the MALs, shop manuals, manufacturer's parts books, portions of the federal stock catalog and a cardex inventory control. Stock record cards for each part are filed by stock number showing the location of the item in the pack-up kit. Use of the pack-up kits makes it possible to secure for movement or to set up for full-scale operation in a matter of hours.

When a part is required to make repairs to battalion equipment, the mechanic tells the men in the repair parts section of his need. Using Manufacturers' Catalogs they identify the part by federal stock number. If the part is physically in the assembly they issue it, if not, they make out a requisition which is sent to the nearest supply activity. The assembly usually provides 50% of the parts required.

In addition to USN equipment, each battalion has equipment and tools not included in MAL repair parts support. This includes power saws and wrenches, blueprint machines, electric ranges, movie projectors, laundry extractor washers, shop machinery, radiac equipment, communications equipment, etc. Most of this equipment has repair parts boxes furnished with the original equipment. When a part is used from the repair parts box, a DD Form 1348 is prepared for a replacement. The company commander or staff officer responsible for the equipment must anticipate the needs for repair parts to support this equipment. Quite often parts for such equipment are not stocked at supply activities. Therefore, procurement may take many months. Details of repair parts management may be found in NAVFAC P-404.

SECTION G - EMBARKATION

1. Battalion Mobility. The mobility of the fleet and fleet support units is a decisive factor during contingency operations or in wartime. As fleet units, the battalion must be constantly ready to pick up, move, and build on the shortest possible notice. One project may be cancelled in favor of another at any stage. The battalions attain the needed mobility by practiced foresight and by "operating out of a box." Movement techniques are developed, practiced, and tested in peacetime for immediate application in war.

In other than full mobilization in peacetime, an MCB usually spends about five months between deployments at its home port. This time is devoted to leave, overhaul of equipment, technical and combat training, and outfitting for the next deployment. The ACB does not deploy as a battalion, but various tactical teams spend from four to eight months embarked aboard amphibious ships. Upon return to the homeport they overhaul equipment, attend training courses, participate in local training exercises and prepare for another deployment.

In wartime, the MCB usually redeploys directly from one advanced base to another. The MCB may land at the site of an advanced base on or soon after D-Day. The ACB tactical teams secure operations soon after the assault phase is ended. In some cases, their mission may be completed early enough that they can reembark aboard the task force to conduct operations at another location.

To enable the battalion or its detachments to make these movements the battalion usually:

a. Maintains a running analysis of potential cargo.

b. Uses multi-purpose containers for control and movement of the battalion allowance.

c. Prepacks, paints and marks these containers.

d. Updates a series of deployment check lists.

The battalion is usually advised of the specific requirements for each deployment through movement orders and operation orders. Movement orders establish ready dates and designate the vessel or vessels that will lift the battalion. The movement order is addressed to the battalion and to the ships involved. Long-distance liaison between the battalion and the ships provide the basis for detailed planning of lifts and the early establishment of firm movement dates.

The Commanding Officer establishes liaison with all commands concerned, initiates comprehensive planning and preparations with the battalion, assigns specific responsibilities, and sets target dates for each phase of the movement. As chief staff officer, the executive officer coordinates all arrangements relying upon the embarkation officer to carry out the details of battalion movement.

2. Routine Movement. During routine movements to standard deployment sites, the battalion gear is usually loaded aboard MSTS or commercial shipping by stevedores of the base to which the battalion is deployed. Personnel are routinely transported by Military Airlift Command. The battalion Supply Officer acts as liaison between the base, the transportation agency and the battalion.

Soon after the movement order is received indicating the ship and her approximate date of arrival, the Supply Officer contacts the ship to indicate the number of personnel and the amount of cargo to be lifted and to obtain the ships troop regulations, berthing data and diagrams, requirements for troop watches and work details, information on chaplain, medical, dental, messing and recreational facilities and any other information needed to plan the move. If traveling on MSTS shipping, the Voyage Staff Instructions of the 4650 series provide a great deal of information needed by embarked units.

The Supply Officer works out the details of cargo assembly and movement to the embarkation area with the port authorities of the base. Usually the battalion responsibility is limited to the packing and delivery of allowance items to the port authority for shipment.

Each company commander and custodial officer is responsible for the preservation, listing, packing, weighing, and final marking of all assigned material. Each category of cargo and personal hold baggage must be turned in to the S-4 at the designated cargo assembly (transit) areas on schedule. A standard packing list for each container turned in for shipment is prepared by the custodial officer in quadruplicate. This packing list is a standard NAVSANDA Form 225 and shows the shipping activity, container number, itemized description of contents in sufficient detail to permit replacement of each item, the length, width, height, cube and weight of the container, date packed, the packer, the custodial officer and a general description of the contents for listing on the Ocean Manifest. The original is sent to the Embarkation Officer, one copy is placed inside the container, one copy is placed in a waterproof envelope attached to the outside of the container and one copy is retained by the custodial officer.

An Ocean Manifest, NAVSANDA Form 1121 must be prepared for all ocean shipments. These documents are prepared by port authorities or by the battalion Supply Officer in accordance with Joint Shipping Instructions and OPNAV Instructions of the 4620 series.

After battalion gear is assembled by the Supply Officer it is turned over to the port authorities of the base who are responsible for seeing that it is outloaded. Usually project material is handled entirely by the base, and the battalion does not assume custody until arrival at the new site.

Specific arrangements must be made by the battalion for the embarkation of personnel. This includes:

a. Provision for the return of library books, recreational gear, hobby shop equipment, and related items.

b. Settlement of mess, pay, BOQ, and other personal affairs.

c. Check of immunization and health records.

d. Marking and turning in hold baggage.

e. Announcement of regulations governing uniforms, carry-on baggage, mail, cameras, radios and customs requirements.

f. Assignment of duties to individuals while embarked.

g. Inspection of troop compartments by company commanders and platoon leaders.

h. Pre-assignment of berths.

i. Distribution of boarding plan.

j. Boarding of messmen, galley personnel, initial watchstanders, and MAA guides.

k. Loading hold baggage.

l. Loading personnel records and other carry-on gear.

m. Final messing ashore.

n. Final clean-up and inspection of all areas being evacuated.

o. Orderly and exactly timed movement of troops and carry-on gear to shipside.

p. Checking of boarding list as each man comes aboard ship.

q. Swift and orderly movement of troops to assigned compartments and berths.

r. Delivery of passenger list to the Captain or Master of the ship.

3. Contingency Planning. The specific tasks involved in battalion movement depend on the support, harbor and cargo-handling facilities available at the sites of embarkation, debarkation, and construction operations. The battalion must be prepared to adjust rapidly to the specific requirements of each deployment and to each phase of each deployment.

The Commanding Officer is advised of planned deployments as far in advance as possible. However, the battalion must always be ready for sudden changes in plans, for premature roll-up of one deployment, and for swift redeployment of a higher priority operation. In planning for rapid mount out under emergency conditions the battalion will discover that little outside help will be available. The move will probably be in support of a Marine expeditionary force and thus shipping space will be shared with Marine units. The battalion will be required to prepare loading plans for allowance gear and all project material and actually load these items aboard. The embarkation plan may indicate that certain ships are to be combat loaded. Because shipping is scarce, the battalion may find that only the highest priority items can be embarked in the initial loads and that much of the battalion gear will be loaded on follow-on shipping. These factors greatly increase the embarkation work for a SEABEE unit. Therefore, a specially trained officer assistant is detailed to the S-4. This battalion officer should have been trained at the Embarkation School run by PHIBLANT or PHIBPAC and be familiar with the battalion allowance, Marine Corps embarkation forms, Ships Loading Characteristics Pamphlets (SLCP) and loading plans.

Embarkation planning consists of planning for surface mount-out and movement of personnel, equipment and supplies to a forward area. Since actual embarkation depends upon many factors such as ship availability, debarkation conditions and overland movement, the battalion must be responsive to changes that may occur on short notice. The required flexibility can only be accomplished by proper pre-planning, practice and a knowledge of potential difficulties.

It is convenient to divide supplies and equipage into the same 10 classes of supply used by the Army and Marines in embarkation and field operations:

a. Class I supplies are subsistence including combat rations.

b. Class II supplies are all unit allowance items not included in other classes, such as clothing, individual equipment and weapons, tentage, organizational tool sets and kits, hand tools, administrative and housekeeping supplies and equipment.

c. Class III supplies are all the fuels, oils and lubricants.

d. Class IV supplies are construction materials including installed equipment and all fortification/barrier materials.

e. Class V supplies are all types of ammunition.

f. Class VI supplies are personal demand items (non-military sales items).

g. Class VII supplies are major end items that are ready for their intended use, such as mobile machine shops and vehicles.

h. Class VIII supplies are all medical materials including medical peculiar repair parts.

i. Class IX supplies are all other repair parts and components to include kits, assemblies and subassemblies, repairable and non-repairable required for maintenance support of all equipment.

j. Class X supplies are material to support non-military programs such as agricultur-

al and economic development.

The embarkation forms are designed to accommodate change. The most important forms are:

a. The Vehicle Summary and Priority Table (VS&PT) which includes a list, in priority sequence, of all automotive and construction equipment, the dimensions, weight and cube of each vehicle as well as the weight and cube of any cargo it carries.

b. The Cargo and Loading Analysis (C&LA) which is the overall control of bulk and prepared cargo (See Figure 60).

c. The Unit Personnel and Tonnage Table (UP&TT) shown in Figure 61 which gives a summary of the entire load by classes of supply, indicating personnel, personal equipment and operational equipment such as causeways and hoisting gear, etc. These forms are very helpful in controlling the embarkation phase but in order to properly place each item in proper unloading sequence a diagram of the ships configuration must be drawn to scale and templates of vehicles and loads moved about until the optimum solution is reached for each compartment and for each ship. This is then drawn up and becomes the loading plan (See Figure 62).

FIGURE 60
SAMPLE CARGO & LOADING ANALYSIS (C&LA)

UNIT: MCB TWO
SHIP: USS NOBLE (APA-218)

UP&TT LINE NO.	DESCRIPTION	NUMBER & TYPE CONTAINER	NUMBER ROUNDS RATIONS ETC.	STANDARD CARGO		NUMBER OF HEAVY LIFTS			MOBILE LOADED		WHERE STOWED
				CU FT.	WT. (LBS)	NO LIFTS	SQ FT	HT.	CU FT	WT (LBS)	
1	RATIONS	5&1 TYPE	870 CASES	957	26,970						
5	MECH TOOLS	2 BOXES	———						6	400	V-8
5	OFFICE SUPPLIES	5 BOXES	———						61	1890	V-8
5	REPAIR PARTS	2 PALLETS	———						108	2400	V-3
29	AMMO	A-124	1 BOX	1	75						AMMO MAG.
	PAGE TOTALS			958	27,045				175	4690	
	GRAND TOTALS										

FIGURE 61
SAMPLE UNIT PERSONNEL AND TONNAGE TABLE (UP&TT)

OFFICERS	CAPT	CDR	LCDR (2)	LT (1)	LTJG(2)	ENS	WO	TOTAL OFF	5
ENLISTED	CPO2	PO1 & PO2 32		OTHER ENL 139				TOTAL ENL	173
								TOTAL PERS	178

SUPPLY CLASS	LINE No.	DESCRIPTION	CU.FT.	WT. (lbs)
		CARGO (less vehicles)		
	1	RATIONS	982	27695
I	2	WATER		
	3	TOTAL CLASS I	982	27695
II , IV MEDICAL AND DENTAL		**AUTHORIZED ALLOWANCES**		
	4	TROOP SPACE CARGO	965	43920
	5	OTHER CARGO	4057	87414
	6	TOTAL CLASS II ,IV, MEDICAL & DENTAL (less veh.)	5022	131334
IIA & IVA	7	AVIATION MATERIAL		
	17	ENGINEER		
	18	GENERAL SUPPLY (QM)(less exchange supplies)		
	19	MOTOR TRANSPORT		
IV	20	ORDNANCE		
	21	CHEMICAL (less inflammable agents)		
	22	TRANSPORTATION CORPS		
	23	TOTAL CLASS IV		
IIA & IVA	24	AVIATION MATERIAL		
		POL		
	25	GASOLINE & KEROSENE		
	26	OTHER POL		
	27	TOTAL CLASS III		
IIIA	28	AIRCRAFT FUELS AND LUBRICANTS		
		AMMUNITION		
	29	SMALL ARMS	25	1875
	30	HIGH EXPLOSIVES		
	31	INFLAMMABLES (pyrotechnics & chemical agents)	639	9902
	32	NUCLEAR WEAPONS		
	33	TOTAL CLASS V	664	11777
VA	34	AIRCRAFT AMMUNITION		
		OTHER SUPPLIES		
	35	EXCHANGE SUPPLIES		
	36	MEDICAL & DENTAL		
	37	TOTAL OTHER SUPPLIES		
	38	TOTAL CARGO (add lines 3,6 ,7, 15 , 23 , 24 ,27, 28 , 33 , 34, and 37) (Sq. Ft. of unitized cargo and heavy lifts from C & LA)	6668	170806

		VEHICLES			
II , IIA & IV	39	TOTAL VEHICLES:	3006	28335	174625

GRAND TOTALS	SHORT TONS	MEASUREMENT TONS			
	172.7	875	3006	35003	345431

UNIT: MCB-TEN EMBARK TEAM No 3 CERTIFIED: _____ COMMANDING

SHIP: USS NOBLE APA-218 DATE _____

117

FIGURE 62
TYPICAL LOADING DIAGRAM

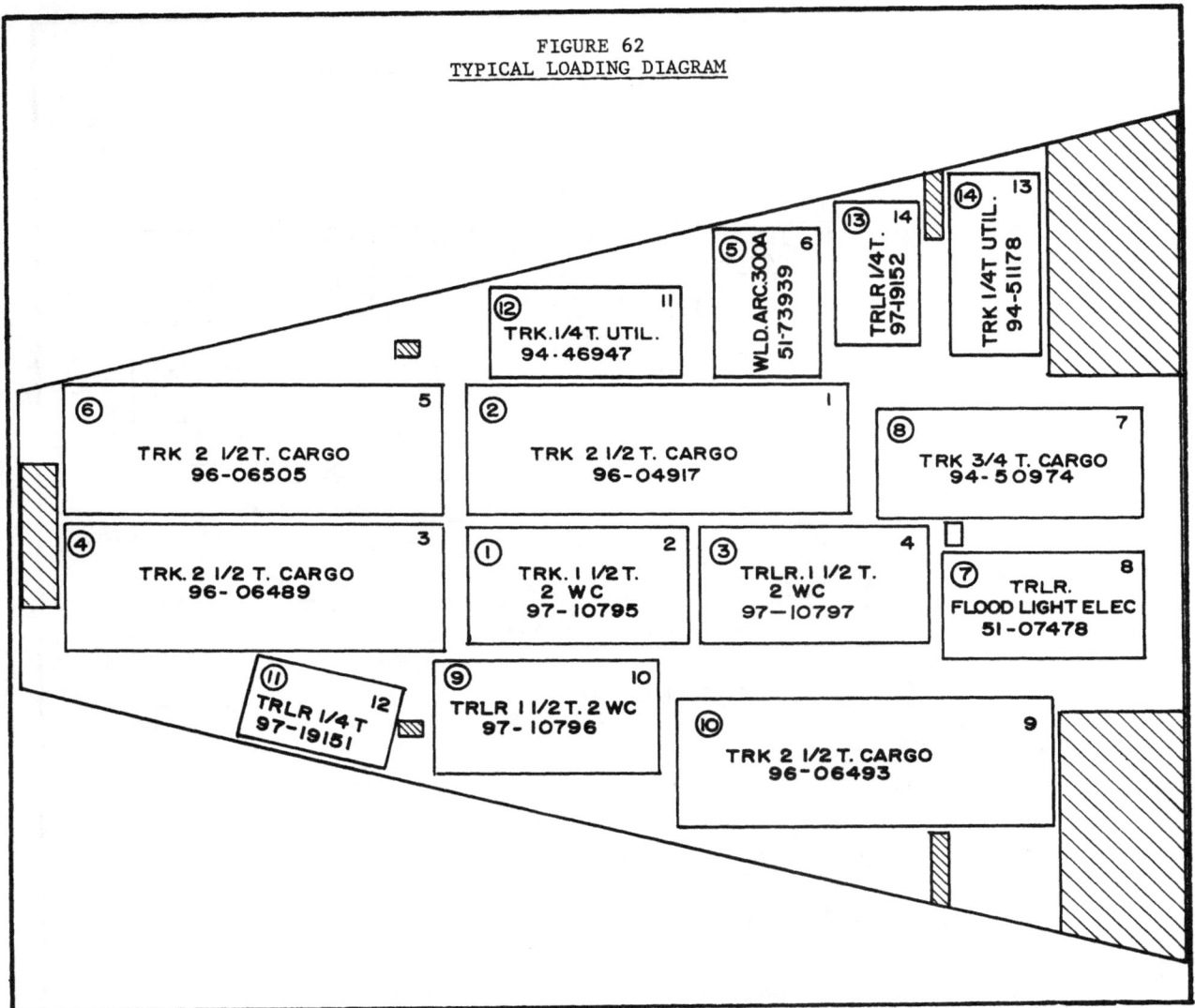

4. <u>Emergency Movement</u>. In the case where the battalion must mount out on an emergency basis, all the functions listed under routine movement must be done plus many others, and usually in a very short time. Job sites must be secured, project materials stored and projects turned over. Loading plans must be prepared, Class I, III and V supplies and often an entire camp must be procured, stocks of clothing and small stores, barber supplies, etc., must be obtained. In addition to packing battalion tools and preparing the equipment for shipment, the battalion may be required to take-off, requisition, and carry some or all of the building materials needed for the deployment. Excess personnel baggage must be collected and transferred for shipment back to homeport. Dunnage and shoring must be estimated and purchased. Unpaid battalion bills must be transferred to the base for payment when due. Many other tasks must be performed including perhaps the actual loading of the ship.

In order to handle these many tasks the battalions have developed standard embarkation plans and have completed cargo loading and analysis sheets and loading diagrams for hypothetical conditions. Also the concept of a Mount Out Control Center (MOCC) has been used to reduce the possibility of overlooking any details. The MOCC is designed to provide the battalion Commanding Officer with up-to-date information on his unit's preparedness to mount-out and accomplish a contingency task. It is essentially a set of file folders which contain detailed check-off lists for staff and company officers. They should be kept in a conference room which contains the necessary equipment to be used as a command post (e.g., tables, chairs, telephone, blackboard, file cabinets, plotting boards and security so that classified information may be displayed.

The file folders contain routine step-by-step procedures, personnel to be contacted, holiday routine procedures and critical deficiencies.

Usually a folder is set up for these categories:

a. Class I and III supplies
b. Class V supplies
c. Tool kits and central tool room
d. Ordnance gear
e. Communications gear
f. NBC equipage
g. Camp components
h. Equipment and repair parts
i. Medical and dental gear
j. Other equipment and supplies
k. Mount-Out tasks assigned to companies

Mount-out exercises are conducted by each battalion in order to locate and correct problem areas in the battalion's plans and to indictate to supporting commands the scope and timing of assistance required.

A discussion of the packaging of the allowance for rapid movement is included in Annex H. Figure 63 shows a typical emergency mount-out by an MCB.

5. <u>New Concepts in Mobility</u>. The military action in Vietnam resulted in many lessons learned for all of the U.S. Armed Forces including the Naval Construction Forces. One of the major change as to both tactics and hardware concerned mobility. The Army founded their Air Calvary Division; the Seagoing Navy developed new concepts of river warfare. The Naval Construction Forces were also profoundly affected. While development of major enclaves was still proceeding rapidly, the scene of battle began to shift from the coast line to the interior of Vietnam. The pattern here was new but undoubtedly the pattern of the future. The NCFs were called upon to support Army and Marine operations in an underdeveloped country in remote areas inaccessible by fixed wing air craft or cross-country vehicles. SEABEES met the challenge by taking to the air in helicopters with men, material and equipment. An entire series of special light weight (under 10,000 pounds) equipment was developed. It included dozers, graders and dumptrucks. This equipment was air-lifted to remote sites by CH-46, 47, 53 and 54 helicopters belonging to the Army and Marines. It soon became evident that to properly initiate and support remote areas with a short time for construction the SEABEES would need organic helicopter support. Organic helicopter support is now planned to be attached directly to Contingency Area Brigade or Regiment Commanders.

FIGURE 63
EMERGENCY MOUNT OUT

SECURING PROJECTS

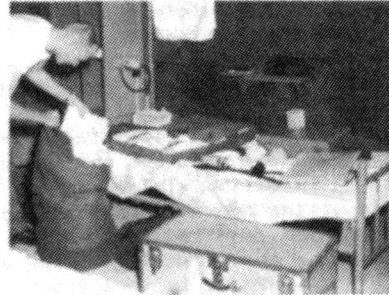

PACKING UP PERSONAL GEAR

TRAILERS ARE LOADED
BY EACH OFFICE

CONSUMABLES ARE DELIVERED
BY SUPPLY

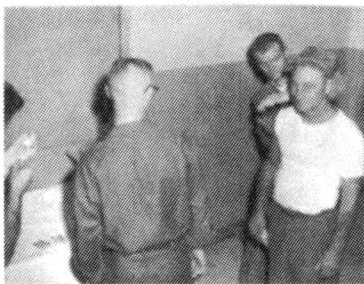

THE CREW GETS SHOTS

THE ADVANCE PARTY
MOVES BY AIR

FIGURE 63 (Cont'd)
EMERGENCY MOUNT OUT

CRAWLER EQUIPMENT
IS MOVED EARLY

SUPPLIES ARE STAGED
ON THE PIER

VEHICLES ARE PREPARED
FOR SEA

HEAVY EQUIPMENT IS
HOISTED BY CRANE

AUTO EQUIPMENT IS
DRIVEN ABOARD

EQUIPMENT IS COMBAT LOADED
IN AN LSD

Chapter VII

COMPANY COMMANDERS

SECTION A -
THE ROLE OF THE COMPANY OFFICER

1. Command Responsibilities. Junior officers in construction battalions are confronted with a different combination of problems than shipboard officers. Peacetime construction projects, in particular, require that Naval procedures be combined with management practices not far removed from those used in the construction industry. The Naval officer must uphold his assigned task as efficiently and economically as possible. Faced with a constant task of personnel management, he has a more difficult duty than a civil engineer whose personnel responsibilities end at quitting time. Every officer has encountered the Division Officer's Guide at some point in his career. Although this guide was originally intended for shipboard use, most of its general details apply to any Naval unit; and it serves as a useful source for management and personnel procedures. An officer is, in effect, a manager, and must be aware of his real managerial responsibilities. Early in his career, a Naval officer discovers that an officer seldom turns out an end-product directly, and that results are achieved through other persons. A company commander is seldom a first-line supervisor. The Navy has evolved a system of developing on-the-job supervisors -- the petty officers. These are men who have demonstrated proficiency in leadership as well as skill in a trade. When petty officers are properly trained, it may only be necessary to indicate goals and to establish deadlines.

Immediate supervision of other lower enlisted personnel is the responsibility of all petty officers. Proper delegation of authority is necessary to keep a unit, whatever its size, in operational order. The majority of battalion petty officers have earned their ratings in construction skills rather than seamanship, but their basic function of leadership in the Naval organization remains the same. However, as the Division Officer's Guide points out, reliance on petty officers to do their work can only be assured through proper supervision by their commissioned officers.

In both MCBs and ACBs, the function of officers as commanders is basically two-fold. They must operate the land-based equivalent of a taut ship, which involves administration of personnel, and they must be equally able to fulfill their engineering assignment as administrators of a construction activity. For an MCB, construction activity can be taken in the usual sense of a building project; for an ACB the meaning, although more specialized, still applies.

The objectives of any assignment can be understood through intelligent analysis of what is expected. Once an officer understands the part he and his men are to play in a project, he is responsible for communicating to the petty officers the nature of the task and other information they should have at the beginning. The petty officers in turn, pass the word to the men. If each man can be made to feel that he is important to the project, a set of common goals will be established throughout the chain of command. Enthusiasm, unlike discipline, cannot be imposed. In a formal sense, authority is delegated through the chain of command. However, on the working level it must be delegated in such a way that each man, including the construction apprentice, is aware of his individual responsibility and is prepared to execute it. If fixing responsibility for work and equipment is not practiced in fact as well as in theory, individual initiative among the men, and maintenance of equipment will show the effects. The corollary to this delegation of authority is supervision, which must not be neglected merely because men have been informed of what is expected of them. Supervision at all levels must be exercised constantly to make certain that the expectations are fulfilled.

Evaluation must be a constant process for the company officer. He in turn, is being constantly evaluated by his superiors. The evaluations must be made of the entire scope of activity which has become his responsibility. The key to evaluation is inspection. Formal inspections are necessary, and must be timed at intervals frequent enough to keep men alert, but no so frequent as to interfere with the work. Casual attention, when it can be given unobtrusively, is the best way of keeping informed about personnel and job progress. The intelligent officer knows enough to refrain from merely prying, which will demoralize the operation, but he also learns to keep an eye on his men in a way that promotes progress and good feeling. These arts are not learned from instruction books; they are acquired in the field.

Each company commander commands his company in accordance with the policies of the Commanding Officer. The general duties and responsibilities of company commanders are comparable to those prescribed for heads of departments in Chapter 9, U.S. Navy Regulations. Company Commanders are assigned to their billets by the Commanding Officer.

Each company commander performs the following functions:

a. Exercises command through his platoon leaders.

b. Organizes and trains his company for construction (or construction support), for combat and for disaster control operations.

c. Executes efficiently all work assigned to his company by proper authority.

d. Maintains his company in a state of readiness to rapidly redeploy to meet emergency situations.

The Company Commander is responsible for company administration involving the following functions:

a. Morale and welfare of men assigned.

b. Training and state of readiness of his company.

c. Economical use of material and funds.

d. Safety.

e. Recreation.

f. Discipline.

g. Inspections.

h. Formations

i. Directives, correspondence and reports.

Junior Officers will find additional guidance in Annex J.

2. _Personnel Administration_. The company officer works with his petty officers as they strive to reach their goals. He fosters an atmosphere which promotes such enthusiasm, interest, and pride in their work that they may well exceed their own expectation. He permits freedom of action so they too, may exercise initiative and leadership. Yet the limits must be outlined, defining the bounds which are established by his decisions as well as by those set by other authorities.

The company commander must be flexible and willing to forego the luxury of blowing his top. He must be able to see sincere petty officers and men make mistakes and charge the cost as a worthwhile investment in the development of his most valuable resources -- people. To criticize too much is to discourage the individual's willingness to try again. Battalion officers will always find that SEABEES are the most ingenious and resourceful men in the world, and when the pressure is on, they can produce in any situation because they are not afraid to tackle something new or strange to them. The overwhelming success of the SEABEE Teams in the counterinsurgency effort is but one of the latest of a long string of such examples stretching back to those early days on Guadalcanal.

The company officer promotes an "atmosphere of approval." This is the spirit prevailing in which each man feels free to exercise his own best judgment, rather than working in fear of making a mistake. To accomplish this requires patience and understanding.

The greatest productivity with highest morale results from properly channeling each man's own drives. A man works hardest when he has a stake in the outcome. Thus the best re-sults will come from sincere attention to each man's wants. Special attention should be paid to the petty officers by getting them in on planning and encouraging real participation. This may take time, but it pays rich dividends in better plans and especially better-executed plans. One of a petty officer's greatest sources of satisfaction is being "in the know." Therefore they must be kept informed.

Every company officer must look after the needs of his men. He should:

a. Encourage preparation to qualify for advancement in rating.

b. See that they have as good a place to live as circumstances permit.

c. Get them good liberty.

d. Make the food as good as possible.

e. Pay attention to safety and minimize hazards.

f. Encourage cohesive groups and promote teamwork.

g. Build up each man's self-esteem and sense of importance and give him the satisfaction of a job well done.

h. Let each man know you are interested in him as a fellow human being by showing a sincere interest in the solution of his problems, but without undue familiarity.

i. Provide high quality leadership by setting a good example and showing impartiality.

Personnel administration will consume many hours of the company commander's time. Its importance cannot be overemphasized. It includes:

a. Billeting of all company enlisted men and inspection of the company's berthing area.

b. Assignment and rotation of messmen, watchstanders, and working parties drawn from the company.

c. Coordination of liberty, leave, special requests and privileges granted to his men.

d. Regular inspection of company personnel in ranks.

e. Routine clothing, bag, and locker inspections of company personnel.

f. Performance of company personnel in formations, parades, reviews, and ceremonies.

g. Keeping up the company watch quarter and station bill.

h. Keeping up the company scoreboard of Advancement in Rating.

Many functions shown under the S-1 and S-2 are actually carried out by the company commanders, and further, since they are responsible for the readiness and morale of their men, they must be familiar enough with enlisted personnel administration and the battalion training program to be able to monitor them and to be able to advise and assist their men.

Company administration should include some type of individual data card, the practical factors sheet, a record of training and a watch quarter and station bill. The individual data card may be tailored to each battalion's needs or standard forms such as the Personnel Accounting Card, NAVPERS 500 or the Division Officers Personnel Record Form, NAVPERS 2840 (shown in Figure 64) may be adapted. The practical factors sheet, NAVPERS 760 along with a record of courses attended can be kept in a training folder maintained for each man. In some cases the company commander will display the status of training for each man on a training scoreboard maintained in the company office.

The watch quarter and station bill is a roster of all personnel on board by name and rate. It is drawn up according to the company's military organization but it also shows each man's job assignment, berthing space, tasks for drills such as fire, storm, etc., as well as any other duties assigned with the company.

3. Leadership. Not all officers recognize the full scope of their responsibilities. Some merely solve problems as they arise. Some confuse paperwork with leadership. The effective officer recognizes where problems exist or will develop and he states these problems in terms that are understandable. When the problem is clear, the solution is usually obvious. He predicts the future, using the present and the past as a guide.

Leadership is an art which an officer must develop himself and also stimulate and cultivate among his men. Not every man will qualify for petty officer ratings, but it is the responsibility of the company commander to single out those men who have such ability. Effective leadership is encouraged first of all by being a good officer. Both the Division Officer's Guide and the Armed Forces Officer, are helpful toward that end. Equally important, an officer learns his calling through professional association with senior officers and his contacts with the older chief petty officers in his command. The kind of cooperation and work an officer gets from his men is determined by the quality of his leadership.

No officer has time to make every decision; yet a good officer does this, in effect, by explaining his policies and establishing procedures. In many enterprises, such policies and procedures are not set forth in writing because each man is expected to glean them from the words and actions of his superiors and co-workers as he "gets the feel of the job." However, this informal approach often leads to misunderstanding and misdirected effort, and sound policies and effective procedures are frequently lost in the transition. In a construction battalion where there is a continual turnover of personnel, it takes each man many months to formulate his understanding of battalion policies. Therefore, all instructions and regulations which are applicable to any number of persons over any extended period of time should be issued in written form. It is the responsibility of the petty officers to explain such instructions to the men, but unless this material exists in written form, it is difficult to obtain compliance.

Policies are guides for action, indicating what the commander considers important and what priority he places on it. When policies are well thought out and clearly stated, they give subordinates a good idea of how their superior looks at the overall situation. Procedures are specific ways of handling certain repetitive situations. Since an officer does not have the time to handle these matters, they are reduced to standard procedure which can be carried out without his decision. Thus, only irregular situations are referred to the superior officer. This type of procedure is management by exception. It is the only way that an officer can devote the necessary time to the special cases which need his attention and yet be sure that routine operations are performed as scheduled. Most regulations and directives are written for this purpose. They are helpful devices at all levels of command, down to the shop and crew. From the viewpoint of the petty officer and constructionman, the use of management by exception can be more beneficial than many officers realize. If an officer states his policies and acts accordingly, everyone knows where he stands. Adequate dissemination and consistent implementation of policies and procedures are probably the biggest factors in promoting high morale.

4. Task or Project Responsibility. The operations of the battalions are based on the prime contractor-subcontractor principle. Skills are concentrated in various companies as described in Chapter I. This arrangement concentrates highly specialized supervision and serves to clarify relationships, fix technical responsibilities, simplify planning, improve scheduling, pool equipment, and utilize trades to their best advantage. Each project or task assigned to the battalion is converted into job orders or task assignments by the operations officer. The special trades platoons are assigned subcontractor responsibilities for all work in their specialties. Tasks are assigned on the basis of workload and job priority. Any company commander, however, may be designated as prime contractor, depending on the workload or the nature of the work. The company commander selected may in turn assign the task to one of his platoon leaders as project officer, but the company commander remains responsible as the prime contractor.

Each company commander who is assigned prime contractor responsibilities for a task must study the plans and estimates for the job and recommend changes to the S-3 before the schedule is submitted to and approved by the Commanding Officer. His thorough analysis of each assigned project should include a review of drawings and specifications together with available reconnaissance reports, topographic maps, hydrographic surveys, and area

FIGURE 64
PERSONNEL DATA CARD

REPORTED ON BOARD	FIRST ENLISTED	EXPIRATION OF ENLISTMENT	DATE OF PRESENT RATE
RATE DESIRED	ELIGIBLE NEXT RATE	DATE OF BIRTH	RELIGION

GCT	ARI	MECH	CLER	READ	MK MECH	MK ELECT	RAD APTT	EDUCATION	LANGUAGE

DEPENDENTS OR NEXT OF KIN

RELATIONSHIP: CHILDREN   ADDRESS

SPECIAL QUALIFICATIONS
OR INTERESTS

SPORTS

ENTERTAINER   AWARDS (MEDALS)

		MARKS				PREVIOUS SERVICE AT SEA	
		DATE	MK	DATE	MK	SHIP	DUTY - GQ STATION
BONDS							
VOTING							
U.C.M.J.							
CODE OF CONDUCT							
RE-ENLIST							
FULL BAG							
LOCKER INSP.							

DISCIPLINARY RECORD

LAST NAME	FIRST	INITIAL	RATE	SERIAL NUMBER	DUTY STATUS					
					LEAVE	SCHOOL	SICK LIST			

DIVISION OFFICER'S PERSONNEL RECORD FORM, NAVPERS 2840 (New 6-61)

FIRST AID	SWIM	ARTIFICIAL RESPIRATION	TEL TALK	FIRE FIGHTER	DRIVER	TRUCK DRIVER	BUS DRIVER	COX	BOAT ENGINEER			

LEAVE

FROM	TO	NO. DAYS	LEFT	FROM	TO	NO. DAYS	DAYS LEFT	FROM	TO	NO. DAYS	DAYS LEFT

USAFI COURSES

COURSE	DATE COMPL.	SCORE	INITIAL	COURSE	DATE COMPL.	SCORE	INITIAL

NAVY SCHOOL

COURSE	DATE COMPL.	SCORE	INITIAL	COURSE	DATE COMPL.	SCORE	INITIAL

U.S.N. TRAINING COURSE PROGRESS

COURSE	PROGRESS TEST RECORD							FINAL MARK	DATE COMPL.	PRACTICAL FACTORS	DATE RATED

studies. He must look at his job in relation to other facilities that are existing or proposed and determine what construction plant is needed at the job site for both prime and subcontract work, what communications facilities are needed, what access roads are to be constructed, what personnel support facilities such as messing, heads, etc. are to be used, where water supplies and fuel supplies are located and what facilities are needed to supply power. He must review the requirements for jobsite defense or security of equipment, tools and materials and integrate these with the requirements for base and camp security. He must provide a plan for emergency evacuation of the job site caused by mount-out or by disaster control requirements for the base or camp. He must look at the factors which affect the efficiency and thus the productivity of men and equipment, such as proposed phasing of the work, hauling distances, weather, length of the proposed work day, specification requirements, and lost time due to travel to and from work and to meals. He must determine the safety measures required and special training needed by members of the crew in preparation for the tasks. After checking the details of each project, he must:

a. Review the Stage II schedule established by the S-3 and develop a detailed Stage III schedule which indicates crew sizes, individual assignment to crews, crew schedules and equipment schedules and the date various construction materials should be delivered on the job site.

b. Lay out the job site reserving space for all temporary facilities and material stockpiles.

c. Instruct project petty officers and crew leaders in project layout and schedule.

d. Arrange for and schedule subcontractor support.

e. Organize, direct and inspect all phases of the project.

f. Collect required labor time cards and make progress reports to the Operations Officer.

When the company commander is performing subcontractural services he must:

a. Check the time schedules and all other details of the work scheduled by all prime contractors.

b. Carefully schedule all work assigned to him.

c. Maintain liaison with prime contractors to coordinate all matters affecting his work.

d. Check on the status of all equipment and materials required.

e. Organize, supervise, and inspect all work performed by his men in conjunction with the prime contractor.

f. Order organized crews or squads to report to prime contractors.

g. Retain administrative control over all members of his platoons and technical control over their work.

Most work loads can be efficiently handled through the careful scheduling of operations related to, but not specifically part of, the end product. During normal operating time as well as during delays caused by late receipt of material due to such factors as weather or enemy action, the following aspects and assignments should be given detailed consideration. If certain tasks could be done during opportune periods, most overloads at peak construction times will be largely alleviated. Selection of these tasks should be based upon the completion of critical activities that will help to shorten the overall project time. For example:

a. Shop work, prefabrication and stockpiling.

b. Field and site preparatory work.

c. Authorized "fill-in" work, non-productive assignments, and general work party duties.

d. Rough-in and finish work.

e. Maintenance work.

f. Regular and special training and special details.

g. Overtime or additional shifts.

5. Company Commander Assistants. The company commander may or may not have another officer assigned. If he does he usually assigns him as company executive officer and delegates authority to him in order to prepare him to assume responsibility for his own company.

Each company commander normally selects the senior chief petty officer assigned to his company as company chief. The chief acts as administrative assistant to the company commander and supervises the company office. A non-rated man with some typing capability is selected as company clerk. He types memorandums and muster reports, files correspondence, acts as mail orderly and assists in keeping the watch quarter and station bill current. He may be used in the disaster control or combat situation as the Company Commander's driver, messenger or radio operator.

Platoon leaders act in accordance with the policies of their company commanders. The general duties and responsibilities of the platoon leader are the same as those prescribed for the division officer in Article 1044, U.S. Navy Regulations and the Division Officer's Guide. Additional military duties are outlined in the Landing Party Manual.

The platoon leader in construction companies organizes and trains his units for construction, and for military and disaster control operations. He acts as prime contractor or subcontractor on work assigned on battalion job orders and is responsible to his company commander for the timely and efficient completion of all work assigned. All platoon leaders command their platoons as rifle or automatic weapon units in combat operations. They take charge of platoon administration, and are responsible for the performance of special and collateral duties assigned.

The squad leader carries out the orders of the platoon leader. He is responsible for the discipline, appearance, training control and conduct of his men and for the condition and care of their tools and equipment. In combat he is responsible for fire discipline, fire control and maneuver of the three fire teams in his squad.

The fire team leader is responsible for directing the men in his work/fire team in construction, defensive combat and disaster control operations. Since his unit is always kept intact, no matter what the assignment, the attitude of his men, the condition of their tools and equipment and their performance in all areas are his primary concern and directly reflect his leadership ability.

SECTION B - THE HEADQUARTERS COMPANY

1. Mission. Headquarters company is an administrative unit for enlisted men assigned to the service departments. All enlisted men as-signed to battalion headquarters, the operations department, the service departments and the MAA force are under the Headquarters Company Commander for military administration. However, department heads direct the functional and technical duties of the officers and men assigned to their departments. The service departments provide the same basic administrative and logistic support services in construction and military situations with the exception that certain of these personnel will compose reserve machine gun and rocket launcher sections and rifle platoons. (See Figure 65).

2. The Company Commander. The Headquarters Company Commander is responsible for the usual detail of company command. His billet reflects the administrative nature of headquarters company, differing from letter company assignments in that it involves no direct construction mission. Normal company command responsibilities include, but are not limited to, assignment, billeting, inspection and training of personnel and initiation of action for their transfer or replacement.

FIGURE 65
ORGANIZATION OF THE HEADQUARTERS COMPANY

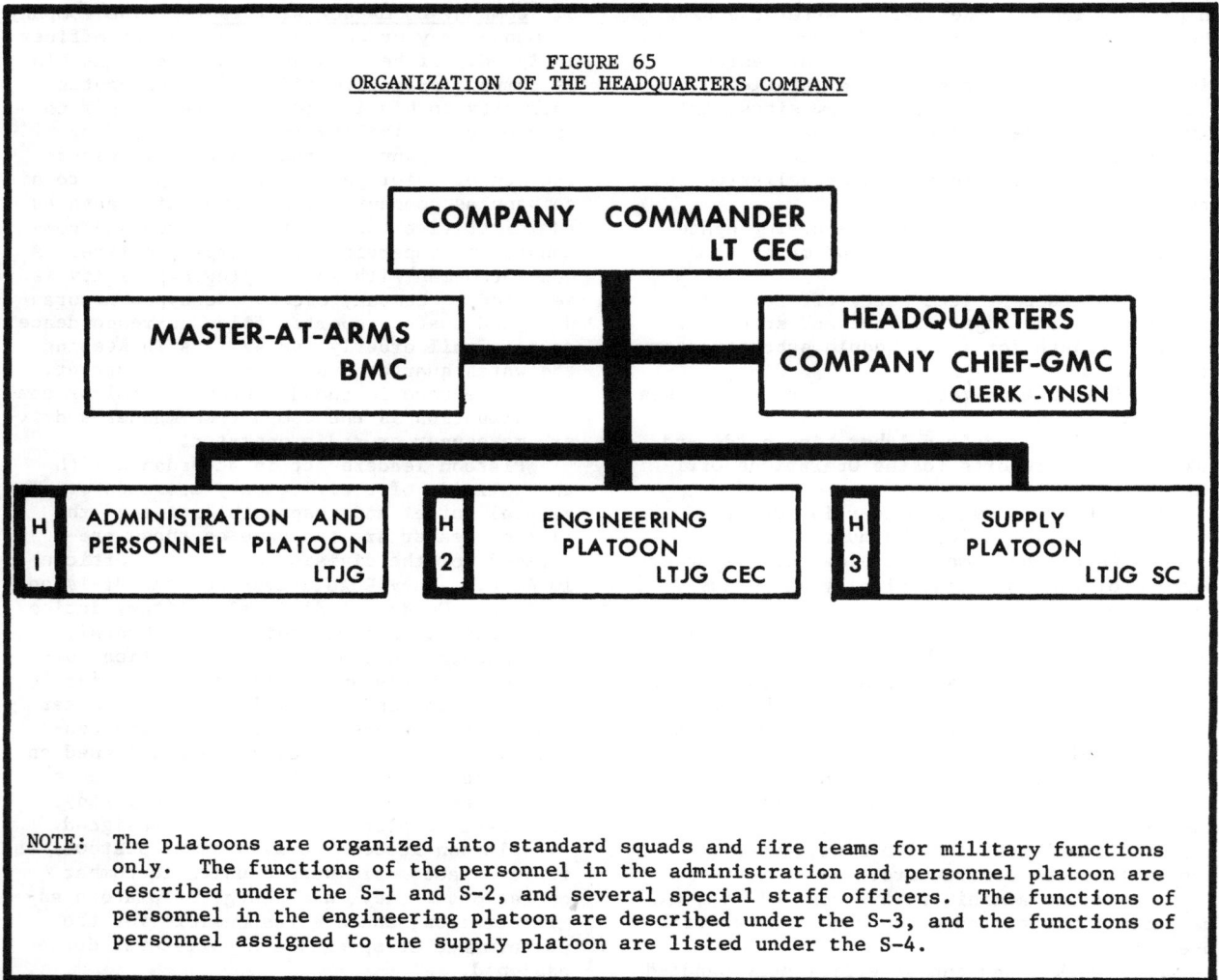

```
                    ┌─────────────────────────┐
                    │  COMPANY  COMMANDER      │
                    │         LT CEC           │
                    └─────────────────────────┘
      ┌──────────────────┐         ┌──────────────────────────┐
      │ MASTER-AT-ARMS   │         │    HEADQUARTERS           │
      │      BMC         │         │ COMPANY CHIEF-GMC         │
      │                  │         │    CLERK -YNSN            │
      └──────────────────┘         └──────────────────────────┘

┌──┬─────────────────────┐  ┌──┬─────────────────┐  ┌──┬─────────────────┐
│H │ ADMINISTRATION AND  │  │H │  ENGINEERING    │  │H │    SUPPLY       │
│1 │ PERSONNEL PLATOON    │  │2 │   PLATOON       │  │3 │   PLATOON       │
│  │      LTJG            │  │  │   LTJG CEC      │  │  │   LTJG SC       │
└──┴─────────────────────┘  └──┴─────────────────┘  └──┴─────────────────┘
```

NOTE: The platoons are organized into standard squads and fire teams for military functions only. The functions of the personnel in the administration and personnel platoon are described under the S-1 and S-2, and several special staff officers. The functions of personnel in the engineering platoon are described under the S-3, and the functions of personnel assigned to the supply platoon are listed under the S-4.

When the battalion is operating a separate camp in the field, the headquarters company commander is responsible for the security and defense of the command post area and acts as a reserve force for the companies on the defense perimeter.

Although the headquarters company commander has many duties, several of them overlap. Most of these duties require direct supervision of enlisted men. Ordinarily all these men work in or near the battalion camp. This grouping of related duties under a single experienced officer reduces the duplication of effort and time-consuming coordination that would be needed if these duties were parcelled out among several officers. Some of these other staff duties may, however, be assigned to other officers at the discretion of the Commanding Officer. For example, the headquarters company commander may serve as the S-1 (Administration and Personnel), although usually the duties of the S-1 require full-time attention of one officer for a 500 to 700 man battalion. Many other special staff functions easily relate to his primary billet, for example, Brig Officer, Camp Operations Officer, Fire Marshal, etc.

3. Company Organization. The Headquarters company consists of a company headquarters, an S-1/S-2 platoon, an S-3 platoon and an S-4 platoon.

The company headequarters may consist of the senior enlisted man in the company, a company clerk, and any other petty officer designated by the company commander.

H-1 Administrative/Personnel Platoon. The Personnel Officer is the platoon commander. He has the senior petty officer in the S-1 department as his platoon chief.

The platoon chief serves as mustering petty officer. The platoon commander is responsible for training and directing the platoon only for military functions such as muster, inspection, etc., and for combat and disaster control operations. He assists with advancement in rate training as does his platoon chief. The duties and functions of the men in this platoon are described under functions of the S-1.

H-2 Engineering Platoon. The Engineering Officer is the platoon leader for this platoon. He is assisted by a platoon chief who is generally the senior petty officer in the platoon who serves as mustering petty officer. The platoon commander's duties are similar to those of the S-1/S-2 platoon commander. The duties and functions of the men in this platoon are described under the functions of the S-3.

H-3 Supply Platoon. The Disbursing Officer serves as platoon commander for this platoon. He also is assisted by a platoon chief who is the senior petty officer in the platoon and who serves as mustering petty officer. His responsibilities are similar to those described above for the H-1 platoon commander.

The duties and functions of the men of this platoon are described under the functions of the S-4.

SECTION C - THE MCB EQUIPMENT COMPANY

1. Mission. In the MCB, the "A" Company operates and maintains automotive and construction equipment. This company serves as a subcontractor for earth-moving, grading, excavation, paving, hauling, pile-driving, well drilling, heavy lifting, blasting and demolition, and may also serve as a prime contractor on earthmoving projects, paving projects, etc. The company operates the transportation pool, a machine shop and the equipment maintenance shops. (See Figure 66). In the combat situation this company would form as three or more rifle platoons.

2. The Company Commander. The "A" company commander serves as equipment superintendent and as a rifle company commander. Except during combat emergencies his principal responsibility is to ensure the accomplishment of the construction work assigned to his company and to assure that all automotive and construction equipment assigned to the battalion is properly utilized and maintained. He is responsible for the usual detail of company command listed previously. He must be familiar with qualification for advancement in rating, service school curricula, and Navy Training Courses for equipment operator (EO), construction mechanic (CM), machinery repairman (MR), and constructionman (CN). As company commander, he directs "A" company military and disaster control training and operations and he may also serve as Emergency Recovery Section Commander. His duties include serving the battalion as a staff member. He normally is appointed Motor Vehicle Accident Investigator. He may be appointed to courts or boards as the Commanding Officer desires.

3. Company Headquarters. "A" company consists of company headquarters, two (or more) equipment operations platoons, and an equipment repair platoon. When required, additional men may be assigned to the existing platoons or organized as extra platoons. Platoons must be prepared to operate on a two-shift (10-12-hour) 7 days a week basis for any given emergency. The company headquarters usually consists of a company chief (EQCM), project coordinator (EOCS), maintenance supervisor (CMCS), dispatcher (EOC) and a company clerk. If additional assistance is required, it is supplied from the numbered platoons.

Company Chief. The company chief, who is the senior enlisted man in the company, is the principal assistant to the company commander. His duties are to execute and enforce policies of the company commander; supervise administration, operations and maintenance;

```
                    ┌─────────────────────────────┐
                    │   COMPANY  COMMANDER         │
                    │   LT CEC                     │
                    └─────────────────────────────┘
┌──────────────────────────┐
│ ASSISTANT COMPANY         │        ┌──────────────────────────────────┐
│ COMMANDER    LTJG,CEC     │        │        HEADQUARTERS              │
├──────────────────────────┤        │  COMPANY CHIEF      -EQCM        │
│ DISPATCHER        EOC     │        │  PROJECT  COORDINATOR -EOCS      │
└──────────────────────────┘        │  MAINT. SUPERVISOR  -CMCS        │
                                     │           CLERK     -CN          │
                                     └──────────────────────────────────┘
```

| A | EQUIPMENT OPERATIONS PLATOON | | A | EQUIPMENT REPAIR PLATOON |
| A 2 / A 1 | EOC / EOC | | 3 | CMC |

EQUIPMENT OPERATIONS PLATOON — EOC

- MOTOR POOL
- CONSTRUCTION EQUIPMENT
- CONCRETE BATCH PLANT
- ASPHALT PLANT
- QUARRY

EQUIPMENT REPAIR PLATOON — CMC

- AUTOMOTIVE SHOP
- CONSTRUCTION EQUIPMENT SHOP
- MACHINE SHOP
- ELECTRICAL SHOP
- CHASSIS,BODY AND FENDER SHOP
- TOOL ROOM
- SERVICE GROUP

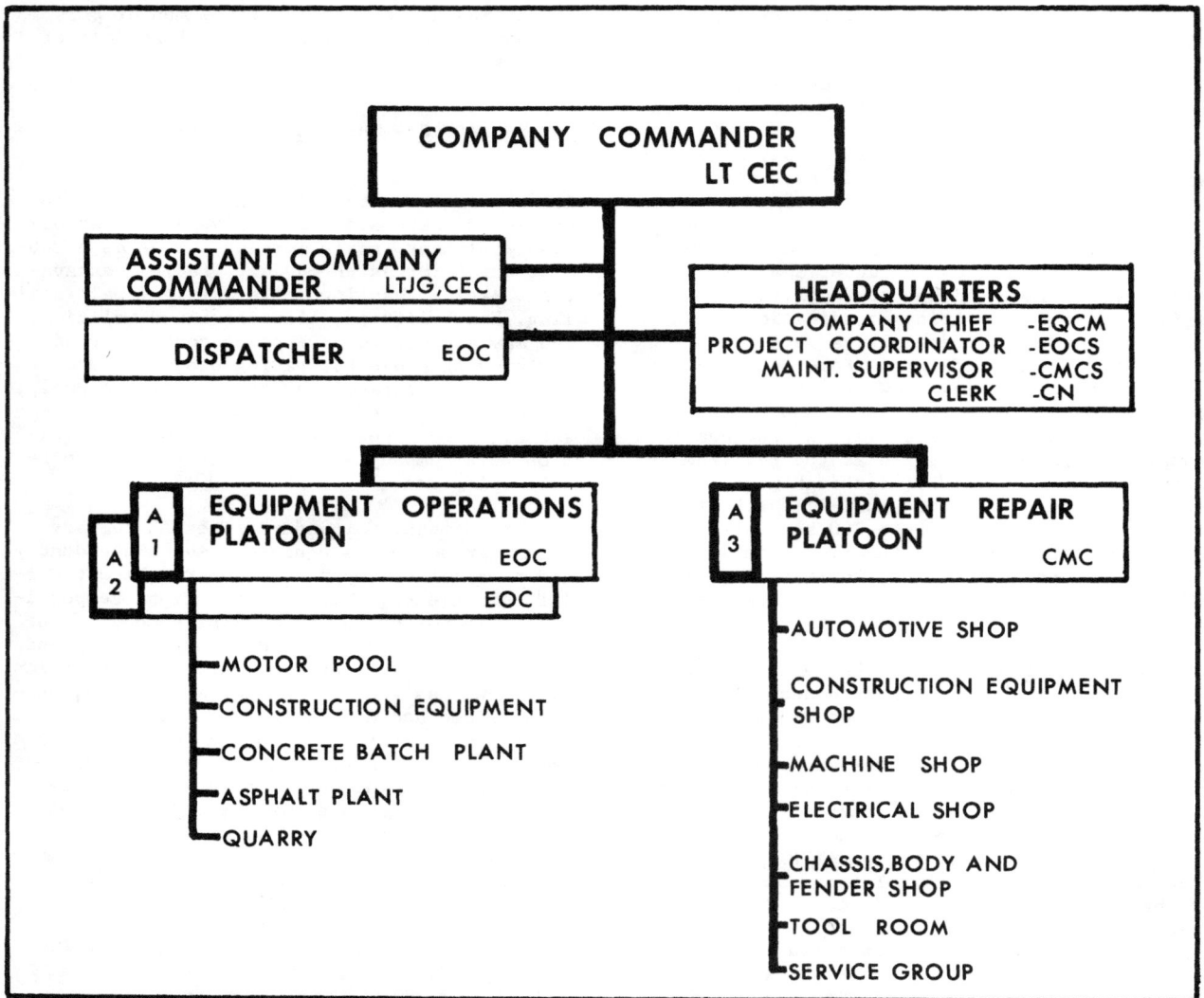

study and make recommendations regarding proper utilization of all equipment, training requirements, and safety practices required to operate and maintain all equipment with maximum degree of efficiency. He is responsible for the preparation of company correspondence and the maintenance of an up-to-date disaster control bill, fire bill and a company organization chart indicating personnel assignments and those on TAD, at sickbay, in the brig, etc. He maintains a company library consisting of publications pertaining to transportation and maintenance of equipment, such as Type Commander instructions, Technical Manuals (TM) and manufacturers' catalogues.

"A" Company Project Coordinator. Usually a senior chief equipment operator (EOCS) is assigned to project petty officers. He coordinates equipment requirements for the projects, reviews project estimates and makes necessary recommendations. He enforces the poli-

cies of his company commander with regard to equipment operations.

Maintenance Supervisor. Usually a senior chief mechanic (CMCS), is given the principal duty of ensuring proper maintenance and repair of all automotive, construction, materials handling and weight handling equipment. He also enforces company commander policies concerning preventative maintenance and cost control programs for all equipment assigned to the MCB. He arranges for repairs, overhaul and services provided by various depots and bases (5th echelon maintenance). He performs the functions required in the administration of the cost control program for all automotive and construction equipment assigned to the battalion in accordance with current COMCBLANT/ COMCBPAC and NAVFAC Instructions. He provides training and instruction necessary to achieve the highest degree of proficiency in the repair and maintenance of all equipment. He makes arrangements with the battalion supply officer for the supply of fuel, lubricants and solvents. Further guidance will be found in NAVFAC P-404.

The Dispatcher. The Dispatcher is usually an EOC. When necessary, the dispatcher may be assisted by one or more assistant dispatchers and a records clerk. His prime obligation is to keep informed of the location, assignment and availability of all automotive, construction and weight-handling equipment assigned to the battalion and to maintain the associated vehicle operations records. He is usually designated by the command to examine the qualifications of all personnel who operate battalion equipment and records the limitations of their U.S. Government Operator Permits. He also keeps a ready reference list of all men qualified to operate each type of equipment. He releases equipment on Class A, B, and C assignments as directed. The dispatcher supervises the efforts of his assistants and clerk in receiving, studying and logging of daily availability reports and in the implementation of requests for equipment and operators to meet battalion and project requirements.

4. The Equipment Operations Platoon. The A-1 and A-2 platoons are identical in mission and organization. The platoon leader is a junior officer of the Civil Engineer Corps. The platoon chief is usually an equipment operator chief (EOC) assisted by a right guide. The platoon leader usually serves as a subcontractor on almost every type project and is responsible for all work assigned to his platoon. He must be familiar with the qualifications for advancement in rating, service school curriculums and Navy training courses

for equipment operators. The operation of an equipment pool includes the systematic servicing, inspection, and preventive maintenance of vehicles and equipment; detection and correction of causes of failure; dispatching, correct operation, recovery, and evacuation of vehicles and equipment; and the maintenance of records. The platoon leader is responsible for the 1st echelon maintenance and operation of his equipment, and for the training of operators and drivers.

The platoon leader is responsible for submitting work or training reports as required and for continually striving to increase the effective use of the equipment by taking action or making recommendations as necessary. The equipment operations platoon leader trains and directs his platoon in defensive combat, combat support and disaster control operations. He may operate a batch plant, in which case he would locate suitable materials and exploit them by setting up and operating processing facilities at key locations. His responsibilities would include the stockpiling of critical material, controlling and recording mixtures, scheduling hauls with precise timing, material and quality control, and keeping proper issue records.

Men assigned to A-1 and A-2 platoon billets are trained as heavy equipment operators (trucks, cranes, tractors, graders, paving machines, rock crushers, concrete mixers). Others are trained in job specialties in connection with the operation of such facilities as quarries, sand and gravel pits, and batch and asphalt plants. Their construction assignments include earthmoving, grading, excavating, batching, hauling, pile driving, well drilling, heavy lifting, blasting and demolishing. The men are permanently assigned to squads and

FIGURE 67
EARTHWORK

One of the Equipment Operations Platoon Leaders would serve as prime contractor for this runway extension.

131

FIGURE 68
SAMPLE SHOP REPAIR ORDER

MAVFAC 9-11200/7A(REV.9-67)
SUPERSEDES NAVDOCKS 1948
S/N-0105-004-1000

SHOP REPAIR ORDER

(1) SRO NUMBER: A-0001
(2) JOB ORDER NUMBER:
USN NUMBER: 94-0001

(3) EQUIPMENT DESCRIPTION: TRUCK, ½ T Pickup
(4) MAKE: FORD
(5) MODEL: F-100
(6) YEAR: 66
(7) EQUIP. CODE: 6313
(8) DOD ALPHA: G
PAGE 1 OF 1

(9) ACTIVITY: MCB-4
(10) PHONE NUMBER:
(11) LAST "A" TYPE PM: 29,800
(12) LAST "B" TYPE PM: 27,602
(13) LAST "C" TYPE PM: 31,900
(14) ACCU. MILES/HRS.: 34,000

PM GROUP (15): 25
PM TYPE DUE (16): A
PM DATE DUE (17): 3/3

TO BE COMPLETED UPON EQUIPMENT AVAILABILITY FOR MAINTENANCE REPAIR

(18) DOWNTIME — IN: DATE 3/3 TIME 0800 — OUT: DATE TIME
(19) PRESENT METER READING: 34,000

(20) WORK GENERATION (Check applicable box): 1 - SCHEDULED ☐ 2 - INTERIM ☐
(21) SCHED. FOR REPLACE (Approx. No. Months): 3 ☐ 6 ☐ 12 ☐
(22) WORK PERFORMANCE: 1 - OWN ACT EQUIP. ☐ 2 - CUST EQUIP ☐
(23) REPAIRED BY: 1 - OTHER GOVT./SHOP ☐ 2 - COMM CONT ☐

TOTAL HOURS:

(24) QTY.	(25) DESCRIPTION	(26) MFG. PART NO.	(27) AMOUNT	(28) WORK CTR.	(29) LCC	STD	ACT.	(30) MAN-HRS	(31) WORK DESCRIPTION	(32) MECH. INIT.
1	TIRE		16.05						LUBE CHASSIS (SERVICE AIR CLEANER AND BATTERY) [1]	JQ
							.5		CHANGE MOTOR OIL AND FILTER CARTRIDGE [2]	JQ
1	MUFFLER		6.45						REPLACE BATTERY (CLEAN TERMINAL AND BOX) [3]	
1	POINTS AND COND.		3.00				.7		ADJUST BRAKES CHECK FLUID MSTR CYL. [4]	JES
6	PLUGS		3.00				1.0		REPLACE MUFFLER	JES
							1.2		TUNE ENGINE	JQ
							1.0		REPAIR HOOD RELEASE ASSY	JES

(34) TOTAL MATERIAL $: 28.50
(35) TOT. LABOR HRS.: 4.4

(33) PARTS ON ORDER (Reqn. Numbers):

(36) WORK AUTHORIZED (Inspector's Signature): John Doe DATE 3/3
(37) WORK APPROVED (Supervisor's Signature): C. M. Wright DATE 3/3

(38) CONTRACTUAL SERVICE REQUEST (Receipt of above order and equipment is hereby acknowledged. Permission to exceed work specified above MUST BE AUTHORIZED by requesting activity).

CONTRACTING FIRM

SIGNATURE   DATE

FOR CUSTOMER JOB ESTIMATING

(39) LABOR (MAINT.) _____ HRS. = $
(40) LABOR (OPER.) _____ HRS. = $
(41) MATERIAL $28.50
(42) OTHER $
(43) TOTAL COST $28.50

MATERIAL RECORD

132

work crew/fire teams, but are equipped for construction according to the requirements of any assigned work. On different deployments, the platoon may operate as an excavation group, paving group or concrete production crew; it may be split in order to accomplish more than one assignment. Equipment is drawn as necessary from the equipment pool, when possible, and is retained for the entire deployment on a Class B assignment. Equipment operators assigned to the construction equipment pool operate equipment released by the dispatcher and perform operator maintenance services on all battalion construction equipment.

5. The Equipment Repair Platoon. The Platoon Leader of the repair platoon is usually a warrant officer, Civil Engineer Corps. The platoon chief of the A-3 platoon is usually a Construction Mechanic Chief (CMC) assisted by a right guide. The platoon leader is responsible for shop and field maintenance and repair of all automotive, construction and weight-handling equipment assigned to, or serviced by, the battalion. He must be familiar with the qualification for advancement in rating for construction mechanics (CM) and machinery repairmen (MR). As Master Mechanic he assigns, routes and supervises all work via Shop Repair Order (SRO), applying standard allowable hours for each item of work. Figure 68 is a sample SRO. He supervises the work of the records clerk and the shop and field equipment inspectors, and assigns all work and approves work requests, expediting priority work and balancing the workload of his men. He schedules inspections and reports excessive maintenance to the equipment operations platoon leaders, making recommendations when equipment should be rebuilt or replaced. He instructs his personnel in the procedures used in the NAVFAC Preventive Maintenance and Cost Control Program. He trains and directs his platoon in defensive combat, combat support and disaster control operations.

Personnel assigned to the A-3 platoon are trained in gasoline and diesel engine repair, chassis, body and fender repair, lubrication and inspection. They draw repair parts and related supplies on requisitions from the repair parts section of the Supply Department, returning replaced parts, assemblies and accessories for disposal. Inspectors and mechanics draw tools as needed from the platoon tool room. Specific duties of key members of the platoon are listed below:

Records Clerk. The Records Clerk assists the equipment repair platoon leader and performs the clerical work involved in the Preventive Maintenance and Cost Control Program. His specific duties include:

a. Maintaining the Preventive Maintenance (PM) Scheduling File.

b. Submitting SROs for equipment that requires PM service inspection and notifying the dispatcher 48 hours in advance of the scheduled times for each vehicle.

c. Keeping daily time records relating to the NAVFAC Cost Control Program.

d. Preparing monthly Equipment Maintenance Control Reports and other reports as required.

e. Maintaining Equipment History Jackets.

Inspectors. Inspectors are assigned to assist the equipment repair platoon leader in his responsibility for inspection of all automotive and construction equipment serviced by the platoon. A shop automotive inspector, a shop equipment inspector and a field equipment inspector are often designated. Inspectors must be seasoned mechanics able to determine readily the nature and extent of necessary repairs on any type of equipment and exercise independent judgment as to whether trouble reports require immediate attention or can be delayed until the next scheduled PM service inspection.

a. Performs inspections as scheduled, completing appropriate record forms when required, noting deficiencies and clearly describing on the SRO any work needed.

b. Checks the file of trouble reports on the equipment prior to starting his inspection

c. Uses available testing equipment.

d. Makes minor adjustments incidental to the inspection.

e. Delivers the initialed SRO to the platoon leader for assignment and scheduling of needed repairs.

f. Road or field tests equipment following repair or overhaul.

g. Releases equipment for return to service after final inspection.

Repair Shops. The construction mechanics assigned to the automotive and construction equipment repair shops perform all work as specified on the SRO and authorized work request. Because certain repair work can be accomplished in the field with less down time involved, field repair crews of construction mechanics are formed, who perform as much of the authorized work as possible without removing the piece of equipment from the job site.

Mechanics Tool Room. The mechanics tool room serves as the central point for issue, storage, inspection, maintenance and repair of all mechanics tools under an approved custody control system. Shop equipment is held on sub-custody by the equipment repair platoon leader. Kits and tools needed continuously are issued to individuals on custody receipts. Other tools are issued on tool chits for particular job assignments. The tool room personnel perform tool repair within their capability, and request assistance of other battalion shops when necessary.

Machine Shop. The machinery repairmen (MR) are assigned to operate the machine shop trailer which contains lathes, drill presses, grinders and other machine tools. They are an extremely valuable asset because they have the capability of manufacturing or repairing equipment parts, tools or machine parts needed

FIGURE 69
MECHANICS REPAIRING A JEEP

Members of the Equipment Repair Platoon
Overhaul a Jeep during an Amphibious
Exercise.

to perform the work required by the SROs.

Chassis, Body, Fender and Radiator Shop.
The work assigned to the construction mechanics of this shop via SROs or work requests include the repairing or rebuilding of chassis, fenders and other body components; the repairing, rebuilding and testing of radiators; the repair of crane booms, bulldozer blades and other steel components and performing welding and brazing tasks.

Electrical Shop. Manned by construction mechanics the electrical shop repairs, rebuilds, cleans, adjusts, and tests all automotive electrical parts and accessories except batteries. This includes generators, starters, voltage regulators, etc. Assignments are authorized on SROs or work requests.

Service Group. The mechanics of the following service activities perform all work as required on SROs and work requests:

    a. Battery Shop
    b. Tire Shop
    c. Lubrication racks
    d. Field lubrication
    e. Fueling station

Men assigned to the battery shop build, maintain, and recharge wet cell batteries, mix electrolyte, disassemble and reassemble used batteries and keep a supply of spare (fully charged) batteries for all types of equipment used by the battalion. The construction mechanics assigned to the tire shop provide a repair and replacement service for all pneumatic-tired equipment in the battalion. The mechanics assigned to the lubrication racks maintain adequate stocks of all lubricants required by the battalion and lubricate automotive and mobile construction equipment as required under the Preventive Maintenance Program. The construction mechanics responsible for field lubrication regularly check the lubrication of all construction equipment stationed in the field and lubricate equipment as required under the Preventive Maintenance Program.

The equipment operators assigned to the fueling station maintain adequate stocks of all liquid fuel and solvents required by the battalion, keep records and submit reports on fuel and solvent inventories and issues. They operate all battalion fuel storage and dispensing equipment and deliver and dispense fuels to the automotive and construction equipment pools, to all gasoline and diesel equipment in the field, and to oil-heating plants and diesel-electric generators.

FIGURE 70
SAFETY PAYS OFF IN THE TIRE SHOP

This man was saved from serious injury when the tire he was inflating blew out because he was using the protective cage shown behind him.

134

SECTION D –
THE MCB SPECIAL CONSTRUCTION COMPANY

1. Mission. In the MCB, the "B" Company serves as prime contractor on exterior water, sewer, power, and communications projects and on fuel systems; subcontractor for structural steel fabrication, sheet metal and millwork, air conditioning and refrigeration, etc. This company can operate a sheet metal and steel shop, a paint shop, a carpenter shop, plumbing shop and an electrical shop. "B" company also performs camp maintenance. This company provides three rifle platoons for battalion defense. Company functions are illustrated in Figure 71.

FIGURE 71
FUNCTIONS OF THE MCB SPECIAL CONSTRUCTION COMPANY

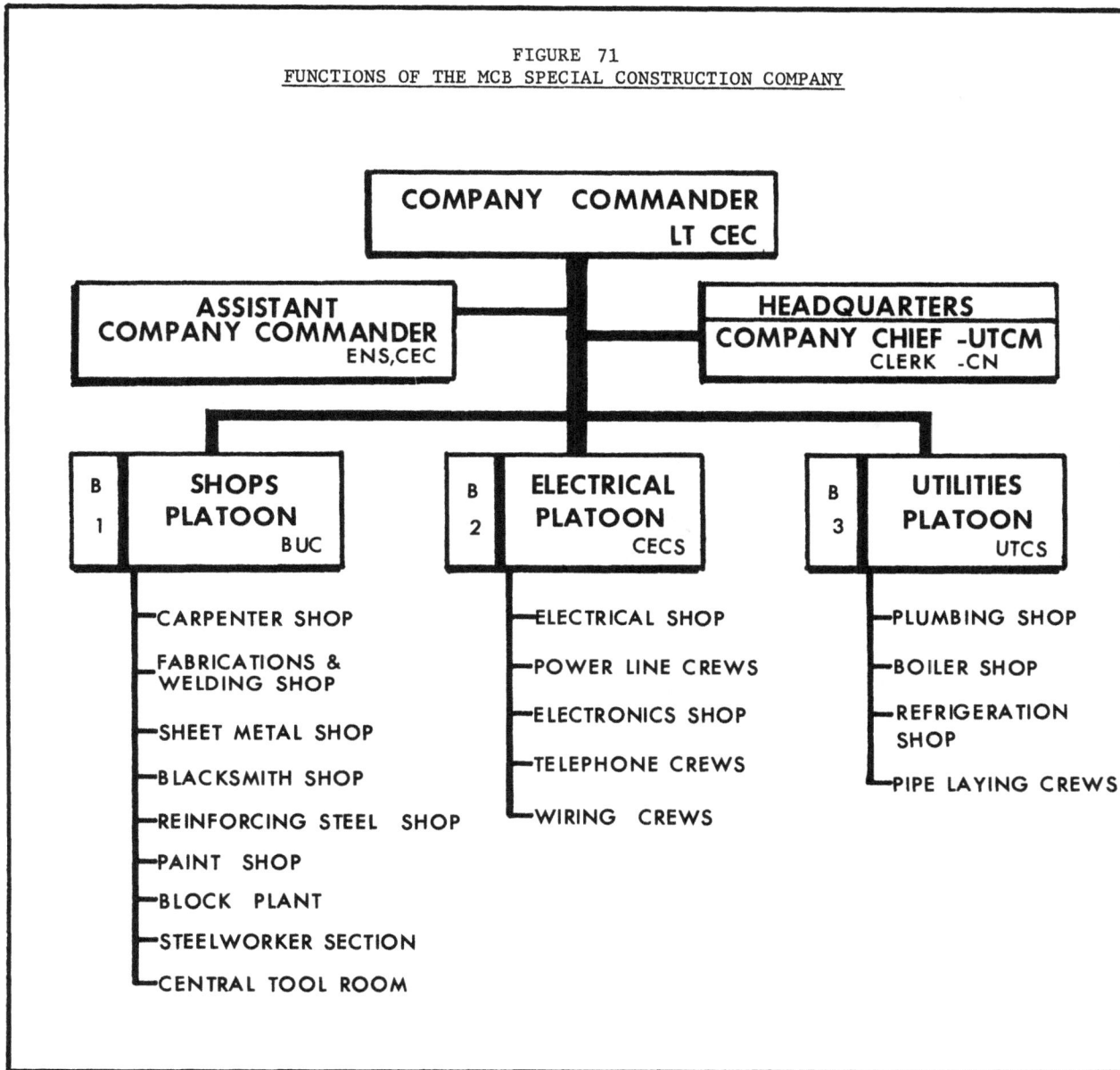

COMPANY COMMANDER
LT CEC

ASSISTANT COMPANY COMMANDER
ENS, CEC

HEADQUARTERS
COMPANY CHIEF -UTCM
CLERK -CN

B1 SHOPS PLATOON
BUC
- CARPENTER SHOP
- FABRICATIONS & WELDING SHOP
- SHEET METAL SHOP
- BLACKSMITH SHOP
- REINFORCING STEEL SHOP
- PAINT SHOP
- BLOCK PLANT
- STEELWORKER SECTION
- CENTRAL TOOL ROOM

B2 ELECTRICAL PLATOON
CECS
- ELECTRICAL SHOP
- POWER LINE CREWS
- ELECTRONICS SHOP
- TELEPHONE CREWS
- WIRING CREWS

B3 UTILITIES PLATOON
UTCS
- PLUMBING SHOP
- BOILER SHOP
- REFRIGERATION SHOP
- PIPE LAYING CREWS

2. The Company Commander. The "B" Company Commander serves as Special Construction and Rifle Company Commander. He is responsible for all work assigned to B-1 Shops Platoon, B-2 Electrical Platoon, and the B-3 Utilities Platoon. Except during combat emergencies, his principal task is to ensure that the shops provide timely support to the projects and that utility systems are integrated with project progress. He is responsible for the usual details of company command listed in Section A of this Chapter. He must be acquainted with qualifications for advancement in rating, service school curricula, and Navy training courses for CE, UT, SW, BU, SF, and CN in order to guide personnel in furthering their knowledge and capability. As company commander, he directs "B" Company military and disaster control efforts during training and operations and may serve as Emergency Recovery Section Commander. He may be assigned any staff duties, but is usually assigned as Camp Maintenance Officer because of his control of the battalion shops and personnel whose irregular assignments make them available for performing maintenance tasks. He may also be assigned as Communications Officer because the electronics repair shop is usually under his control.

3. Company Organization. The "B" Company consists of the company headquarters and three platoons - the Shops Platoon, the Electrical Platoon, and the Utilities Platoon. When required, additional men may be assigned to existing platoons or organized as extra platoons. The platoons must be prepared to operate on various shifts in order to support construction projects.

The company headquarters consists of a company chief (UTCM), a company clerk and the platoon leaders acting as technical advisors to the company commander. The company chief is the senior enlisted man in the company. His duties are to execute and enforce the policies of the company commander, supervise company administration, coordinate the efforts of the shops and the subcontractual utility work.

4. The Shops Platoon. The platoon chief is the senior petty officer assigned to the battalion shops. He is usually a senior chief steelworker (SWCS). He is assisted by a mustering petty officer and a right guide. He provides shop services for all projects and camp maintenance. He must train his men to perform all assigned tasks. These tasks include operation of carpenter shop, a concrete block and precast concrete plant, a sheet metal shop, a fabrication and welding shop, a paint shop, reinforcing steel shop, a rigging section and the central tool room. The shops are set up according to the projected workloads for each deployment with platoon members permanently assigned to squads and work crew/fire teams. The platoon leader must coordinate his subcontractual shop work with project petty officers. He may become a project petty offi-

cer on such jobs as tank farms or pipeline installation. He trains and directs his platoon in defensive combat, combat support and disaster control operations. A summary of the duties of the shops is listed below:

Carpenter Shop. The Carpenter Shop is staffed by builders (BU) who operate all woodworking machinery, machine tools, and hand tools in the construction of wedges, hubs, stakes, millwork, cabinet work, windows, doors, frames, stairs, chairs, tables, trusses, mouldings, boxes, crates, signs, wooden scaffolds, wooden forms for concrete, and all related wooden items.

Fabrication and Welding Shop. Manned by steelworkers (SW), this shop is assigned work pertaining to structural steel fabrication such as tanks, scaffolds and other products. They perform structural repairs and welding tasks on machinery, automotive and construction equipment as required.

Sheet Metal Shop. The steelworkers assigned to the sheet metal shop fabricate, install and repair sheet metal and light plate items. They lay out, shear, rivet, weld, braze, trim, and solder sheet metal. They fabricate and install ductwork, flashing, gutters, hoods and other sheet metal items. They also are called upon to repair metal-clad buildings, sheet metal galley equipment, metal boxes, etc.

Blacksmith Shop. Steelworkers (SW) are assigned to the blacksmith shop. They operate furnaces and forges to heat-treat, case-harden, forge, and forge-weld metals. The blacksmith shop fabricates and repairs metal shapes, structural members, hooks, shackles, brackets, and forged machine parts. In addition, they make repair, sharpen, and temper tools and tool bits.

Reinforcing Steel Shop. The steelworkers assigned to the reinforcing steel shop cut, bend, prefabricate and weld reinforcing steel.

Paint Shop. The builders assigned to this shop operate the battalion paint locker and perform painting details of all kinds as directed, including painting of automotive and construction equipment.

Concrete Block and Pre-Cast Concrete Plant. This plant is operated by builders who manufacture concrete and cinder block, concrete lintels, cast stone, pre-cast concrete joists, beams, columns, splash blocks, floor and roof slabs and other prefabricated concrete units.

Steelworker Section. Members of the steelworker section are organized according to the projected workload for each deployment. The men are specifically trained and capable of erecting and dismantling butler buildings, tank farms, Bailey bridges, steel towers, cranes, craneways, derricks, and other steel structures They maintain the rigging loft, rig "A" frames, gin poles, booms, blocks and tackle and prefabricate nets and slings.

5. The Electrical Platoon. The platoon leader is a CEC Officer. The platoon chief is the senior construction electrician (CECS or CEC)

FIGURE 72
PREFABRICATING BENTS FOR A BUTLER BUILDING

A Member of the "B" Company Fabrication and Welding Shop is Welding the Column Section of a Rigid Frame Bent to the Haunch Section. Note that his Rifle is Within Easy Reach.

They must be able to operate power plants, generators and telephone systems.

6. The Utilities Platoon. The Platoon Leader is a CEC Officer. The Platoon Chief is usually a senior chief utilitiesman (UTCS) and is assisted by a platoon right guide. He provides exterior and interior utilities subcontractural work for projects and camp. He may be assigned as project petty officer on such tasks as water supply or sewage disposal systems. He is responsible for technical, military and disaster control training of his platoon and he directs their efforts in these areas. Members of the utilities platoon are trained to install, maintain, operate and repair:
    a. Steam boilers
    b. Pumps
    c. Compressors
    d. Water supply systems
    e. Purification and distillation plants
    f. Pipelines
    g. Plumbing
    h. Sewage disposal systems
    i. Air conditioning and refrigeration plants.

Members of the platoon are permanently assigned to squads and work crew/fire teams. However, squads and work crew/fire teams are grouped and assigned specialties according to the projected workload for each deployment.

assigned to the battalion assisted by a right guide in the administration and training of the platoon. The platoon leader provides all types of electrical installation as a subcontractor. He may be project officer on such jobs as transmission lines or telephone switchboard installation. He trains and directs his platoon in defensive combat, combat support and disaster control operations.

Members of the Electrical Platoon are permanently assigned to squads and work crew/fire teams. However, squads and work crew/fire teams are grouped and assigned specialties according to the projected workload for each deployment. They are trained to install, maintain and repair:
    a. Transformers
    b. Telephone systems
    c. Interoffice communications systems
    d. Public address systems
    e. Alarm systems
    f. Circuit breakers
    g. Advanced base portable electric power plants
    h. Central electric power stations
    i. Power and telephone lines and poles
    j. Underground conduits
    k. Interior wiring
    l. Electrical fixtures

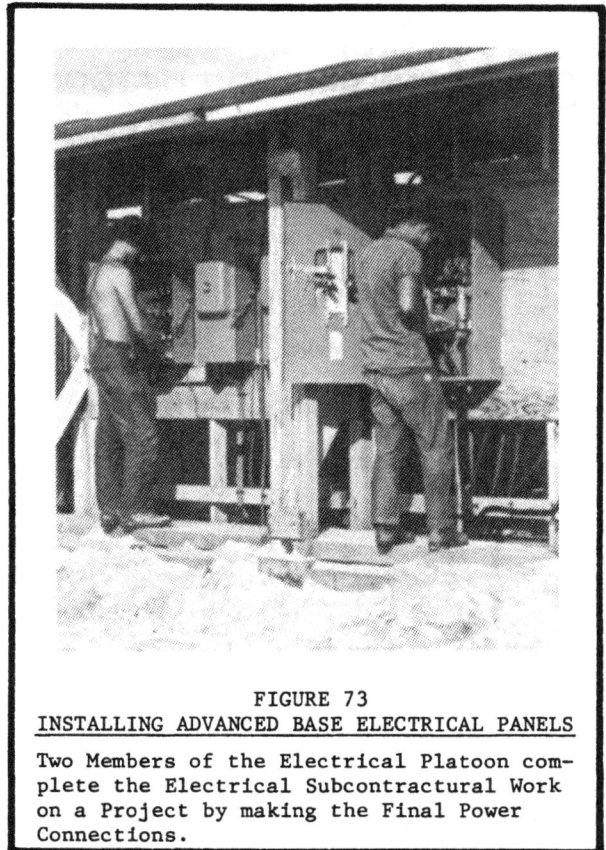

FIGURE 73
INSTALLING ADVANCED BASE ELECTRICAL PANELS

Two Members of the Electrical Platoon complete the Electrical Subcontractural Work on a Project by making the Final Power Connections.

SECTION E –
THE MCB GENERAL CONSTRUCTION COMPANIES

1. **Mission**. In the MCB, the "C" Company and "D" Company serve as prime contractor on all general construction. Members of these companies are all-around construction craftsmen, versatile in supervising or performing rough and finished carpentry, masonry, concrete work, dockbuilding, painting and related con-

struction work. See Figure 74. "C" Company and "D" Company consist of two rifle platoons and one weapons platoon, each. These military assignments are based on the general concept of divisibility. "C" and "D" Companies are the two basic companies to which units or personnel of companies "A" and "B" and headquarters could be attached. "A", "B", and headquarters are basically rifle companies which simplifies the task organization or assignment of these companies to either company "C" or "D" to form a large detachment.

FIGURE 74
FUNCTIONS OF THE MCB GENERAL CONSTRUCTION COMPANIES

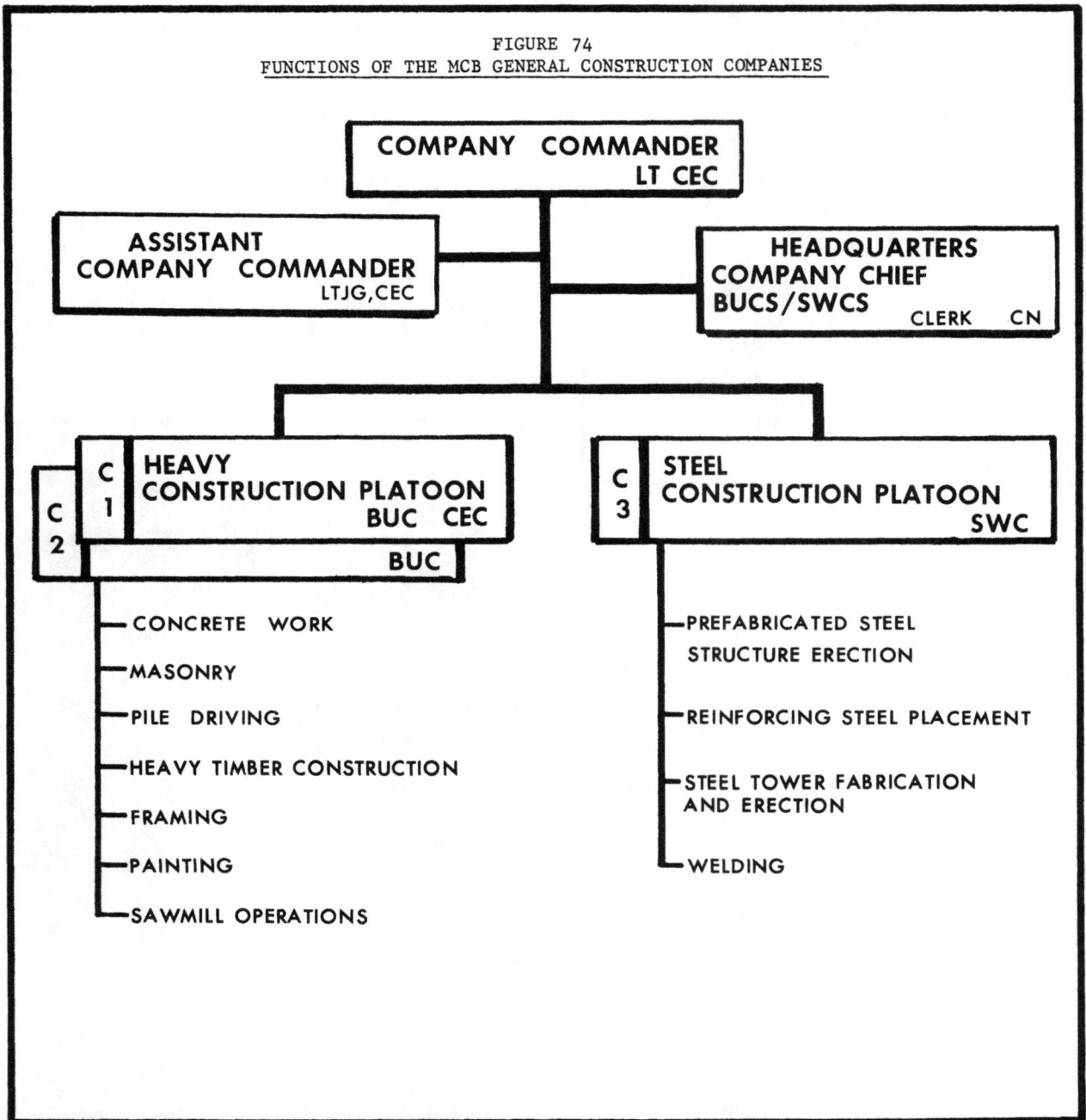

COMPANY COMMANDER LT CEC

ASSISTANT COMPANY COMMANDER LTJG,CEC

HEADQUARTERS COMPANY CHIEF BUCS/SWCS CLERK CN

C 2
C 1 **HEAVY CONSTRUCTION PLATOON BUC CEC** BUC

- CONCRETE WORK
- MASONRY
- PILE DRIVING
- HEAVY TIMBER CONSTRUCTION
- FRAMING
- PAINTING
- SAWMILL OPERATIONS

C 3 **STEEL CONSTRUCTION PLATOON** SWC

- PREFABRICATED STEEL STRUCTURE ERECTION
- REINFORCING STEEL PLACEMENT
- STEEL TOWER FABRICATION AND ERECTION
- WELDING

FIGURE 75
HARBOR DEVELOPMENT

A typical example of a Construction Task that may be assigned to the "C" or "D" Company would be the Improvement of Beach Landing Sites and the Development of Harbor Facilities and Warehouse Space.

quainted with qualifications for advancement in rating, service school curricula and training courses for BU, SW, CN, and any other ratings assigned. He prepares his company by formal training, on-the-job training and drills to perform their assigned construction, combat support and disaster control roles. He may be assigned several staff functions and be appointed to courts and boards by the Commanding Officer.

3. Company Organization. The company usually consists of a company headquarters, two identical Heavy Construction Platoons, and a Steel Construction Platoon.

When the battalion is staffed up to allowance levels, an assistant company commander is assigned. He is responsible for executing and enforcing the policies of the company commander and may be assigned as project officer on some of the projects assigned to the company. In this role, he would perform all the functions of project planning and supervision, but would submit his proposals to the company commander for approval since the company commander retains the responsibility for all projects assigned to his company.

The company headquarters consists of a leading chief, a Senior Chief Builder (BUCS), and a company clerk. In combat, additional messengers and communicators would be assigned to the company headquarters. The company chief is the senior enlisted man in the company. He supervises company administration, including correspondence, time cards, material requests from project petty officers, the company organization chart and the watch quarter and station bill. He may provide technical assistance to project petty officers in such things as material take-off, project scheduling and crew assignments.

2. The Company Commander. The company commander of either company serves as the general construction and rifle company commander. He is responsible for the construction of all projects assigned, including the work done by subcontractors. In combat, he is assigned a defense sector and must see that the machine guns and rocket launchers of his weapon platoons are properly positioned and integrated with the battalion defense plan. He is responsible for the usual details of company command described in Section A. He must be ac-

FIGURE 76
HANGAR CONSTRUCTION

The Construction of a Hangar and Maintenance Shop is an Example of a Large Project that may be assigned to the Heavy Construction Platoons.

139

4. **The Heavy Construction Platoon**. The platoon leader of the C-1 and C-2 platoon is usually a CEC Officer assisted by a Chief Builder (BUC) and a right guide. He is assigned as project officer for one or more projects and is responsible for coordinating all work crews assigned. He must be familiar with qualifications for advancement for BU, and other men assigned to his platoon. He trains and directs his platoon in defensive combat, combat support and disaster control operations. Members of the platoon are permanently assigned to squads and work crew/fire teams. However, squads and work crew/fire teams may be grouped and assigned specialties according to the projected workload for each deployment. Members of the platoon are trained to:

a. Build, erect, maintain, and repair wooden and concrete structures.

b. Perform pile-driving and dock building operations.

c. Perform logging and sawmill operations.

d. Mix and apply paints.

e. Build concrete forms.

f. Lay brick, stone, concrete block and building tiles.

g. Batch, mix and place concrete for all types of structures (including underwater installations.

5. **The Steel Construction Platoon**. The platoon leader is usually a CEC Warrant Officer assisted by a steelworker chief (SWC) and a right guide. He is assigned as project petty officer for one or more projects and is responsible for coordinating all work crews. Part of his platoon may be utilized as a subcontractor force. He must be familiar with the technical, military and disaster control requirements for the SW and other ratings assigned. He trains and directs his platoon in all phases of these operations and in defensive combat, combat support and disaster control operations. Members of the platoon are assigned as they are in the heavy construction platoon. They are trained to:

a. Place reinforcing steel.

b. Erect and dismantle steel structures.

c. Erect advanced base prefabricated structures.

FIGURE 77
ELEPHANT HUT ASSEMBLY

The Construction of the Large 40 x 100 Quonset Hut is Assigned to the Steelworkers of the Steel Construction Platoons.

SECTION F - THE ACB EQUIPMENT COMPANY

1. **Mission**. In the ACB, the "A" company is responsible for the operation, custody and maintenance of automotive vehicles, construction equipment, propulsion units, bottom-laid amphibious assault bulk fuel systems, communications equipment and winches. This company operates the battalion's field phone system, the motor pool and equipment repair shops. It performs beach salvage and is capable of performing limited earth-moving tasks during amphibious assaults. Field maintenance for all Naval Beach Group Staff automotive and construction equipment is done by the ACB equipment company.

2. **The Company Commander**. The company commander is usually a lieutenant in the Civil Engineer Corps who is responsible for the direction of his company in the accomplishment of assigned tasks and for the maintenance and repair of equipment in his custody plus repair of marine equipment assigned to the pontoon and limited construction companies. He is responsible for the maintenance of safe working conditions in the repair shops and compounds, assignment and training of personnel and the control of funds allotted to his company. He must be familiar with advancement requirements and schools for EO, CM, MR, EN, BU, CE, SW and SF. During amphibious operations he usually becomes the team commander of the bottom-laid fuel system installation team.

3. **Company Organization**. The company is divided into three platoons, as shown in Figure 78, the Equipment Operations Platoon, the Equipment Maintenance Platoon and the Shops Platoon. The company commander also has a company executive officer, a company chief and a company clerk. The company executive officer directs the training programs and

FIGURE 78
FUNCTIONS OF THE ACB EQUIPMENT COMPANY

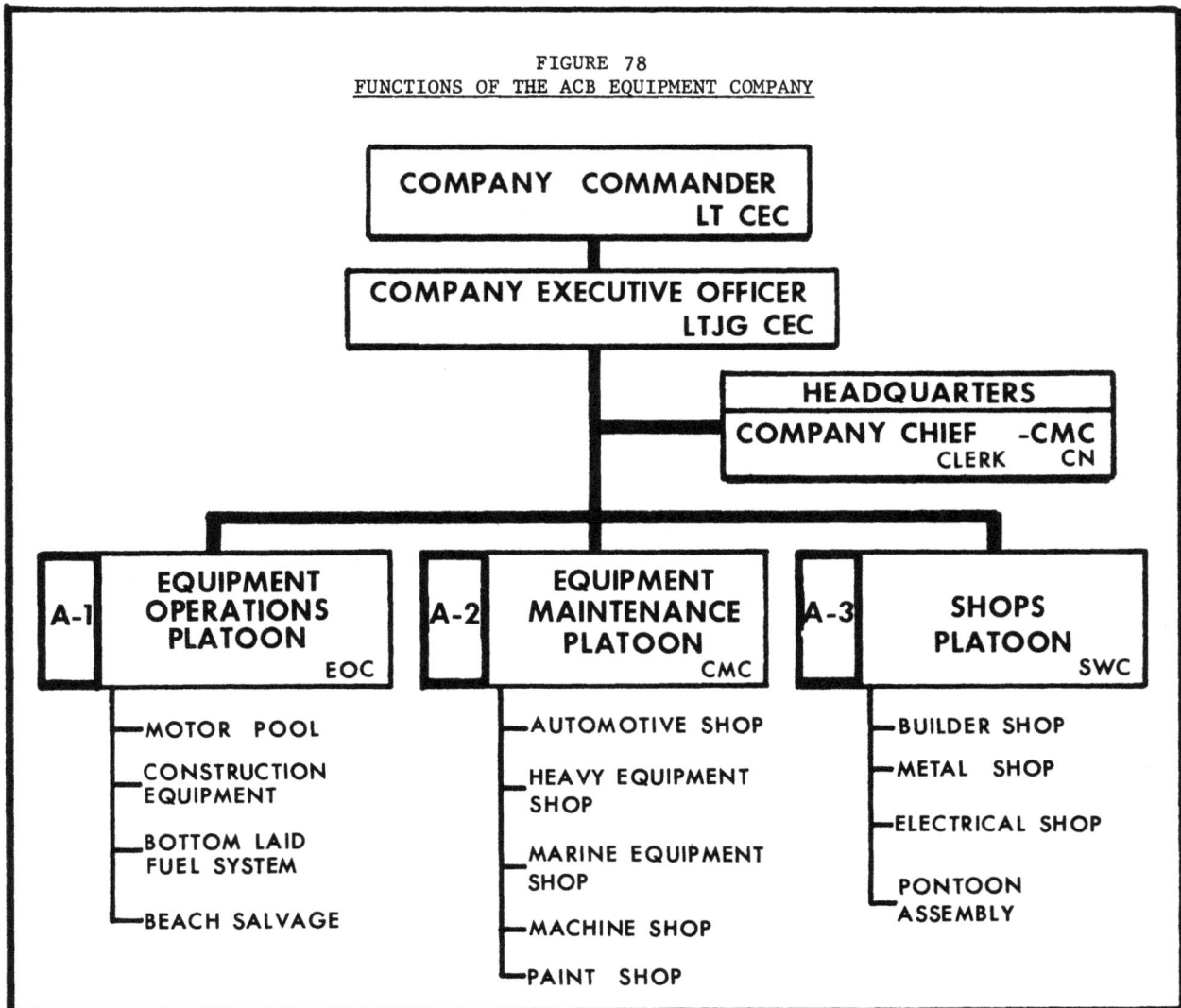

141

carries out other assignments given to him by the company commander. The company chief is the senior enlisted man in the company. He is responsible for coordinating the activities of the shops, enforcement of the company commander's policies and maintaining the company roster and the watch, quarter and station bill.

A-1 The Equipment Operations Platoon. The platoon leader of the A-1 platoon is usually an equipment operator chief (EOC) who is responsible for providing the men and equipment that are required to perform limited earth-moving, materials handling, beach salvage and transportation tasks. He trains and directs his men in the operation of an equipment pool and the installation of the bottom-laid fuel system. He is particularly concerned with the preparation of vehicles from embarkation through the surf.

A-2 The Equipment Maintenance Platoon. The platoon leader of the A-2 platoon is usually a construction mechanic chief (CMC) who is responsible for maintenance and repair not only of automotive and construction equipment, but also marine equipment such as outboard propulsion units, boat engines, and winches. He ensures that equipment is properly maintained in accordance with NAVFAC criteria.

A-3 The Shops Platoon. The platoon leader is usually a steelworker chief (SWC) who provides the same basic services performed by the MCB "B" company. In addition, he is responsible for the assembly and overhaul of pontoon structures. When deployed he may be given the responsibility for camp maintenance.

SECTION G - THE ACB PONTOON COMPANIES

1. Mission. The pontoon or operating companies "B" and "C", are responsible for providing causeways, rush rolls, lighterage and transfer barges, warping tugs, causeway tender boats and buoyant amphibious assault bulk fuel systems, along with all the necessary anchoring, hoisting and lashing gear, to support an amphibious assault. They must be able to maintain, repair and operate pontoon structures including causeways, barges, tugs and floating cranes. They must be able to perform 1st echelon maintenance on propulsion units, boat engines and winches.

2. The Company Commander. The company commander may be a CEC officer or a line officer. He is responsible for all floating capabilities of the battalion and for the accomplishment of tasks assigned to his personnel. He is responsible for the training of his men and must be familiar with advancement requirements, schools, practical factors and training courses for BM, EN, SM and EM. During amphibious operations he is the causeway installation commander and often acts as team leader for the primary causeway installation team and maintains responsibility for the other causeway teams when they are in the same area.

3. Company Organization. The company is divided into three platoons as shown in Figure 79, the Warping Tug Platoon, the Causeway Platoon and the Barge Platoon. The company commander is assisted by a company executive officer and a company headquarters consisting of a company chief, a supply petty officer and a company clerk. The company executive officer directs the training program and carries out the duties given to him by the company commander. Often he is the buoyant fuel system commander during amphibious operations. The company chief is the senior enlisted man in the company and should be a boatswain's mate chief (BMC). He is responsible for tabulating personnel assignments within the company and for keeping the watch quarter and station bill current. He enforces the company commander's policies and coordinates the activities of the platoons.

B-1 The Warping Tug Platoon. The platoon leader of this platoon is normally a boatswain's mate chief (BMC) who is responsible for the loading, re-assembly and operation of warping tugs and floating cranes and for the installation of the buoyant fuel system. He provides seaward salvage of broached craft in the surf zone, salvage of sunken gear, mobility to the causeway pier and subsequent placement of causeway anchors using his warping tugs. His men operate the specially configured LCU or LCM which lays the buoyant fuel system. In amphibious operations he would usually act as the team leader of the AABFS (buoyant) team.

B-2 The Causeway Platoon. Normally a BMC is the platoon leader for the B-2 platoon. He is responsible for first echelon maintenance of all causeway sections and associated lifting, lashing and mooring gear. His task during an amphibious assault is to get the causeway sections into the water, attached end to end, form the causeway pier, bring it into the beach on the range markers and anchor it in place.

B-3 The Barge Platoon. The platoon leader, a BMC, is responsible for rigging and preparing barges for lighterage and if necessary for transfer. The transfer barge has a crawler crane chained to the deck and is used to transfer cargo from one boat to another type of craft. This is used when an obstruction, such as a reef, prevents one boat from making the entire trip. Tender craft to help install the causeway and fuel system are also provided by this platoon.

SECTION H -
THE ACB LIMITED CONSTRUCTION COMPANY

1. Mission. When the ACB is at full strength and operating under mobilization conditions, many Group VIII personnel who are assigned to the battalion are formed into a fourth company.

This company, usually designated as "D" Company is capable of fulfilling general construction missions. It would be landed on the beach to build egress roads, small airfields, helicopter landing areas, etc., as directed by the Naval Beach Group Commander. When the "D" company is not formed, personnel normally assigned to "A" company perform the limited construction tasks.

FIGURE 79
FUNCTIONS OF THE ACB PONTOON COMPANIES

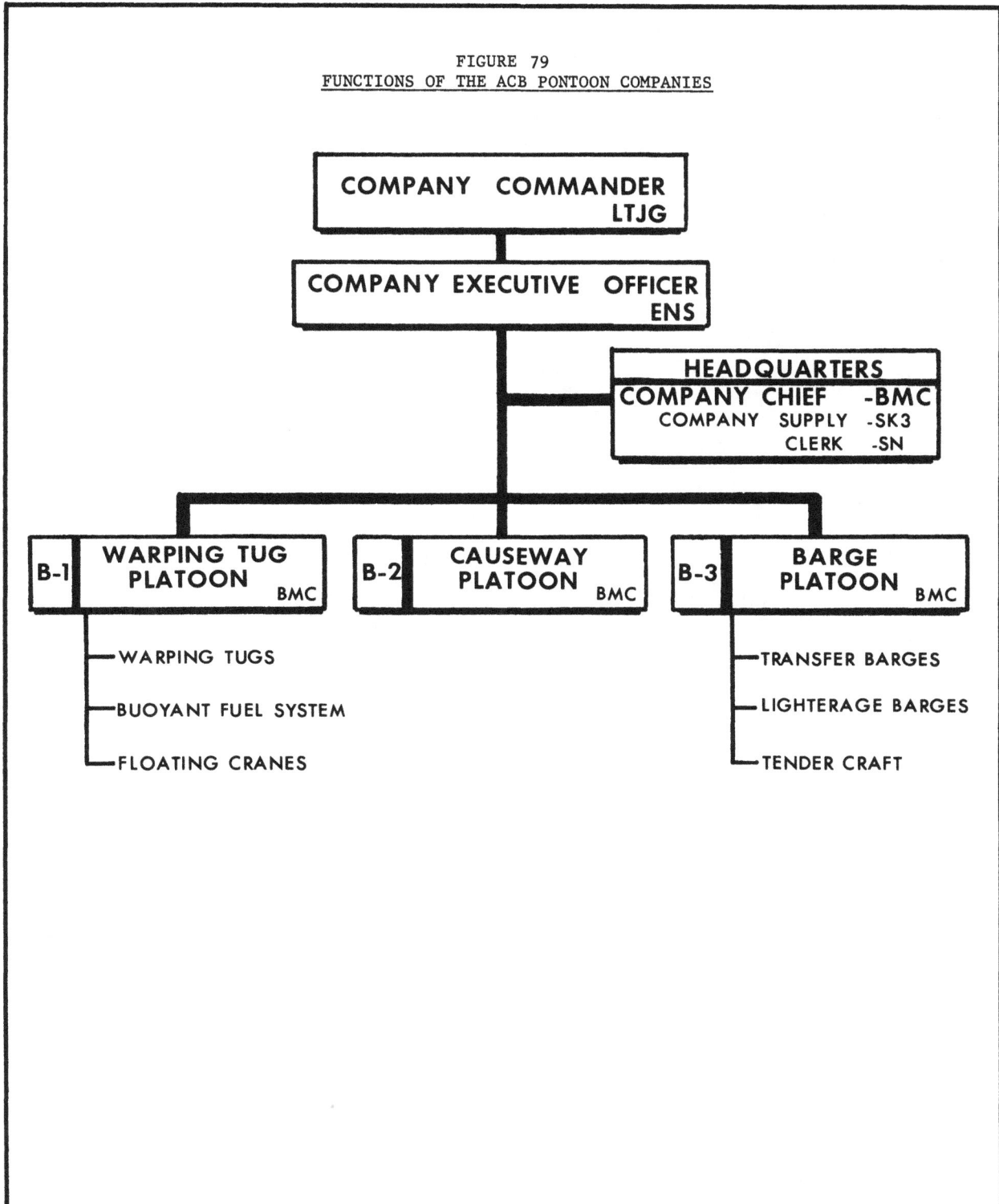

```
                    ┌─────────────────────────────┐
                    │   COMPANY  COMMANDER         │
                    │          LTJG                │
                    └─────────────────────────────┘
                                  │
                    ┌─────────────────────────────┐
                    │ COMPANY EXECUTIVE  OFFICER   │
                    │           ENS                │
                    └─────────────────────────────┘
                                  │
                         ┌────────────────────────────────┐
                         │        HEADQUARTERS            │
                         │ COMPANY CHIEF      -BMC        │
                         │   COMPANY  SUPPLY  -SK3        │
                         │        CLERK   -SN             │
                         └────────────────────────────────┘

   ┌───┬──────────────────┐  ┌───┬──────────────┐  ┌───┬──────────────┐
   │B-1│ WARPING TUG      │  │B-2│ CAUSEWAY     │  │B-3│ BARGE        │
   │   │ PLATOON      BMC │  │   │ PLATOON  BMC │  │   │ PLATOON  BMC │
   └───┴──────────────────┘  └───┴──────────────┘  └───┴──────────────┘
        │                                                │
        ├─ WARPING TUGS                                  ├─ TRANSFER BARGES
        │                                                │
        ├─ BUOYANT FUEL SYSTEM                           ├─ LIGHTERAGE BARGES
        │                                                │
        └─ FLOATING CRANES                               └─ TENDER CRAFT
```

143

FIGURE 80
BACKLOADING AFTER AN EXERCISE

In this Picture of an MCB Backloading over the ACB Causeway Pier, much of the Floating
Equipment of the ACB Operating Company can be seen. On the left is a Warping Tug, then
the Causeway Pier in the Center. To the right is an LCU Specially Configured with a
Hose Reel to Install the Buoyant Fuel System and Another Warping Tug is in the Foreground.

2. Company Organization. The "D" company
would be organized into a company headquarters
and two platoons as shown in Figure 81. The
equipment platoon would assume the responsibility of beach salvage and beach improvement
along with construction of roads, landing
fields and storage areas and may be given responsibility for maintenance of a section of
the main supply route.

The construction platoon would consist of
personnel in the EA, BU, UT, CE and SW ratings
and would, therefore, have a great deal of
potential to perform small construction tasks

required by the Naval Beach Group Commander.
These might include camp buildings, sewage
systems, water supply, power systems, etc.
Care must be taken in planning the employment
of this company to ensure that the proper
equipment and tools are available or provided
and that construction consumables be procured
and loaded aboard ships of the task force.
The duties of the "D" company chief and platoon leaders would be essentially the same
as the duties of petty officers with similar
missions. These duties have been listed
previously.

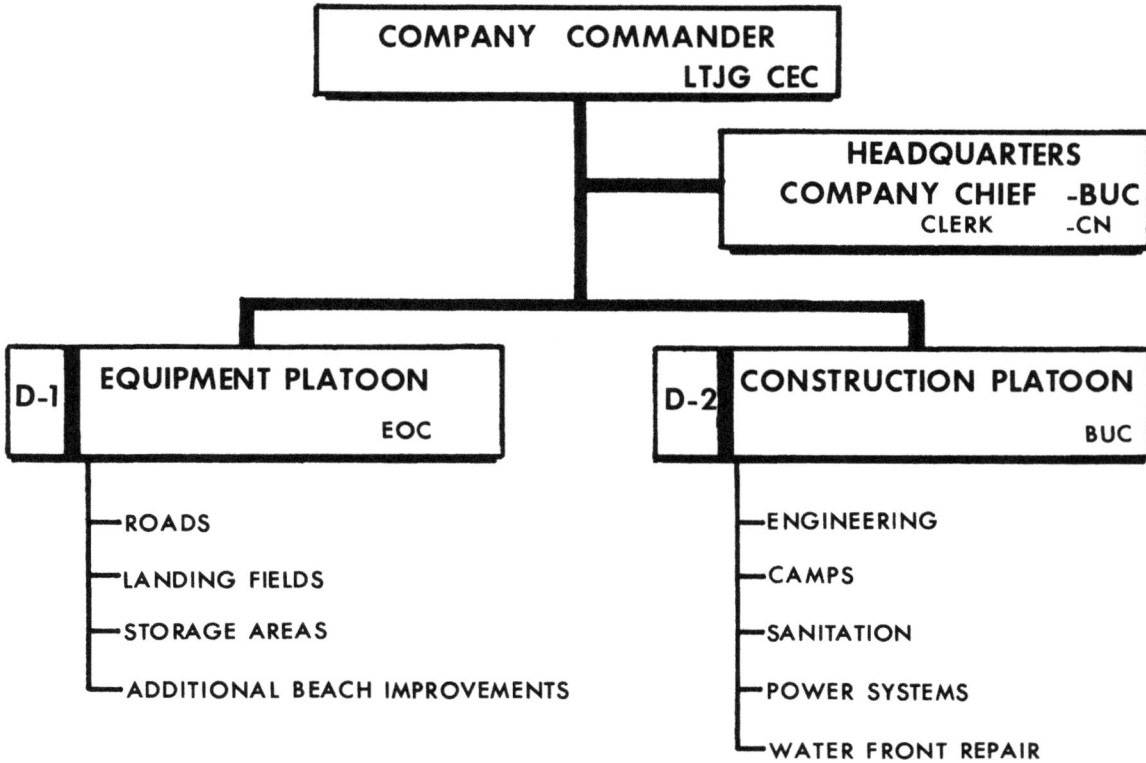

FIGURE 81
FUNCTIONS OF THE ACB LIMITED CONSTRUCTION COMPANY

Chapter VIII

THE BATTALION CAMP AND SECURITY

SECTION A - CAMP ESTABLISHMENT

1. _Camp Planning_. The length of time that a battalion will be assigned to a particular area and the available existing facilities determine what arrangements must be made to make the camp area habitable. During peacetime, the battalion will usually be berthed in established SEABEE camps or in the barracks of the base to which they are deployed. In this case planning is limited to determining what barracks the personnel will be assigned and which spaces will be used for purposes such as offices, storage, sick bay, etc. All the existing facilities must be inspected and arrangements made to effect the necessary repairs. At an established SEABEE camp, the battalion may decide on the desirability of making changes or adding new buildings, utilities or facilities. Such factors as the sources of labor, materials and funds must be weighed against the anticipated advantages gained from the proposed changes. The final decision rests with the Commanding Officer of the activity having responsibility for the camp facilities.

Under emergency or wartime conditions, the battalion is usually deployed to new locations where bases are being built or expanded, and thus, the battalion supplies its own housing and shop facilities. The battalion allowance contains enough tents for this purpose, however, prefabricated steel or wooden structures may be substituted either because of the climatic conditions or the anticipated length of stay. When a new camp is to be constructed, decisions must be reached as to the location of quarters, showers, heads, storage spaces, mess hall, shops and other facilities. The NAVFAC P-140 contains a series of drawings of various camp configurations and some of the details for construction. These drawings should be used as a guide and site-adapted by the S-3 (See Figure 85).

2. _Camp Facilities_. Whether permanent structures or the more austere tent camp is to be used, the basic facilities must be provided for the conduct of battalion activities. (A typical workday routine for peacetime operations is shown in Figure 82.

Separate berthing facilities are normally provided for officers, chiefs and other enlisted personnel. If the base already has living quarters, the battalion makes use of them. If no quarters are available, the battalion must erect its own. Officers are assigned one or two to a room during peacetime

if facilities permit. During time of emergency, they may be assigned to tents or to quarters similar to barracks with several officers in the same tent or room. Chiefs generally live in a barracks-type building, but they may have separate rooms. Enlisted men also live in barracks. If the battalion uses tents, all personnel will be quartered in them. Living conditions in tents are the same for all personnel, except that more enlisted men are quartered in one tent than are officers. Battalions vary on the procedure for assigning men to particular quarters and to a bunk within the quarters. Assignment to quarters is usually made by the company commander. This should reflect the man's squad assignment. It is very desirable to maintain squad integrity in berth assignments. The bunk assignment is

FIGURE 82
TYPICAL DAILY ROUTINE
FOR TRAINING DEPLOYMENT

0530	REVEILLE
0530	BREAKFAST
0640	OFFICERS CALL
0645	QUARTERS
0700	TURN TO - SICK CALL
1130	DINNER
1230	TURN TO
1645	SET THE WATCH
1700	SECURE WORK
1700	OPEN CLUBS
1730	SUPPER
1800	OPEN RECREATION FACILITIES
1900	DUTY SECTION MUSTER
1930	EVENING MOVIE
2100	SECURE RECREATION FACILITIES
2130	SECURE CLUBS
2200	TAPS

FIGURE 83
CONSTRUCTION OF STRONG-BACK TENTS

A 16 x 32 Squad Tent plus Wooden Framing,
A Stove and Electrical Wiring Constitute a
1360 Assembly. There are 40 such Assemblies in the P-25 Functional Component.

made by the senior petty officer in the quarters. This man is the squad leader or senior
fire team leader and is designated as hut (or
tent) captain. The hut captain is responsible
for seeing that the regulations affecting his
barracks are carried out. He inspects for
cleanliness, appearance, security and safety.
Bunks are tagged to show the name, rate, and
company of the occupant. The OOD or another
designated officer makes a daily inspection of
living quarters. Weekly, semimonthly or
monthly inspections are usually made by the
Commanding Officer. Any criticism by the inspecting officer is passed on to the company
commander whose men are quartered in the barracks and also to the hut captain.

Messing for battalion members may range
from subsisting in a permanent mess at the
homeport or at an established base to eating
"C" rations during the initial stages of an
assault. It is usually desirable for the battalion to operate its own galley when deployed, and it is staffed and equipped to do so.
This procedure allows the Commanding Officer
more flexibility in getting the construction
task accomplished because he can control the
meal hours and set the uniform requirements
for the battaion galley. Local working

conditions may make it desirable to subsist
some personnel at other messes nearer the work
site, especially for the noon meal. The two
supply officers concerned must make the necessary arrangements. U.S. Navy Regulations permit chief petty officers to have a separate
mess. At some locations the chiefs subsist
entirely from food prepared in the general
mess. Their food may be served in a separate
section of the mess hall or at a different
location. However, the Commanding Officer may
authorize a closed mess for the chiefs which
would be operated by them in accordance with
regulations established by BUPERS. As with
the CPOs, the officers may subsist from the
general mess or set up a closed mess. Normally when the base facilities are inadequate or
when the battalion is operating its own camp,
the battalion is authorized to establish a
wardroom mess. In such cases, food is usually
purchased from the general mess and is served
to the officers by the battalion stewards.
The stewards bring the meals from the general
mess at prescribed hours. Regulations prohibit
their being assigned to the commissary section
or serving food to enlisted men. Under the di-
rectiom of the Mess Caterer, the stewards serve
food in the officers' mess, keep the messing
space in proper condition for the serving of
meals and maintain officers' quarters in a
habitable condition.

The battalion will usually set up or utilize offices, shops, warehouses, equipment
compounds and tool rooms that are separate and
distinct from any other unit. Other camp facilities such as a dispensary, chapel, library,
post office or theater may be provided by
others or set up and run by battalion personnel. The battalion has the personnel to provide most of these services, and when operating alone or at a distance from existing facilities, may choose to do so. Although it
may cost some extra manhours to provide these
services, the convenience of having them available in the camp gives the men a sense of
unity, and provides a boost to morale. Utilizing the standard personnel allowance, Figure
84 indicates the personnel that would be assigned to camp operations. Some savings in
manpower can be realized by using existing
facilities at an established base, but many of
the battalion personnel would have to be assigned to the base to offset the increased
workload generated by the battalion.

3. Camp Maintenance. When in quarters supplied by the base, the battalion usually reports any trouble, in accordance with established procedures, to the base public works
department. However, the battalion usually
performs maintenance for its own area even
when occupying established camps. Minor maintenance consists of repairs to barracks, shops,
telephone systems, light systems and so on.
The necessity for repairs is ascertained by
means of inspections or is reported by the
occupants of a shop or barracks. Minor

FIGURE 84		
SAMPLE MANPOWER REQUIREMENTS		
FOR CAMP FUNCTIONS		
GALLEY		46
Cooks	14 CS + 2 SN/SA	
Messmen	28 CN/CA	
Messdeck MAA	2 Group VIII	
SHIPS SERVICE		8
Laundry	2 SH + 2 SN/SA	
Retail Store	1 SH + 1 SN/SA	
Barber	2 SH	
DISPENSARY		7
Dental	2 DT	
Medical	5 HM	
ADMIN PERSONNEL OFFICE		13
Admin	8 YN + 2 SN/SA	
Pers	2 PN + 1 SN/SA	
SUPPLY OFFICE		18
Supply	12 SK + 3 SN/SA	
Disbursing	3 DK	
MAA OFFICE		8
MAA Force	4 BM + 4 Group VIII	
CAMP OPERATION		26
Maintenance	2 Group VIII	
Watches		
(24 hr/7 days)	22 Group VIII	
Fireman	2 Group VIII	
ARMORY		5
	4 GM + 1 SN/SA	
POST OFFICE		2
	2 PC	
WARDROOM		6
	6 SD	
COMMUNICATIONS SHOP		2
	2 ET	
SPECIAL SERVICES		5
	5 Group VIII	
COMPANY CLERKS		5
	CN/CA	
CHAPLAIN'S ASSISTANT		1
	SN/SA	
TOTAL =	27% of 563 or	152

repairs to windows, doors, roofs and floors may be approved by a designated officer, such as the headquarters company commander. Major repairs or alterations require the approval of the battalion Commanding Officer. In some cases approval for major work must be obtained from COMCBLANT/COMCBPAC and when approved, may be included as part of the construction task. The maintenance is usually performed by several petty officers in the shops company, assisted by regular members of various shops. When a considerable expenditure of man-days is involved, the project is assigned to one of the construction companies. Such facilities as generators, boilers, wells and distillation units are operated and maintained by specialists from the shop company assigned to service watches.

At an established camp, facilities to maintain camp sanitation are usually available. These facilities consist of sewerage systems and refuse disposal systems. If the influx of battalion personnel makes additional sewerage and disposal systems necessary, the battalion may expand existing facilities. Concurrence of the base public works officer and the base sanitation officer are necessary. At new locations, the battalion must construct its own disposal systems. The length of time that the battalion will be in the location and the nature of the installation determine the type of facilities that are most practical.

Cleanliness is one of the major points checked during the various inspections. Battalion directives establish the procedure to be followed to keep living quarters, shops, heads and showers clean. A cleaning detail is normally assigned for this purpose. This detail may consist of one or more men from each barracks who clean their own quarters, or it may consist of a few designated men who clean all barracks. Each hut captain sees that his barracks is ready for the cleaning detail. Trash is placed in cans or other receptacles; excess gear, such as seabags and suitcases, is stored in assigned spaces, and clothing is stowed in lockers. The cleaning detail is also assigned to clean heads, showers, and perhaps other spaces. The chief master-at-arms inspects the work of the cleaning detail to ensure that the work meets the prescribed standards. The officer directly in charge of a shop or an office is charged with the cleanliness and upkeep of that space. The men assigned to this officer generally do the cleaning for shop and office spaces.

The headquarters company commander is usually assigned the task of coordinating camp operations. As such, he drafts directives governing the functions related to camp operation such as refuse disposal, pest control, operation of utilities and traffic control. He also coordinates the activities of the men that are responsible to carry out these functions.

FIGURE 85
ADVANCED BASE DRAWING OF A SEABEE CAMP

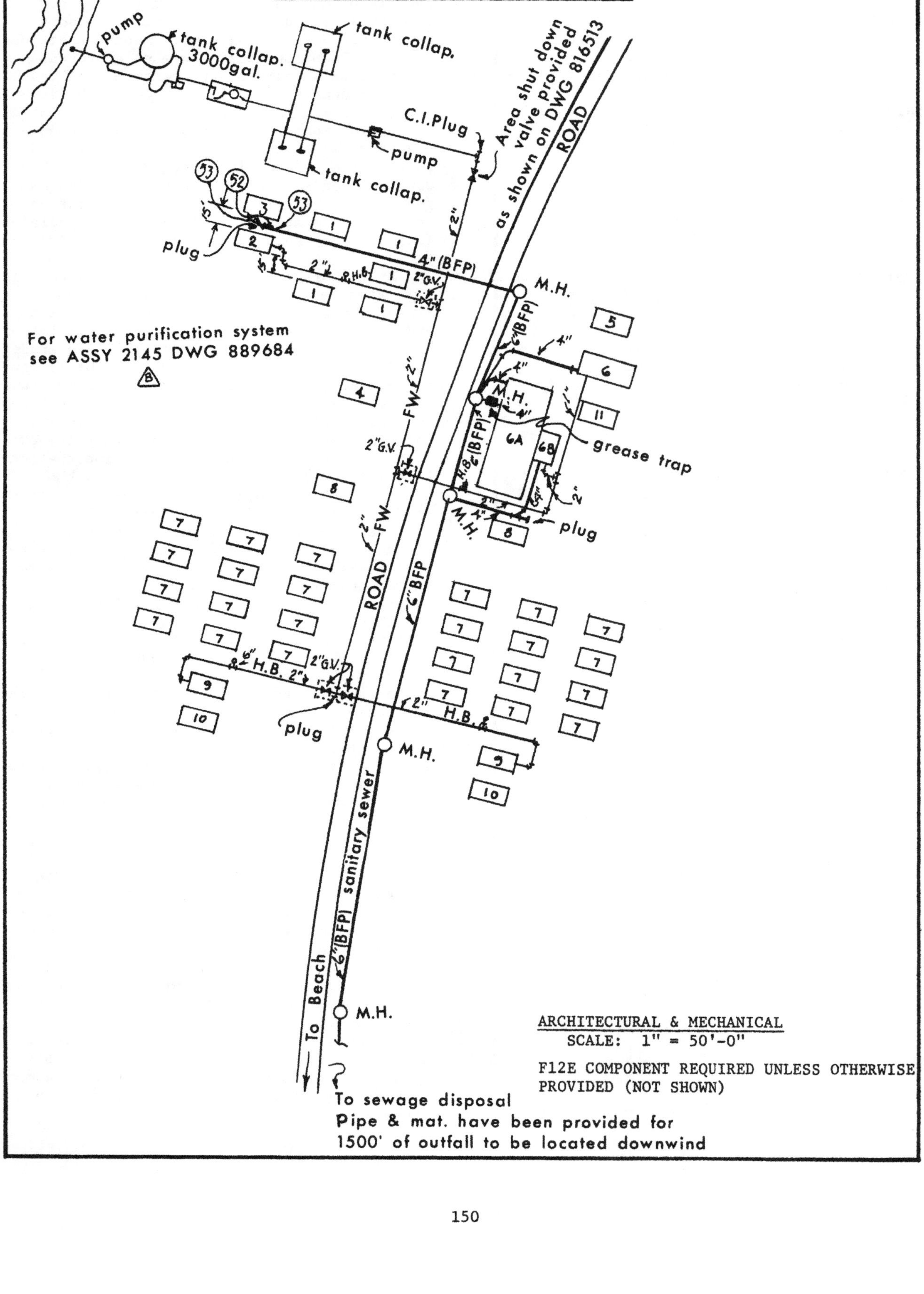

pump
tank collap. 3000gal.
tank collap.
C.I.Plug
pump
tank collap.
Area shut down valve provided as shown on DWG 816513
ROAD
53
52
53
plug
4" (BFP)
2"
H.B.
2" G.V.
M.H.
6" (BFP)
4"
For water purification system see ASSY 2145 DWG 889684
FW
2" G.V.
grease trap
6A
6B
plug
ROAD
6" BFP
6"
H.B. 2"
2" G.V.
plug
M.H.
sanitary sewer
6" (BFP)
To Beach
M.H.
To sewage disposal
Pipe & mat. have been provided for 1500' of outfall to be located downwind
ARCHITECTURAL & MECHANICAL
SCALE: 1" = 50'-0"
F12E COMPONENT REQUIRED UNLESS OTHERWISE PROVIDED (NOT SHOWN)

FIGURE 86
CAMP FACILITIES

TENT CONSTRUCTION

TEMPORARY EQUIPMENT REPAIR SHOPS

CHOW HALL

E.M.CLUB

OVERALL OF TENT CAMP

FIGURE 86
CAMP FACILITIES (Cont'd)

VOLLEY BALL

STORAGE YARD

MORTAR PITS

CAMP DEFENSE

FIGURE 86
CAMP FACILITIES (Cont'd)

QUONSET HUT BERTHING

IMPROVED SHOP SPACES

BATTALION LIBRARY

PERMANENT CAMP FACILITIES

SECTION B - THE BATTALION WATCH

1. Functions of the Watch. While the battalion sleeps, someone must remain alert to sound the alarm in case of fire or disorder. When men are away from the job, camp or shop, someone must be posted at each location to protect government property. At headquarters someone must remain on duty to receive and act on important messages at whatever hour they arrive. Moreover, certain essential equipment, such as boilers, generators and switchboards, must be operated through the night. Every key activity, such as the motor pool, supply department and camp operation crew, must remain on a standby status ready to perform occasional after-hours services or to spring into action in an emergency. The fire department and the medical dispensary must be prepared for any contingency. Also, there must always be an alert and thoroughly trained guard force to preserve order and to protect the lives and property of the battalion. All these functions are performed by the Battalion Watch.

The make-up of the watch depends on each battalion's specific needs. The need varies sharply from time to time, especially from peace to war. It may range in size and composition from a minimum of several personnel to assure that urgent matters which arise outside regular work hours receive immediate attention, to a 50% alert status in a perimeter defense to enable the battalion to maintain an around-the-clock, around-the-calendar alert. This burden of day-in-day-out, round-the-clock security and of odd-hour work is distributed equitably among the officers and men of the battalion by means of the watch system. Owing to the nature of the watch, its members must be trained to expect the unexpected and to act deliberately and decisively in the face of any emergency. Only by continuous planning, painstaking preparation and equitable rotation can each officer and man know his duty and be ready to discharge these serious responsibilities immediately. A typical battalion watch structure is indicated in Figure 87.

2. Watch Administration. As with all activities of the battalion, the Commanding Officer has ultimate responsibility for the battalion watch. U.S. Navy Regulations, (Article 1002) require that he establish any watches necessary for the safety and proper operation of the battalion. He approves the battalion watch plan which is published in instructions and notices of the 1601 series. He also requires that the battalion log be submitted to him periodically for approval. To assure that the watch is properly administered, he may appoint a Security Officer and a Senior Watch Officer.

The Executive Officer ensures that necessary security measures and safety precautions are understood and observed. He submits recommendations for changes to the battalion watch plan to the Commanding Officer for approval; he supervises preparation and publication of daily watch lists and he approves entries in the Night Order Book and the OOD Instruction Book.

The Battalion S-2 usually serves as Security Officer. As such, he is a special staff assistant to the Executive Officer on matters relating to the watch. The Security Officer studies the security needs of the battalion. He views and reviews every battalion activity from the standpoint of internal security, external security and physical protection, and he evaluates the effectiveness of the battalion watch plan with regard to safeguards against enemy attack, infiltration, guerrilla activity, sabotage, theft, lawlessness or any incident which would upset normal battalion activity. He helps the Executive Officer prepare battalion watch instructions and notices, the OOD instructions and the special orders for each watch billet. He works with the Classified Material Control Officer in planning safeguards against espionage, sabotage and unauthorized disclosure of classified matter. He correlates the watch plan with the battalion's disaster control plan, and when in a combat situation, the perimeter defense plan The watch must be alert for such emergencies and be ready to execute any prescribed plan at once. These include the:

a. Fire Bill
b. Storm Bill
c. NBC Warfare Plan
d. Air Raid Defense Plan
e. Civil Disturbance and Riot Plan
f. Camp Defense Plan

The Security Officer effects liaison with local Marine, Army, Air Force and Coast Guard Security Officers; ONI and FBI Offices; Fire and Police Departments; and Civil Defense Agencies when appropriate.

Collateral duty as Senior Watch Officer may be assigned to the senior CEC Officer eligible for OOD duty. He prepares the OOD duty roster monthly and instructs officers in order to qualify them for duty as OOD. He makes certain that members of the watch understand and are able to comply with the watch plan and their general and special orders. He periodically examines entries in the Log for adequacy and form, and whenever he notes a relaxation in standards of watch performance, recommends that remedial training be scheduled.

The S-1 prepares the JOOD duty roster and coordinates assignment of enlisted men to specific watch billets. He prepares the daily watch list from lists submitted by the companies and sees that it is published in the Plan of the Day. He makes current rosters of all duty sections available to the OOD, JOOD, and duty MAA.

Each company commander and department head designates the men eligible to fill the watch billets assigned to his company. He assigns

FIGURE 87
THE BATTALION WATCH STRUCTURE

```
                    ┌─────────────────────────┐
                    │   OFFICER OF THE DAY     │
                    └─────────────────────────┘
                  ┌─────────────────────────────┐
                  │  JUNIOR OFFICER OF THE DAY   │
                  └─────────────────────────────┘

┌─────────────────────────┐        ┌─────────────────────────────┐
│  SERGEANT OF THE GUARD   │        │  PETTY OFFICER OF THE WATCH  │
└─────────────────────────┘        └─────────────────────────────┘

  ┌─────────────────────────────┐    ┌─────────────────────────┐
  │   CORPORAL OF THE GUARD      │    │  COMMUNICATIONS WATCH    │
  │       FIRST RELIEF           │    └─────────────────────────┘
  └─────────────────────────────┘    ┌─────────────────────────┐
                                      │      DUTY DRIVER         │
    ┌─────────────────────────┐      └─────────────────────────┘
    │    PERIMETER POSTS       │      ┌─────────────────────────┐
    └─────────────────────────┘      │  B Co. SERVICE WATCHES   │
    ┌─────────────────────────┐      └─────────────────────────┘
    │ INTERNAL FIRE and SECURITY│     ┌─────────────────────────┐
    └─────────────────────────┘      │ HDQS Co. SERVICE WATCHES │
    ┌─────────────────────────┐      └─────────────────────────┘
    │      ROVING PATROL       │      ┌─────────────────────────┐
    └─────────────────────────┘      │     DUTY DISPATCHER      │
                                      └─────────────────────────┘
                                      ┌─────────────────────────┐
                                      │    DUTY FIREFIGHTER      │
                                      └─────────────────────────┘
```

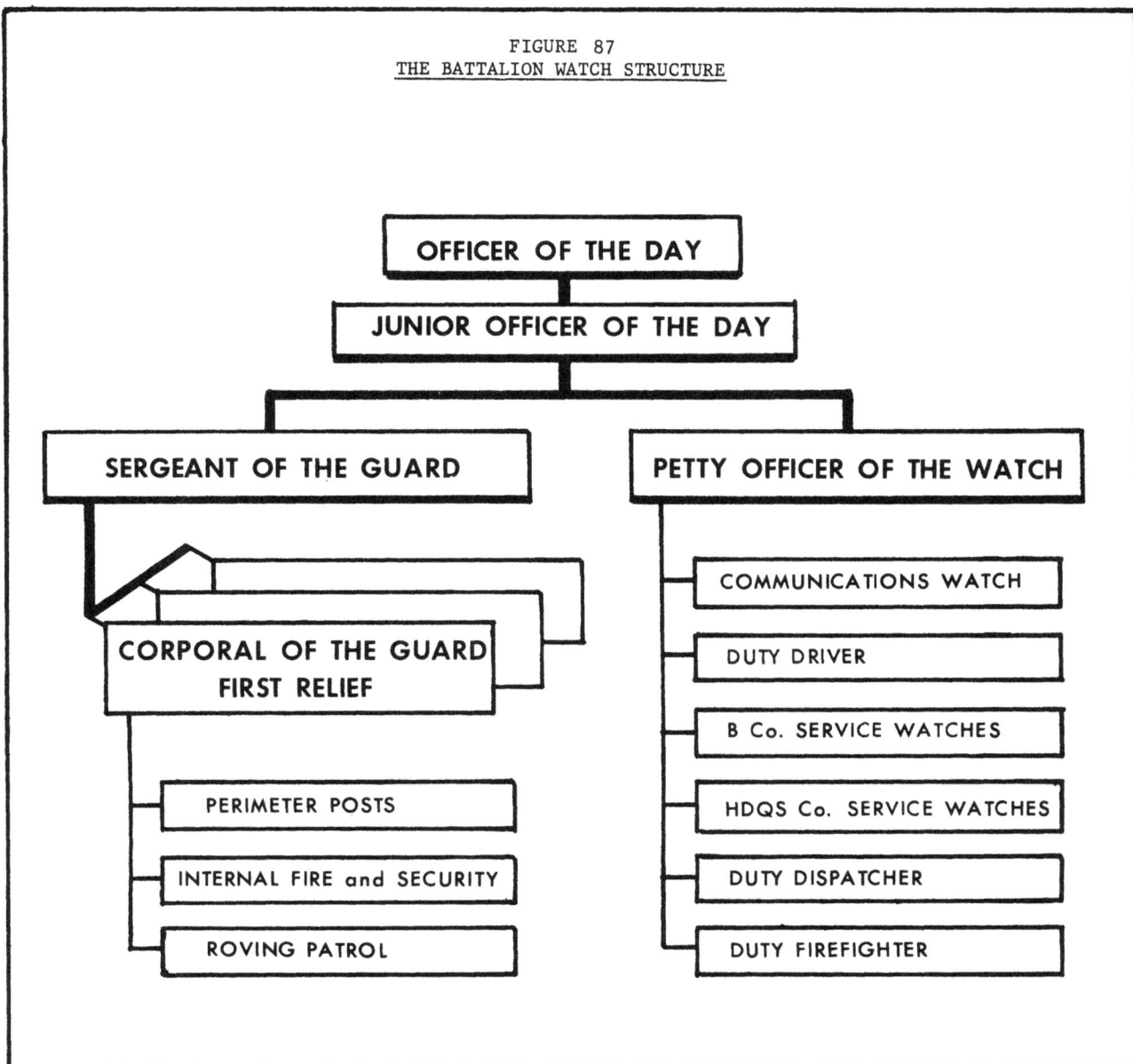

every man under his command to a duty section maintaining an equal distribution of critical ratings and an equal distribution of manpower among the duty sections so that roughly the same number of men will be on duty at any given time. Company commanders submit the names of watchstanders to the S-1 and submit other necessary data to the Executive Officer for entry in the Night Order Book.

Watch instructions include: the general requirements for the watch; duties of the OOD, JOOD, Sergeant of the Guard, Corporal of the Guard, Service Watches, duty MAA, and other watch-standers; the method of assigning men to duty sections; and policy concerning the watch. Watch instructions must be correlated with other directives and must be consistent with directives and must be consistent with directives and orders from higher authority.

The OOD Instruction Book is a ready reference on command policy. This book usually contains the special orders for each watch billet; a complete set of watch notices and instructions; copies of or references to pertinent orders and directives; statements of the Commanding Officer's policy on emergency leave, extension of leave, use of vehicles, granting of liberty or special privileges to men in the duty section, etc.; a list of persons and agencies to be notified in an emergency; the telephone numbers and addresses of battalion officers; and other information likely to be needed by the OOD. Plans for after-hours operations are submitted to the Executive

Officer, and when approved they are placed in the Night Order Book. Information on events or conditions expected during the OODs tour are also described in this book, e.g., weather forecast, persons due to arrive or depart, ship arrivals and departures and activities of nearby units. The Night Order Book is given to the OOD before the close of business on work days, and he relays necessary information or orders to the JOOD and other watchstanders.

3. The Officer of the Day. Every qualified officer or warrant officer assigned to the battalion (except the Chaplain, the Medical Officer, and the Dental Offficer) is eligible for duty as OOD. Tours of duty may be exchanged only upon approval of the Senior Watch Officer. If the OOD becomes incapable of completing his tour, a replacement is designated by the Senior Watch Officer. The OOD tour of duty is 24 hours, and he must remain on duty until properly relieved. On work days he performs his regular duties during work hours unless otherwise specified. Should his regular duties conflict with his duties as OOD, his OOD duties take precedence. During his tour the OOD must remain in the battalion area and is on call at all times. Toward the close of the working day, he reports personally to the Executive Officer to receive last minute instructions and the Night Order Book. The OOD is relieved at battalion headquarters at the time appointed in the battalion watch plan. The change of duty generally occurs in the morning. On work days the change of duty generally takes place in the presence of the Executive Officer. On other days the retiring OOD relays his reports directly to the new OOD.

The OODs office is usually located at battalion headquarters adjacent to the MAA Office. This office is the permanent location of the Battalion Log and the full-time station of the JOOD. The OODs office is the information and communications center of the watch and must be manned at all times by a responsible member of the watch. Sleeping accommodations must be provided in the battalion area for use by the OOD. He may use his regular quarters if they are located in the battalion area but in any case, his quarters must have an around-the-clock telephone connection with the OODs Office.

During his tour of duty the battalion OOD directly represents the Commanding Officer. After working hours the OOD is responsible for all battalion business. At such times, he is charged with the safety, security, discipline and efficiency of the battalion. To accomplish his mission, he must keep informed on all current matters affecting battalion operations, facilities, equipment and personnel. He must understand the battalion's relationship to other commands and know where and how to contact the battalion Commanding Officer or Executive Officer. He promptly notifies them of all matters which may affect the security or the reputation of the battalion. In the exercise of his duties the OOD has full

FIGURE 88
SABOTAGE OF PROJECT MATERIAL

In Forward Areas the Battalion Watch May Have to Cope with Enemy Attempts to Sabotage Project Material. The Greater the Need for Security, the Fewer the Number of Men who can be used for Productive Labor.

authority over all persons in the battalion area who are subject to orders of the Commanding Officer (except the Executive Officer). The OOD is responsible for compliance with all current orders, instructions and regulations issued by competent authority. Thus he must have a working knowledge of, and ready access to, files of all such orders and directives. When about his duties, the OOD appears in the uniform of the day and wears the OOD brassard and carries the weapons and equipment specified in the OODs instructions. He plans his tour so that at any given time he is in the best place to perform his duties. However, he must keep the JOOD informed of his whereabouts and of how he may be contacted.

4. Specific Duties of the OOD. Since the OOD is the senior member of the battalion watch, he is responsible for taking appropriate action and for keeping the Commanding Officer informed. Some specific duties are:

a. He assumes control at the scene of any emergency and takes swift action to prevent loss of life or property. He handles fires, accidents, serious injuries and deaths in accordance with battalion directives. If his estimate of the situation so dictates, he orders execution of appropriate sections of the battalion's disaster control plan. The emergency is brought promptly to the attention of the Commanding Officer, Executive Officer or senior CEC Officer on board.

b. He initiates action in response to, and follows up on official communications received after working hours. Urgent matters are relayed to the Commanding Officer.

c. He may grant leave, extension of leave or liberty under emergency conditions but he must comply with battalion directives and fleet policy.

d. He must be ready to execute the guard mounting provisions of the Landing Party Manual and of the battalion's watch plan. At the Duty Section Muster he checks the uniforms and appearance of the guard and makes certain that all members are present and in condition to perform their duties. He makes substitutions or rearrangements as necessary. He ensures that all petty officers, sentries, guards and patrols know, understand and are able to execute their general and special orders. He makes announcements or remarks that are pertinent at the time.

e. He makes inspections, requires reports and takes other action necessary to enforce military law. Whenever a person is apprehended by anyone subject to the authority of the OOD, the OOD must be notified by the most direct means. The Manual for Courts-Martial, empowers the OOD to order enlisted men into arrest, restriction in lieu of arrest or confinement. In taking such action the OOD must rigidly comply with military law and pertinent battalion directives. Whenever he orders a man into confinement, the OOD must consult at once with the Executive Officer or the Commanding Officer.

f. He assures performance of the daily routine as set forth in the plan of the day and is responsible for proper and timely execution by the duty MAA (or bugler) of routine and special calls. He approves special announcements made over the public address system.

g. He is responsible for accommodation of all personnel who report on board after work hours. He sees that check-in and orientation procedures are initiated and coordinates arrangements for transportation, food and berthing. Similar arrangements are provided to personnel who are due to leave the battalion after hours. Required data on arrivals and departures is recorded in the Log.

h. He is responsible for maintaining the Log in accordance with battalion directives.

5. Inspection by the OOD. Inspection is one of the main duties of the OOD. The OODs instructions require him to make certain formal inspections within specified time limits. Whenever feasible, scheduled inspections are shown on a reproduced check list which provides blanks to be filled in by the OOD. As he makes his rounds, he notes the time and results of each scheduled inspection. He also records any other inspections which must be entered in the Log. Moreover, his entire tour as OOD may be regarded as a tour of inspection and should be as fruitful in this respect as possible. Thus, he checks on the security of buildings, supplies and equipment; use of government vehicles; observance of fire regulations; and he observes the appearance, conduct, morale and general military proficiency of battalion personnel.

Specifically, the OOD visits and samples the general mess (if operated by the battalion) during the serving of at least one meal. He notes unsanitary conditions, such as the presence of insects, holes in the screening and improper disposal of refuse. He inspects for cleanliness of the galley, scullery and mess hall. He assesses the quality and quantity of the ration, and checks on the care, speed and courtesy with which food is served. He receives complaints and evaluates the operation of the mess from the standpoing of morale. He requires the JOOD to make similar inspections of all other meals served during his tour. He enters the time and results of mess inspections in the Log.

The OOD inspects each guard post in the manner prescribed in the battalion instructions He makes certain that each member of the guard knows, understands and is able to execute his general and special orders. Surprise inspections are especially valuable in testing the alertness of the guard. Whenever inadequacies are brought to light, the OOD gives on-the-spot instructions and orders and takes whatever follow-up steps are necessary to uphold the standards of the guard and to preserve the security of the battalion.

The OOD also makes scheduled inspections of the battalion area for cleanliness, good order and security. He may be required, in company with the CMAA or duty MAA, to make an inspection during work hours of the berthing spaces of one or more companies. Such inspections focus on compliance with cleaning bills, fire regulations and other pertinent directives If only one or two company areas are inspected per day, the Executive Officer generally specifies which areas will be inspected by the particular OOD. Sometimes the scope of the inspection is enlarged to cover areas outside the berthing spaces. The OOD ensures that a check is made after work hours of all locked files and safes and that a report is turned in as required by battalion directives, and he personally spot-checks such facilities.

The OOD ensures that men going on liberty are in proper uniform and that their appearance reflects credit on the battalion and the Navy. Each day the OOD inspects some of the men as they pick up their liberty cards, and he requires Company Duty Petty Officer or the MAA to inspect all members of the liberty party and to hold individual liberty cards until standards of appearance are met. The OOD generally inspects the battalion Enlisted Men's Club while it is open to ensure that proper uniforms are worn and that good order prevails.

He later checks to see if the Club has secured promptly at closing time.

Immediately after taps, the OOD inspects the battalion berthing spaces to make certain that lights are out and that quiet prevails. Between taps and dawn, the OOD makes at least one tour of the entire battalion area to check for any irregularities. At the end of his tour, the OOD reviews his check list, makes certain that all necessary entries appear in the Log, prepares and delivers any required written reports and makes an oral report to the Executive Officer (on work days) or to his relief (on other days).

6. The Junior Officer of the Day. The Junior Officer of the Day is the direct representative of the OOD. During his tour (usually 24 hours) he supervises the watch and executes orders of the OOD. He performs the duties of Commander of the Guard listed in the Landing Party Manual and acts as the information and communication center of the watch. He receives, logs and relays to the OOD the reports of the Sergeant of the Guard, the Company Petty Duty Officers, the Service Watches and the duty MAA. He keeps informed of the whereabouts of the OOD at all times and informs the OOD on all matters which require the OODs attention. During normal working hours the CMAA performs the JOOD duties. The JOOD has full authority over all enlisted men within the battalion area who are subject to orders of the OOD. He has authority to place men under arrest or under restriction in lieu of arrest in accordance with battalion directives, but upon doing so he must notify the OOD. The JOOD is responsible for the conduct of the watch, for compliance with current directives; and for execution of the orders given to him by the Commanding Officer, the Executive Officer, and the OOD and for upkeep of the battalion Log. Throughout his tour the JOOD must remain in the uniform of the day and wear the JOOD brassard. He is armed and equipped as specified in battalion instructions. Every qualified CPO assigned to the battalion is eligible for duty as JOOD, however, the CMAA is generally exempted from such duty. The S-1 is responsible for the equitable rotation of duty among eligible CPOs of the battalion. He prepares the JOOD duty roster for approval by the Commanding Officer.

The JOOD must be oriented on matters pertaining to battalion operations, facilities, equipment and personnel. He must understand the battalion's relationship to other commands and to the civil community. He must know and be able to explain the general and special orders for each watch billet. Before assuming duty as JOOD, he must study and be ready to execute all provisions of the battalion instructions. The change of JOODs takes place in the OODs office and is recorded in the battalion Log. The OODs office is the full-time station of the JOOD. If the JOOD leaves the OODs office, he designates the Sergeant of the Guard to remain in his place. In such cases the JOOD states his destination and briefs his temporary replacement.

The JOOD receives and logs official calls, messages and mail. Messages from the American Red Cross are delivered to the OOD or Chaplain before any notification of the individual concerned. The JOOD is in charge of afterhours telephone service if the battalion operates a switchboard. In all cases the JOOD has access to a complete and up-to-date roster of battalion personnel, showing the location of berths, crew assignment and off-station addresses and telephone numbers.

The JOOD has direct supervision over the duty driver. He ensures that transportation assigned to the watch is used for official business only. He must be informed about and approve or disapprove all after-hours requests for transportation and he refers questionable requests to the OOD for decision. He arranges transportation for working parties and for other purposes as directed by the OOD.

The JOOD makes such inspections as are ordered by the OOD. These include security watches, service watches, the general mess, and berthing areas. At the end of his tour, the JOOD reports to the OOD on all entries he has made in the Log and informs him of the status of pending matters which require the attention of the new JOOD.

7. The Interior Guard. The main work of the interior guard is to watch over and protect personnel and property in areas under the control of the battalion. The Commanding Officer is responsible for providing an interior guard adequate to cope with any threat to the safety, security or good order of the battalion. The size, specific organization and special orders of the interior guard vary from time to time according to the battalion's actual needs. At a secure base in peacetime, the investment in the interior guard may be at a minimum and may be stood by members of the construction companies as rotating watch assignments. However, at an advanced base in wartime, a major share of the battalion's strength may necessarily be invested in the interior guard system. Men cannot work without sufficient sleep and the battalion might not rest unless it has full confidence in its interior guard. In wartime it might even become necessary to send an extra man or men with a survey crew or bulldozer to watch for the enemy. In these cases the responsibility for security would be vested in a specially designated security company or platoon that should not be required to carry on project work. But in any case, members of the interior guard must understand their general and special orders thoroughly and be ready to execute them swiftly and well. The interior guard system for the battalion is based on the principles stated in the Landing Party Manual. Special orders and specific requirements are published in the battalion watch plan and are issued by the Commanding Officer,

Executive Officer and OOD. Members of the interior guard report for duty in the uniform and with the weapons and equipment specified in the battalion watch plan or as ordered by the OOD.

The Senior Petty Officer of the Guard, whatever his grade, is officially called the Sergeant of the Guard. This billet is usually filled by a petty officer first-class. The duties are described in the Landing Party Manual. The Sergeant of the Guard receives and executes or relays orders of the Commanding Officer, Executive Officer, OOD and JOOD. He makes scheduled reports to the JOOD within prescribed time limits and makes special reports to the JOOD at once.

When the situation requires, a petty officer is designated as Corporal of the Guard for each guard relief. When such is the case, the Corporal of the Guard performs the duties described for his billet in the Landing Party Manual. He visits the OODs office between tours of the guard posts.

8. Camp and Job Sentries. Sufficient guard posts or patrols are established to watch over and protect all areas under the battalion's control. Each post and patrol should be provided with the best available means of communication with the Sergeant of the Guard and JOOD. These posts are usually relieved every four hours. Any qualified member of the duty section is eligible for duty as an interior guard. Every member of the interior guard must understand his general orders which are explained in the Landing Party Manual and in the Bluejackets' Manual. But the mere ability to recite general orders and a standard explanation is not enough. Each guard must be able to relate his general orders to his specific duties. Special orders define the specific duties for each member of the guard and set the physical limits of his post, arms and equipment required, locations of fire alarms and fire fighting equipment, reports to be made and actions to be taken. To the extent possible, special orders for each post should be incorporated in the battalion watch plan. However, requirements may change from day-today or even from relief to relief. Each interior guard should have full opportunity to study the special orders for his post in writing before assuming duty. If this is not possible, his special orders should be explained to him before he is posted. In any case, in the course of their inspections, the OOD and his representative must make certain that each guard understands his general and specific orders.

9. Service Watches. Service Watches are established as necessary (a) to provide standby service for all essential functions of the battalion; (b) to operate essential equipment on an all-night or around-the-clock basis; and (c) to provide specific after-hours services.

Men assigned to service watches must be specially qualified for their watch duties. The officer regularly in charge of the function involved usually designates the men eligible to stand service watches. They are not usually required to stand other watches. After work hours all men on service watches are subject to orders of the OOD and JOOD. Such watches are usually visited by the OOD and the JOOD on their regular tours of inspection.

Typical service watches are:
- a. Boiler watch
- b. Generator watch
- c. Duty Yeoman
- d. Duty Storekeeper
- e. Duty Corpsman
- f. Duty Driver
- g. Duty Fire Fighter
- h. Duty Librarian
- i. Duty Movie Operator
- j. Hobby Shop Watch
- k. E.M. Club Watch
- l. Duty Commissaryman
- m. Duty Armorer
- n. Duty Electrician
- o. Duty Utilitiesman
- p. Communication watch
- q. Duty Dispatcher

FIGURE 89
PERIMETER DEFENSE

While the Interior Guard may Consist of only a few Fire and Security Watches in Secure Areas, Several Platoons may be Required to man Defensive Positions during Wartime Operations.

10. Company Duty Section. Each company is divided into duty sections (usually four) in accordance with the battalion watch plan. Taken together, the company duty sections make up an effective, organized force. They are available for special work details or to handle emergencies arising after normal working hours. In each company the section on duty is headed by the Company Duty Petty Officer who makes certain that men in his company scheduled for duty report for duty on time and in proper uniform. He assigns other duties to members of his duty section as required. He must be prepared to muster his duty section on very short notice in an emergency. The duty section is mustered at times specified in the plan of the day and at any other times directed by the OOD, JOOD or Company Duty Petty Officer. Members of the section on duty are not eligible for liberty nor are they permitted to drink alcoholic beverages during their duty period nor for six hours before the section assumes the duty. They perform their regular duties during work hours, except that the first guard relief leaves work early.

The Company Duty Petty Officer inspects his company area as required in battalion directives and reports the results of such inspections to the JOOD. He orders all necessary measures to secure berthing compartments against rain, wind, impending storms and other weather conditions along with being responsible for proper cleaning of berthing compartments and for policing the company area.

11. The Duty MAA. A duty MAA is designated to perform the MAA functions during non-working hours. On work days he assumes the MAA duty when the CMAA leaves his office for the day. On non-working days the tour of the duty MAA is usually the same as that for the OOD. The duty MAA is subject to orders of the OOD the JOOD and is required to make periodic and special reports to the JOOD. He must remain in the MAA office except for meals and when his duty requires him to be elsewhere, and he assists the JOOD in the execution of the plan of the day and in maintaining good order and discipline in the camp. He takes charge of all men under arrest or in restriction and musters them at the hours indicated in the plan of the day and at such other times as necessary. Before assuming the duty, he obtains a list of all the restricted men and men serving extra duty. The result of his muster is entered in the Log. He supervises the performance of extra duty and hard labor without confinement when these punishments have been awarded at Captain's Mast or by courts-martial. He records the length of time worked and the kind of work performed by each person. Proper execution of punishments is an important link in the chain of military justice. Unless punishments are strictly enforced, military justice loses its effectiveness as a means of maintaining discipline. The duty MAA tours the camp area after taps to ensure the lights are out and that quiet prevails.

12. Shore Patrol. Senior Officers Present instructions normally require that all units produce officers or men for a shore patrol operated by a naval base or area commander. The battalion itself may sometimes be required to set up an independent shore patrol. The general requirements for a shore patrol are specified in U.S. Navy Regulations, (Article 0625) and the U.S. Navy Shore Patrol Manual, (NAVPERS 15106). When a battalion is operating independently, the Commanding Officer determines when and where a shore patrol is necessary. When the battalion is required to provide men for a shore patrol operated by higher authority, the battalion has no control over the operations of the patrol. However, the Commanding Officer of the battalion is responsible for selecting qualified men for such duty. Hence, he is responsible for their training, appearance, conduct and punctuality in reporting for duty. Before reporting to another command for shore patrol duty, men selected must be thoroughly trained in shore patrol procedures and must understand all pertinent directives.

13. The Battalion Log. A Log is maintained by each battalion and is one of the most valuable official records of the battalion. It must be complete and accurate and is often introduced as evidence at trials. The Log is kept for two main purposes; one administrative the other historical. It serves as a consolidated report for the battalion watch. The time and content of all muster reports and of reports from every element of the watch, the time guard reliefs are posted and relieved, the names of the OOD and the JOOD and the times each actually begins and ends his tour of duty are recorded. The Log serves as a record of arrests, confinements, restrictions, accidents, and other events which may be required as evidence for investigations, disciplinary actions and courts-martial. Each battalion usually publishes an instruction giving guidelines for what should be recorded and how to make the entry. Some standard entries are indicated below:

a. Location of the battalion (daily) and the assumption of duty and relief of OOD and JOOD.

b. Weather (at regular intervals and at time of change).

c. Reveille, morning colors, morning quarters, evening colors, guard mounting, posting and changing of guard reliefs and taps.

d. Receipt of official dispatches, telegrams and telephone calls (after work hours).

e. Reports (regular and speical) by security watches, Company Duty Petty Officers, service watches, the duty MAA and the Shore Patrol.

f. Inspections (time and results).

g. Orders under which the battalion is

operating.

h. Movement information on the battalion and its detachments (embarkation, debarkation, change of base or station).

i. Arrival and departure of supply ships.

j. Official visits and reviews and visits by distinguished persons.

k. Citations or medals awarded to units. or individuals.

l. Change of command (battalion, base, next higher command.

m. Reception and transfer of personnel (showing complete reference data).

n. Reenlistments.

o. Names and rates or ranks of personnel departing on or returning from leave and the departure and arrival of certain liberty parties, such as boating or mountain climbing parties.

p. Apprehension, arrest and restriction of personnel.

q. Declaration of deserters.

r. Courts and Boards (name of senior member or president, time of convening and adjournment, action taken and the findings and sentences.

s. Musters of prisoners and restricted men.

t. Confinement and release of prisoners

u. Combat operations, results of enemy activity and battalion action.

v. Accidents, deaths and injuries or casualties and major damage to equipment or other property.

w. Fires, thefts and other disorders or events of special interest.

x. Special night operations and working parties.

Chapter IX
THE STINGER CONCEPT

SECTION A - THE SYSTEM APPROACH

1. <u>Introduction</u>. Today's lesson learned must
become tomorrow's method of operation if we
are to survive as human beings or as an effective SEABEE construction/fighting force. The
full commitment to Vietnam of the entire assets of the SEABEES, during which the force
level more than doubled in a relative short
period of time, demonstrated to all management
levels that there were shortfalls in SEABEES
Management and that there were ways to correct
those shortfalls -- if we so organized our-
selves and the analysis of SEABEE requirements
to ensure consistent analysis and study.

2. <u>Recognition and Action</u>. With this recognition came action. In early 1966, the Commander, NAVFAC, embarked on a series of studies to define what it takes to field a SEABEE
Unit and support its combat construction mission in a theatre of operations. This was an
extremely complex undertaking. Too often we
tend only to see the obvious actions -- and
neglect to consider the back-up actions that
permit the obvious actions to take place.
Like a TV show, we see only the performers,
that is, the SEABEES. But behind those performers are a lot of supporting people and
events -- script writers, producers, camera-
men, typists, etc. (the NCRs, the CBLANT/PAC
Commanders, NAVFAC, CECOS, NAVSCONS, and the
Construction Battalion Centers). And behind
the whole show are necessary activities -- not
directly related but nevertheless absolutely
necessary; the shop that makes the stage props,
the delivery van that brings them to the station, the electric company that produces the
electricity. The TV company has little control over these assets, but it must make sure
these activities are coordinated with the
show, or the show doesn't go on. These kind
of activities we call "interfaces." The
SEABEES have interfaces too, over which the
SEABEES have no control, but which are most
important to the success of the SEABEE mission. Examples are: The Supply System, NAV-
ORD for weapons, MSTS & MAC, The Congress for
our dollars.

3. <u>Interrelationships</u>. The interrelationships
of all these activities is also important because if we change one element of the TV show
(or the SEABEES), all the other supporting
activities are also affected to one degree or
another. For example, if we increase the number of MCBs, then we affect the number of barracks to house them, the number of students at

NAVSCON, the amount of paper and pencils needed by yeomen, etc.
In order to make available the recourses
necessary to support this change we have to
be able to define precisely -- in terms of
people, dollars, facilities, etc. -- the total
resources required to support that chain of
events which develops by a "simple" increase
in the number of NMCBs, PHIBCBs, SEABEE Teams,
etc.

SECTION B - SYSTEM IDENTIFICATION

1. <u>The System Model</u>. All of these requirements and considerations led to the development of a definition of the SEABEE System, in
which the activities of the performers (the
SEABEE Battalions), the back-up people, organizations, and the interfaces were identified and properly related in a diagramatic
way.
A simplified drawing of the system is
shown in Figure 90. Consider that the boxes
are filled with men, equipment, tools and
materials -- all accomplishing their jobs.
Since the boxes are interconnected, and a flow
of activity occurs along each line it follows
that having defined the system, one can study
and analyze the requirements of each and the
relationships between the several boxes in a
systematic way.
To summarize then, STINGER is a SEABEE
System, and it includes more than just the
NMCBs and the PHIBCBs. It includes consideration of the supply systems that provide building materials; it includes the procurement,
training and deployment of men; it includes a
PWRS and an ABFC program; it includes a CBC
system for homeport, training, and material
support, it includes the equipping of the
units. This list could go on at length, but
at least begins to indicate the detail and
extensiveness of the makeup of the System.
Put in more general terms, STINGER identifies and includes all elements of the NCF, and
the Navy offices, commands, and organizations
which provide direct support, guidance or direction in the accomplishment of combat construction by the deployed units of the NCF.
As such, it includes the interfaces with and
between these various organizations, and also
the activities, actions and services performed
in support of NCF units to ensure successful
SEABEE Tactically Installed, Navy-Generated,
Engineering Resource (STINGER) System mission
accomplishment. It includes planning, programming, funding, procurement, outfitting,

FIGURE 90
THE STINGER SYSTEM ENVELOPE

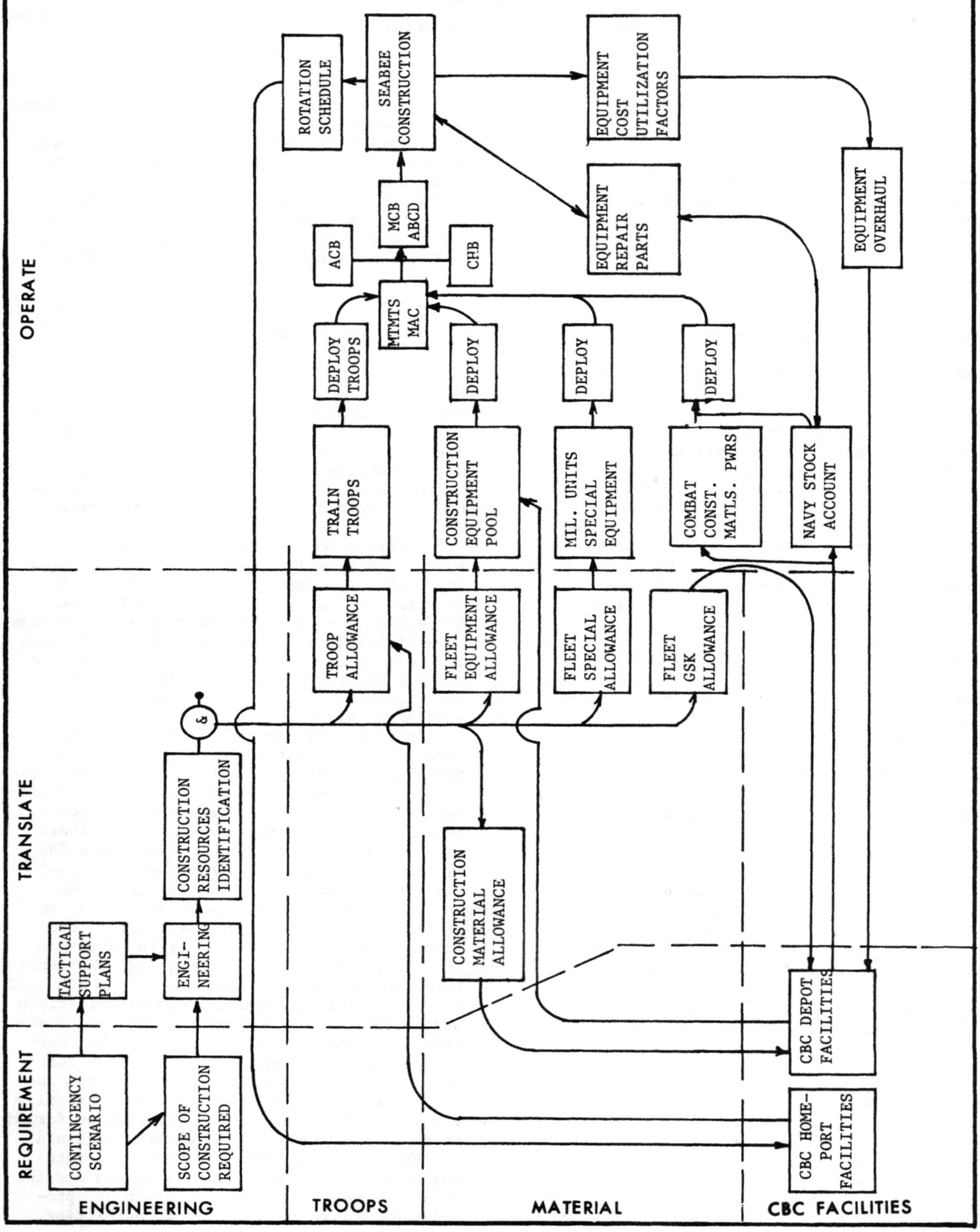

164

command and operation activities integrated with the basic resources used and consumed. It further includes a recognition of the sequential nature of the conduct of actions within the System envelope. Because STINGER is so all-encompassing, it involves the coordination (and to the assigned degree, the direction) of a wide variety of action information, organizational input, data, and the like.

It is important to remember that the STINGER System is a real, viable entity. It is not composed of blocks on a piece of paper interconnected by lines. It consists of organizations and the activities and services rendered by those organizations. Since organizations are composed of people, the System must be a flexible one which can react to changing situations and involvements. The makeup of the System is, therefore, composed so as to permit a flexibility of action that is constantly attuned to changing Navy requirements. A basic goal in the management of the System is the ability to gain and maintain initiative in the response of the NCF to combat construction, vice merely being in a position to respond to need.

2. Computers. Attaining the initiative means the ability to be able to study the system under changing influences and develop requirements rapidly. Computers can't think, but they can be used to store information in vast quantities and provide the means for us to retrieve it rapidly, and assemble the information in many different ways. Computers also provide a means by which we simulate the STINGER operations in the computer. Since we have defined what our System is (Figure 90), we can use Operations Research techniques to represent each of the boxes on the diagram with a computer program.

As noted before, computer programs don't make up the STINGER System; people, commands and organizations do. Computer Simulation Models allow us to study the System and make logical predictions regarding STINGER, but people make it work. The combination of the two, working together, can ensure the effective management of all elements of the STINGER System to ensure integration of the widely varied interests, responsibilities and activities of the System to ensure a well directed application of efforts in building an effective, combat-ready SEABEE Force.

SECTION C - STINGER SYSTEM CAPABILITY

1. Capability of Stinger System Programs. It is considered appropriate to discuss briefly what a simulation involves and why it was selected for STINGER System use. Most operation research problems deal with situations for which an optimum solution is desired. In the mathematical sense, simulation cannot be

classified as an optimizing technique, yet it has become widely used as a method of attacking operations research problems. The reasons for this is that many systems studies are so large and complex that an optimizing technique cannot be formulated within the limits of computer if the system is well understood and expressible mathematically. The usual procedure is to guess at a good solution, simulate this solutuion and examine the results to formulate a guess at a better solution, etc. In this way, the best solution of those tried will be found. The improvement over an existing or contemplated solution can thus be estimated, even though the optimum is not known.

2. Present Programs. The STINGER System programs, at the present time, incorporate the logic and data from the operation of the Naval Construction Forces and simulates these operations by means of a digital computer. Seven subsystem routines are available in the STINGER Simulation which identify required construction to support an MEF (Marine Expeditionary Force), as well as the men, equipment, and materials necessary to place this construction at some specified location within a given time. The Subsystem programs presently in the STINGER Simulation include:
    a. The Combat Construction Support Required by an MEF.
    b. The Navy Construction Force Systems including:
        (1) Material routine
        (2) Manpower routine
        (3) NCF equipment overhaul routine
        (4) Spare parts routine for NCF equipment.
        (5) NCF equipment routine
    c. The home porting routine

3. Program Output. For a given amount of construction described by the number of functional components, tactical support components, and other construction (time phased), the STINGER Simulation will define the amount of work-in-place (as a function of time) that could be accomplished with a specific construction capability.

4. System Program Data Includes:

    a. Number of men and
        (1) Months of service in NMCBs
        (2) Months of service in other duty
        (3) Months of service (total)
    b. Number of NMCBs deployed and months deployed.
    c. Number of NMCBs at home port and months deployed.
    d. Number of NMCBs at home port and months at home port.
    e. Number and type of functional components available and required by MEF.
    f. Number and type of tactical support components available and required by MEF.
    g. Due date for completion of required

construction by components.

h. Man hours of construction distribution of "delay time" in getting material to job.

i. The statistical distribution of MCB equipment failures as a function of operating environment and type of equipment.

j. Cumulative use time on equipment (miles/hours)

k. The probability a part will be required to repair or overhaul equipment as a function of equipment type.

l. The delay time for ordering parts.

m. The time for equipment overhaul as a function of equipment type.

n. The shipping time "stateside to war area" for men, material, parts, etc.

o. The assembly time to take material out of the PWRS and make ready for shipment.

p. The delay time in procuring equipment from the manufacturer.

q. The delay time in procuring materials from the manufacturer.

5. Representative Results

a. The work in place as a function of time.

b. The through-put of an equipment overall shop as a function of time.

c. The average time equipment must wait for overhaul.

d. The men available for direct construction as a function of time.

e. The training requirements to maintain a given combat construction capability.

f. The backlog of work measured in total man days.

g. The amount of work completed, measured in total man days.

h. The man days lost as a result of material delay or the lack of construction assignments, etc.

The following are representative of the kinds of studies that can now be accomplished:

6. Combat Construction Support Required. The objective of this problem assignment is the development of study results which can be utilized to assist the military planner in rapidly analyzing a Scenario or Op Plan in which an MEF is engaged. The identification of such items as the number of functional components and their size, the number of MCBs, and the time from "D" day specific construction is required can be made. Estimates of shipping requirements and cost are developed.

7. Navy Construction Force Systems Study. The program requires information concerning the availability of material required for construction, the number of men required for construction, the attrition of manpower in MCBs, equipment failure data in a MCB, number of spare parts required to keep the NMCB equipment in operation, etc. As a result of the scheduling of initial parameters and their associated performance, the output is the identification of

the total number of NCMBs, the amount of NMCB equipment, the amount of construction material, etc., to perform the required construction within the allowable time under conditions where attrition occurs.

8. SEABEE Officer Qualification Plan. A computer program for this work unit was developed A variety of administrative and system constraints are provided for by the program. Within these constraints, the program determines an optimum allocation rule, based upon retention probabilities. The rule maximizes the number of CEC officers obtaining SEABEE experience by the time they are considered for promotion to LCDR. Using the determined allocation rule, the program simulates CEC officer rotation over a period of time, giving as output the projected SEABEE experience status of the entire Civil Engineer Corps.

9. A Method for Determining Steady State Officer Levels in the Civil Engineer Corps. This is a computer program which calculates steady-state officer levels based on a 30-year cycle. The program calculates the levels based on annual Ensign input, transfer policy, promotion policy, and attrition data. The results are tabulated by grade and years of experience.

10. Trade-Off Study of Equipment for Navy Construction Force. This study examines NMCB equipment and SEABEE Team equipment requirements. It develops a measure of effectiveness whereby NMCB equipment is rated in terms of satisfying construction requirements. For example, if a construction project calls for earth moving to be performed, then a rating system is required to identify the equipment to be used as well as those factors being traded-off (by this selection) in terms of time, cost, and performance if the best equipment for this purpose is not available. Construction jobs and their frequency of occurrence are within the scope of this effort.

11. Civil Engineer Corps Strength Study. The Civil Engineer Corps Strength Study computer program computes, for up to ten years into the future, the predicted number of Civil Engineer Corps officers on active duty. The predicted number is based on the beginning number as modified by promotion, recruitment, transfer, mandatory attrition, retirment policies, and the expected voluntary attrition.

12. Future Effort. Data collection is presently being conducted to determine the number of personnel (officer and enlisted) in the regiments, brigade, NAVFAC, CBPAC Staff, and CBLANT Staff such that growth equations can be developed to determine the effect on these staffs as MCBs are increased or decreased. Also, personnel of the CBCs will be included.

It is planned to include tactical support

components and assemblies in the STINGER System studies as well as the collection and usage of better data to improve the accuracy of the System results for use at all levels of the STINGER System. Identification of construction requirements for a war at sea scenario plan and ACB requirements will also be addressed.

13. <u>STINGER Within the Navy Programming System</u>. Among the uses of STINGER simulation output is the basic justification for NCF units and the resources to support their readiness. This justification is used in forwarding NCF force level/support proposals in the Navy/DOD, Planning, Programming and Budgeting System. Annex I explains the place of STINGER within this system.

Chapter X
STINGER SUPPORT FUNCTIONS

SECTION A –
THE NAVAL FACILITIES ENGINEERING COMMAND

1. <u>Organization and Functions</u>. The Commander, Naval Facilities Engineering Command, as the Chief of Civil Engineers of the Navy, is the principal advisor to the Chief of Naval Operations (CNO) for SEABEES; and, as such, is the focal point for the coordination of all SEABEE operations and support. This involves supporting a well-trained and equipped Naval Construction Force for direct support of the Fleet and Fleet Marine Force or any other agency that is carrying out National Security Policy.

The coordination of this support, because of the diversity of SEABEE units and their employment, and the complex command and support chains involved, requires a system approach. The concept of a systems approach was presented in the preceeding chapter, "The STINGER Concept." The organization within NAVFAC that discharges COMNAVFACENGCOMs SEABEE System responsibilities will now be examined.

The Assistant Commander for Military Readiness (Code 06) has two broad areas of responsibility which relate to the NCF:

a. The formulation of military personnel plans, training plans and policies.

b. Providing for the financial and logistic support of the Naval Construction Forces.

He carries out these two basic responsibilities with the aid of a staff which includes six functional divisions (See Figure 91).

a. The Military Personnel Division (Code 061).

b. The System Plans and Policy Division (Code 062).

c. The System Operations Division (Code 063).

d. The Logistics Division (Code 064).

e. The Functional Component Division (Code 065)

f. The Disaster Control Division (Code 066).

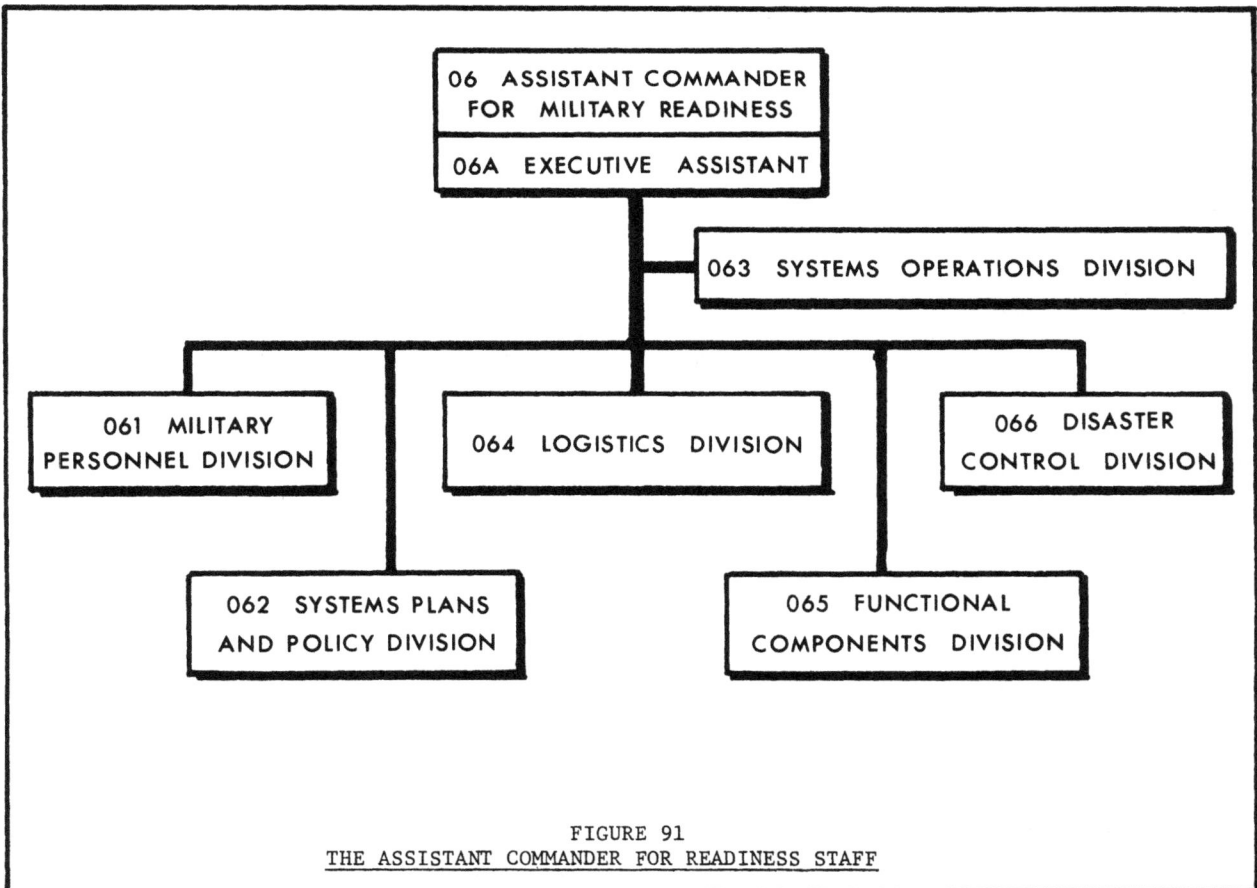

FIGURE 91
THE ASSISTANT COMMANDER FOR READINESS STAFF

2. **Military Personnel Division.** (NAVFAC Code 061) This Division fulfills the Chief of Civil Engineers responsibilities for the procurement, training, and distribution of SEABEE manpower and the determination of billet requirements.

The Division has two Branches, one for the Active Forces and the other for the Reserve Forces. The Reserve Programs Branch administers the Civil Engineer Corps and SEABEE Programs of the Naval Reserve including organization, training, and staffing of SEABEE Reserve Units. The Military Personnel Branch (Active) has two plans, policies, and training sections - one for CEC officers, and one for Enlisted personnel.

The major functions performed by Code 061 are:

a. Maintains liaison and provides advice and information to the Chief of Naval Operations in the conduct of programs such as: Manpower Technical Advisor for qualitative and quantitative CEC and Group VIII billet requirements of the Navy; Designator Sponsor for Code 5100 officers; represents the Selected Reserve Program Sponsor and Specialist Program Sponsor.

b. Maintains liaison and provides advice and information to the Chief of Naval Personnel in the planning, procurement, training, education, promotion, retirement, retention, administration, and assignment of CEC officer personnel and Group VIII enlisted personnel both active and inactive.

3. **Systems Plans and Policy Division.** (NAVFAC Code 062) This Division is charged with the development of SEABEE concepts, requirements, objectives, and plans in consonance with Navy and Marine Corps plans and objectives. Major functions are:

a. Develops STINGER System readiness concepts, requirements and objectives for presentation to higher authority to introduce planning CEC and SEABEE viewpoint into NAVFAC, Navy, DOD and Joint Planning and Programs.

b. Coordinates for the Assistant Commander for Military Readiness RDT&E Projects, including preparation of Navy Planning and Programming System documentation required in connection therewith.

c. Provides a data bank of information within the office of the Assistant Commander for Military Readiness on programs, policies, RDT&E engineering and scientific data, doctrines, and developments of interest to and necessary for the long range development of NAVFAC/STINGER System military readiness.

d. Conducts and/or collaborates in, as assigned, the preparation, review, updating and revision of NAVFAC/STINGER System military readiness elements of the Navy Planning and Programming System. Includes preparation of required documentation dialogue of the System.

e. Maintains appropriate liaison with departmental, field and fleet commands, and elements of the office of the Assistant Commander for Military Readiness to ensure conti-

nuity, cognizance, and for Military Readiness to ensure continuity, cognizance, and concurrentness of Counterinsurgency, Limited War, and Mobilization planning affecting NAVFAC and the STINGER System.

4. **The Systems Operations Division.** (NAVFAC Code 063).

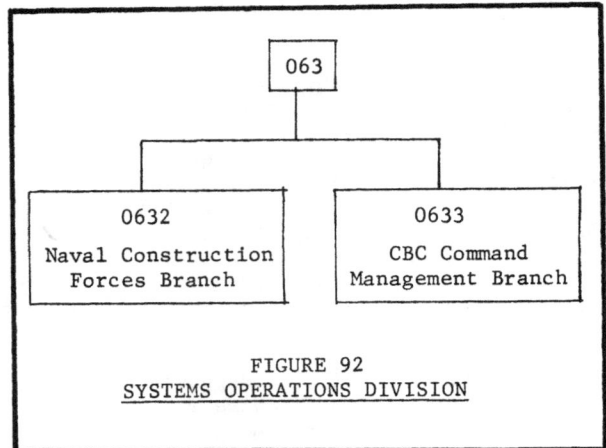

FIGURE 92
SYSTEMS OPERATIONS DIVISION

The Systems Operations Division of Code 06, similar to the Operations Department of an NCF Unit, is the task coordinator for the Assistant Commander for Military Readiness. The major functions of the division are:

a. To coordinate all tasks relating to NAVFAC Program VI for the Assistant Commander for Military Readiness. Represents the Assistant Commander at Program and Budget hearings and conferences.

b. Program manage any special project involved with any phase of the STINGER System which may be assigned.

c. Provide command management of the Construction Battalion Centers.

d. Act as a focal point for the Assistant Commander for Military Readiness for all matters pertaining to operations and support of the Naval Construction Forces.

e. Act as a focal point for the Assistant Commander for Military Readiness for all matters pertaining to operations and support of the Naval Construction Forces.

f. Develop, defend and execute budgets for the active Naval Construction Forces and STINGER System as assigned.

Code 0632 determines and coordinates STINGER System actions required to improve, redirect, extend or expand NCF capabilities and readiness. Included in this is the coordination of STINGER budget information, TSFC, procurement policies, and input to OPNAV/OSD Plans.

He is the focal point for NAVFAC financial support of NCF elements. This support consists of coordinating efforts in two major directions: (1) Construction funds; (2) Operational funds in terms of Initial Outfitting and Operations and Maintenance.

```
                            ┌──────────┐
                            │   0632   │
                            └──────────┘
          ┌──────────────────┬─────────┴──────────┬──────────────────┐
  ┌───────────────┐  ┌───────────────┐   ┌───────────────┐   ┌───────────────┐
  │     06321     │  │     06322     │   │     06323     │   │     06324     │
  │    FORCE      │  │   TECHNICAL   │   │ AMPHIBIOUS &  │   │ AUTOMOTIVE &  │
  │ CAPABILITIES  │  │  COMPONENTS   │   │ SPECIAL EQUIP-│   │ CONSTRUCTION  │
  │   SECTION     │  │   SECTION     │   │ MENT SECTION  │   │  EQUIPMENT    │
  │               │  │               │   │               │   │   SECTION     │
  └───────────────┘  └───────────────┘   └───────────────┘   └───────────────┘
```

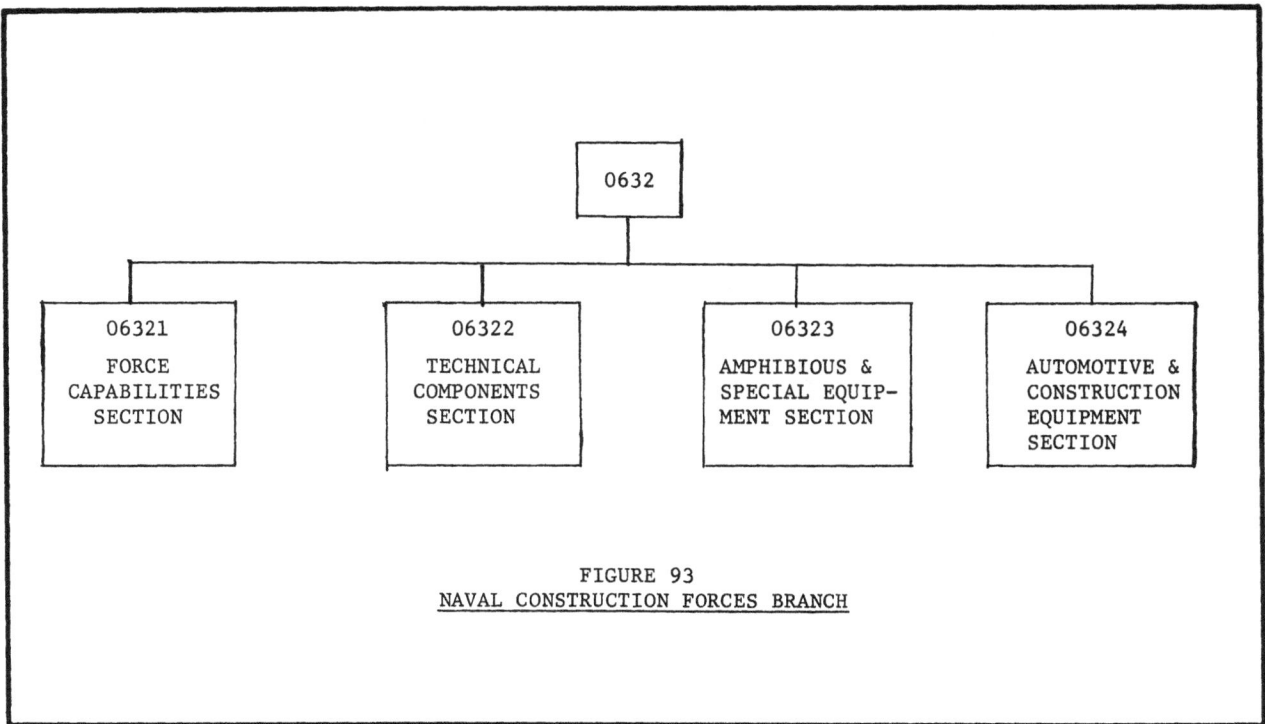

FIGURE 93
NAVAL CONSTRUCTION FORCES BRANCH

a. Construction Funds. Construction material may be generated by several types of funds handled in various ways. The basic fund sources are OP,N- Other Procurement, Navy, and MCON- Military Construction.

(1) Military Construction Funds. Military Construction Dollars are provided to NCF units by NAVFAC through the Navy Component Commander of the locale in which construction is to be accomplished. Construction requirements from the field are submitted to OPNAV through the Fleet Chain of Command. These are submitted to Congress in the form of MILCON Line Items (i.e., particular structures in a specific location).

Upon approval of a line item, funds become available to NAVFAC who then passes them on to the Component Commander. The funds associated with the troop portion, if it is to be accomplished by NCF, is turned over to CBPAC/LANT. CBPAC/LANT in turn distributes these funds for construction material to the Brigades, Regiments, or individual Battalions depending upon the operational organization in the area designated for project accomplishment. A diagrammatic representation of this process is shown in Figure 94. Code 0632 does not fit directly in the MILCON cycle but is responsible for advising NAVFAC program offices and OPNAV offices of MCF MCON usage rates.

(2) Other Procurement, Navy. The second major source of "Combat Construction Support" is Other Procurement, Navy. Code 0632 is responsible for determining NCF requirements in this area. As previously noted, MCON dollars are tied to specific construction line items. The process of translating field requirements into available dollars takes a period of several months. The primary reason for the use of NCF is their ability to provide timely response to and training for continuing construction needs. The time frame for initial contingency operation execution does not lead itself to MCON execution. Further, after establishment of major logistic terminals, field force construction effort turns to support of Marine maneuver battalions constantly on the move in need of rapid Tactical Support Construction response. These needs have generated the OPN funded Tactical Support Functional Component System. TSFCs become the SEABEE ammunition in the overall STINGER Weapons System. TSFCs are ABFCs funded by OP,N dollars; examples are bunkers, bridges, road rock, observation towers, etc. They are treated as a special category of PWRS (Pre-Positioned War Reserve Stock). NAVFAC submits directly into the budget chain requests for OPN dollars necessary to fund TSFCs. Upon receipt of these funds NAVFAC has several alternatives. COMNAVFACENGCOM may (1) procure material directly; (2) give obligational authority to the CBCs; (3) to CBPAC/LANT; or (4) to the NCB/NCR/MCB. In any of these cases, material becomes available to the field in the form of a stockpile of combat construction assemblies ready for unprogrammed use to meet urgent tactical requirements of field commanders. Authority for release of the assemblies has been delegated by CNO to NAVFAC. In the case of the Vietnam contingency, NAVFAC delegated release authority to the 3rd NCB. Release is based upon a tacti-

cal criteria which states in general that the need for the TSFC is urgent, unplanned and based upon tactical necessity. Material listings of TSFCs are in the P-103 (Catalog of ABFCs). The TSFC program was initiated during Vietnam Operations in 1967. OP,N funds will have accounted for over half of all SEABEE project funds by the end of FY 1969. In peacetime TSFCs will be held by Pacific and Atlantic Regimental Staffs for issue to MCBs upon activation of a contingency plan. Thus, MCBs will arrive in the field with all the essentials for immediate construction: men, equipment and material.

(3) Other. Project funds for NCF use may come from many miscellaneous sources. Among them in the past have been: other services' MILCON, other services' O&M, AID/State Department, MAP funds. NAVFAC normally plays no role in obtaining these funds.

b. Operating/Outfitting Funds

(1) Initial Outfitting. Initial outfitting of the equipment and material allowance is generally provided by NAVFAC for all functional components in which it is dominant including the MCBs, PHIBCBs, CBMUs, SEABEE Teams, etc. This is done with OP,N funds. Special items are provided by other Syscoms such as Weapons by NAVORD, and Medical/Dental equipment by BUMED. Initial outfitting for MCBs (not including TSFCs) is about $4.4 million and $12 million for PHIBCBs. OP,N funds also are the source of replacements of major items. Replacement needs are determined by application of the SOAP (Supply Overhaul Assistance Program) and the BEEP (Basic Equipment Evaluation Program). OPN initial outfitting replacement policy, and fund control determinations are made by COMNAVFAC. Code 06323, the Amphibious and Special Equipment Section, and Code 06324. Automotive and Construction Equipment Section handle the detailed operations.

(2) Operations and Maintenance. The Department of Defense has directed that a portion of the Resource Management System be implemented on 1 July 1968. The immediate effects of the forerunner of RMS was to give O&MN control to the Fleet Commanders vice the Systems Commanders as of 1 July 1967. O&MN requirements are now submitted by NCF units through the fleet chain to the Fleet Resources Office (FRO) which is under the Chief of Naval Material. FRO forwards fleet requests for approval and after approval is obtained distributes funds to the Fleets. Individual NCF Commands receive quarterly OPTARS (Operating Targets) from their Commanders. NAVFAC has direct O&MN responsibility only with regard to its own command elements, the CBCs. However, with regard to NCF units, Code 0632 is

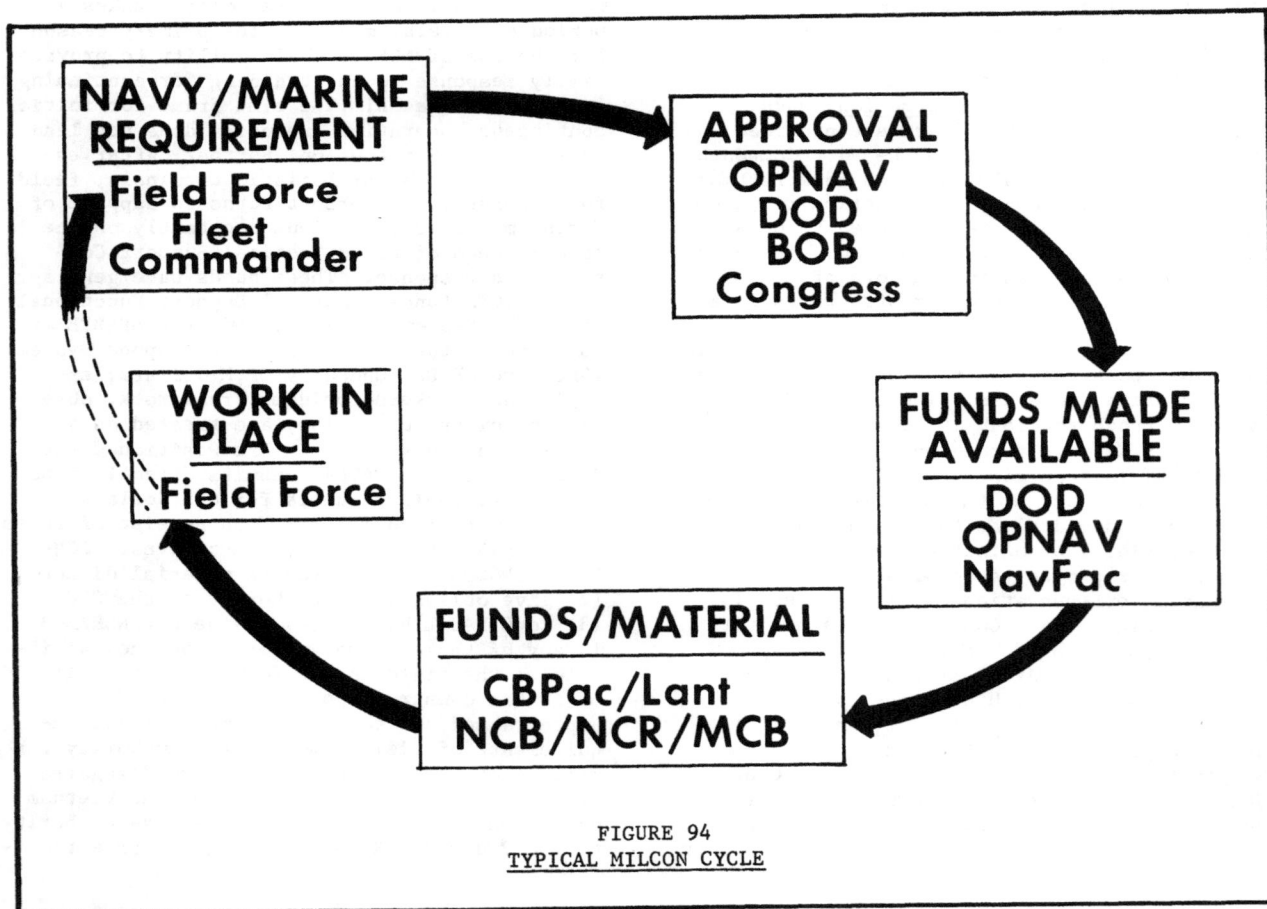

FIGURE 94
TYPICAL MILCON CYCLE

FIGURE 95
TYPICAL OPN CYCLE

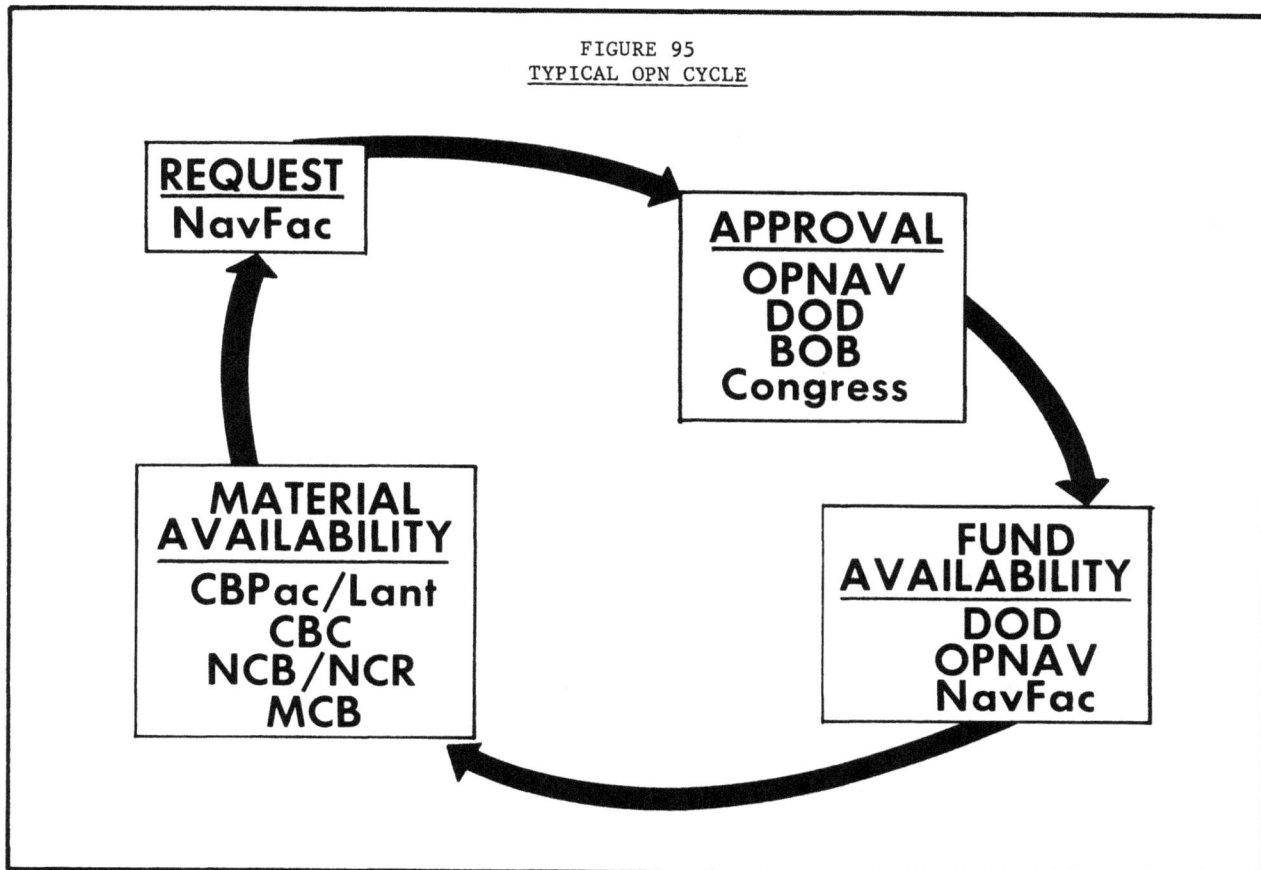

FIGURE 95
TYPICAL OPN CYCLE

cognizant of O&MN requests and defends the NCF O&MN budget throught DOD and Congress. These functions are accomplished to assist the Chief of Civil Engineers in his role of technical advisor to CNO for the Naval Construction Forces. Code 0632 is also responsible for determining NCF requirements to be submitted in Navy Programming documents (See Annex I).

Code 0633 coordinates all NAVFAC actions pertaining to operations of the CBCs including budget preparation and fund assignment, stock management functions, and overall capability to meet NCF support needs.

5. The Logistics Division. (NAVFAC Code 064). The Logistics Division provides logistic support for NAVFAC cognizant material items and establishes policies for supply management for NAVFAC. Major functions of Code 064, both NCF and Non-NCF related, are:

a. Exercises the Command's responsibility as inventory manager for cognizance symbol "2C", Major Construction and Civil Engineering material held in store for any purpose or procured for a planned or mobilization requirement, to support all requirements of the Naval Establishment, including initial outfitting and replacement for operating construction.

b. Manages supply aspects of the PWRS Program to support approved programs.

c. Establishes a program for preservation, maintenance, and quality control to assure the readiness and reliability of Prepositioned War Reserve Stock.

d. Coordinates and implements interservice procurement agreements and purchase assignments for NAVFAC.

e. Administers the Defense Priorities System for the Command and its cognizant field activities.

f. Administers the Defense Materials System for the Command and field activities and controls allotments of materials to prime contractors.

g. Coordinates the development and submission of NAVFAC (peacetime and mobilization) requirements for material within the "Controlled Material Program" as required by the Department of Commerce.

6. The Functional Components Division. (NAVFAC Code 065). The mission of this division is to develop, engineer and determine allowances for NAVFAC dominant functional components. This includes translating conceptual requirements and objectives, technological developments and guidance from higher authority into hardware for support of the Advanced Base Functional Component System.

Major functions performed to accomplish

the division missions are:

a. Engineers and designs Advanced Base Functional Components to upgrade existing capabilities and to meet new requirements.

b. Establishes Advanced Base Functional Component Allowances (including all NCF Units).

c. Coordinates programs to maximize standardization and common characteristics within the Advanced Base Functional Component System.

d. Initiates and monitors Research, Development, Test and Evaluation in connection with Advanced Base Functional Component design and development.

e. Develops, coordinates and monitors re-pair parts policies associated with the ABFCs.

f. Develops, recommends and monitors rotation and replacement policies for functional component material.

g. Coordinates functional component input to assigned programs.

7. The Disaster Recovery Division. (NAVFAC Code 0661) This Division directs, controls and provides guidance in matters pertaining to Disaster Recovery and NBC Warfare Defense and for selected portions of the National Civil Defense Program. This division's direct relationship to the NCF is in the area of establishing disaster recovery organization and training for the NCF.

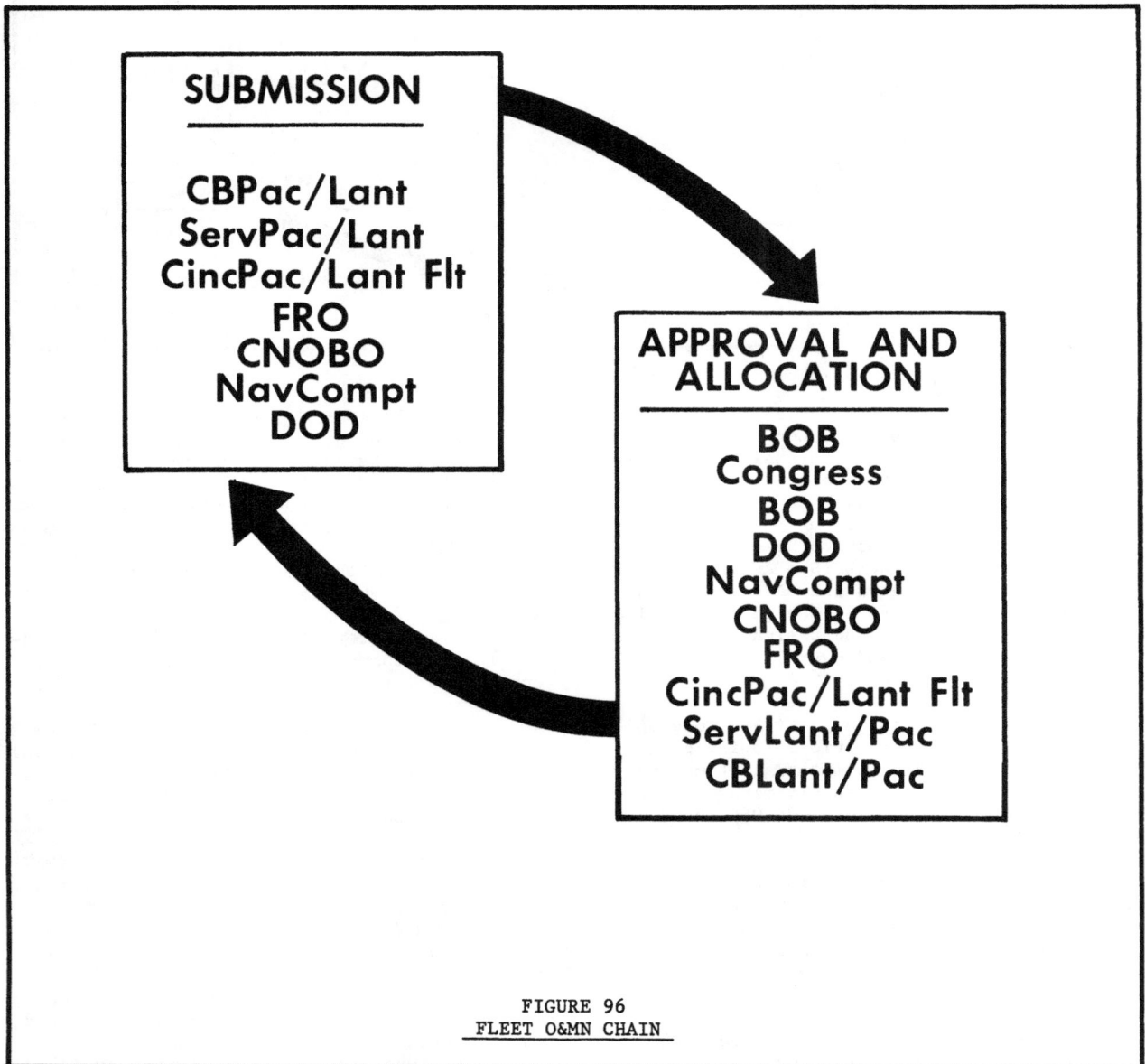

SUBMISSION

CBPac/Lant
ServPac/Lant
CincPac/Lant Flt
FRO
CNOBO
NavCompt
DOD

APPROVAL AND ALLOCATION

BOB
Congress
BOB
DOD
NavCompt
CNOBO
FRO
CincPac/Lant Flt
ServLant/Pac
CBLant/Pac

FIGURE 96
FLEET O&MN CHAIN

SECTION B –
COMMANDER NAVAL CONSTRUCTION BATTALIONS
ATLANTIC AND PACIFIC

1. _Representative of the Service Force Com-
mander_. To assist the Service Force Commander
to effectively fulfill his responsibilities as
Type Commander to the MCBs, a staff has been
established in each fleet. The Commanders,
Naval Construction Battalions, Atlantic and
Pacific have been delegated authority to act
for the Service Force Commander in routine
matters pertaining to the employment, deployment, administration, training, readiness and
operations of the assigned NCF units. The
Amphibious Construction Battalions are not
included as part of the COMCBLANT/COMCBPAC
responsibility. Specific duites of the staff
are:

a. Prepare detailed plans for the deployment and employment of SEABEE units assigned
to the Service Force Commander.

b. Keep the Service Force Commander informed of the state of readiness of NCF units
and their ability to carry out assigned tasks.

c. Recommend appropriate action on
matters of major importance or basic policy.

2. _Evaluation of MCB Readiness_. The COMCB-
LANT/COMCBPAC staffs must keep informed about
the state of readiness of each assigned battalion and assist each battalion to achieve
and maintain a capability to carry out mission
tasks under any emergency conditions. This
evaluation of preparedness includes the train-

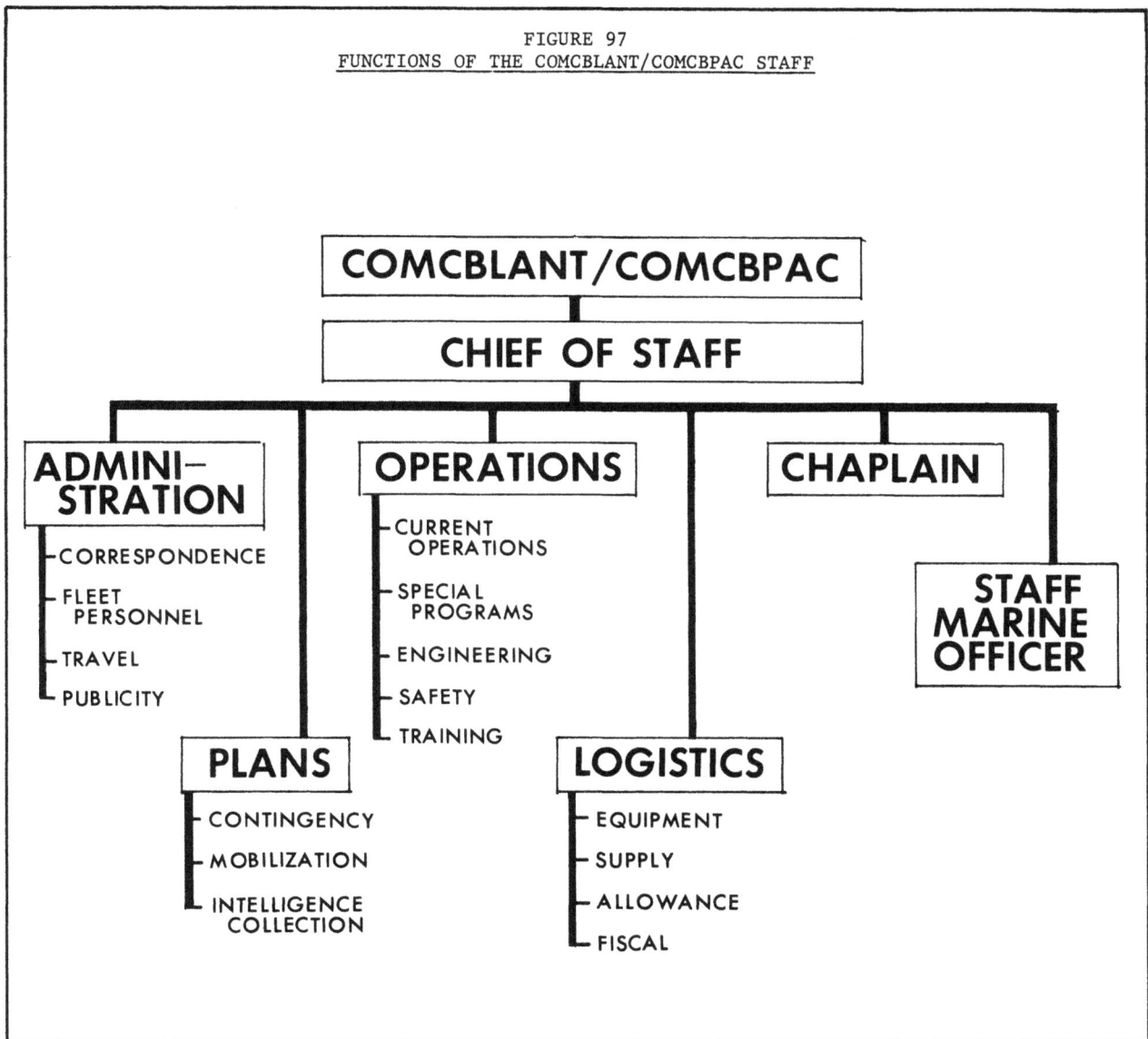

FIGURE 97
FUNCTIONS OF THE COMCBLANT/COMCBPAC STAFF

ing of officers and men and the upkeep of tools and equipment. Since a man's attitude plays a big part in his ability to perform effectively, such things as special services, personnel administration, religious activities and camp administration should be included in any evaluation of readiness. An evaluation system provides an opportunity for COMCBLANT/ COMCBPAC to identify problem areas requiring action by them or the type commander and to provide the battalion Commanding Officer with an appraisal of his units' readiness posture and provide assistance and suggestions for improvements. The evaluation should prove of mutual assistance. Since an evaluation of the actual day-to-day operation of each unit is not possible, COMCBLANT/COMCBPAC use the following methods of evaluation in combination:

a. An annual administrative inspection is conducted at the deployment site. This inspection, which usually lasts about one week, includes a thorough investigation into all phases of battalion administration, training, operations and supply and usually includes some drills to determine the ability of the battalion to respond at any level of command.

b. Informal visits are made by members of the COMCBLANT or COMCBPAC staff at times other than during the administrative inspection.

c. A comparison of various programs conducted by the battalions and of the records achieved, is made by cognizant staff officers. For example, the safety program and the records achieved, is made by cognizant staff officers. For example, the safety program and the records of injury or lost time of each battalion can be compared as can the number of men who qualified with the rifle, or the amount of equipment on the deadline, etc.

d. A year-long study of all the various reports submitted by each battalion is another effective means of evaluation.

The battalion in each fleet which comes out ahead of the others each year is declared the "best of type" and authorized to wear the Battle Efficiency Pennant. This competition for "best of type" stimulates the efforts of the competing units and helps to increase the overall efficiency of the Naval Construction Forces.

3. Support Rendered to the Mobile Construction Battalions. With respect to the MCBs, COMCBLANT/COMCBPAC exercises administrative control, and when in homeport, operational control. The staffs are organized and function in a similar manner. Many duties are delegated to the Naval Construction Regiments at the homeport. Specific functions are shown in Figure 97.

The functions which are of major importance to the battalions are:

a. Issue instructions and notices concerning policy and standard operating procedures.

b. Conduct annual inspections to assist battalions to maintain their ability to carry out wartime task assignments.

c. Review fleet operation plans and maintain continuous liaison with fleet commanders to determine the construction requirements of projects assigned for MCB accomplishment, and publish training guidelines and implementing plans as needed.

d. Collect and evaluate engineer intelligence as reference material for use by battalion officers.

e. Select training projects, assist in developing peacetime deployment workloads and prepare annual deployment schedules.

f. Provide engineering assistance to the battalions.

g. Provide funds for O&M expenses and project support.

h. Provide guidelines for the maintenance of the MCB allowance to carry out emergency assignments, and control the prepositioned equipment and camp components.

i. Coordinate MCB participation in fleet readiness exercises.

j. Make arrangements for movement of battalion personnel, equipment and material to and from the deployment sites.

k. Prepare and publish operation orders and movement orders for each deployment.

Figure 98 shows some of the interactions necessary to prepare for a peacetime deployment. Under emergency conditions these functions would have to be condensed into weeks or days.

4. Home Port Naval Construction Regiments.

a. Mission. Under the direction of the Commander Naval Construction Battalions, insure maximum effectiveness of all units of the Naval Construction Force while at homeport, in achieving the highest possible state of operational and material readiness to meet their Disaster Control, Contingency, and Wartime missions of military and construction support to Naval, Marine and other forces. As a secondary mission, act as a receiving and separating activity.

b. Functions.

(1) Exercise operational control over all units of the Naval Construction Force at homeport.

(2) Exercise various elements of administrative control (as specifically delegated by COMCBPAC/LANT) of all units of the Naval Construction Force at the homeport.

(3) Conduct military training programs and coordinate technical and disaster recovery training programs for all battalions during the homeport deployment.

(4) Coordinates and conduct specialized training for SEABEE Teams and other specialized units. Monitor battalion technical training and assist in obtaining quotas to various Service Schools.

(5) Administer the details of the Naval Construction Force Automotive and Equipment Program (including special tools) in

accordance with policies and directives issued by COMCBLANT/PAC. Determine equipment requirements, coordinate the testing of new equipment, decide on the deposition of used equipment and schedule major overhauls.

(6) Provide liaison with CBC on storage, preservation and shipping of advanced base and mobilization stocks of prepositioning and for the support of the Naval Construction Force.

(7) Control field communications, radiac, weapons, and infantry equipment and training ammunition for units of the Naval Construction Forces homeported at the same activity as the NCR.

(8) Provide management guidance and evaluate effectiveness of military, operational, and material readiness of all homeported units of the Naval Construction Force.

(9) Provide limited logistics support in specifically designated programs to units of the Naval Construction Force, as directed.

(10) Assist CBC in the conducting of active duty military training for reserve MCBs.

(11) Monitor personnel distribution among the NCF units and make recommendations to the Enlisted Personnel Distribution Office (EPDO) to provide sufficient skills to perform each scheduled workload.

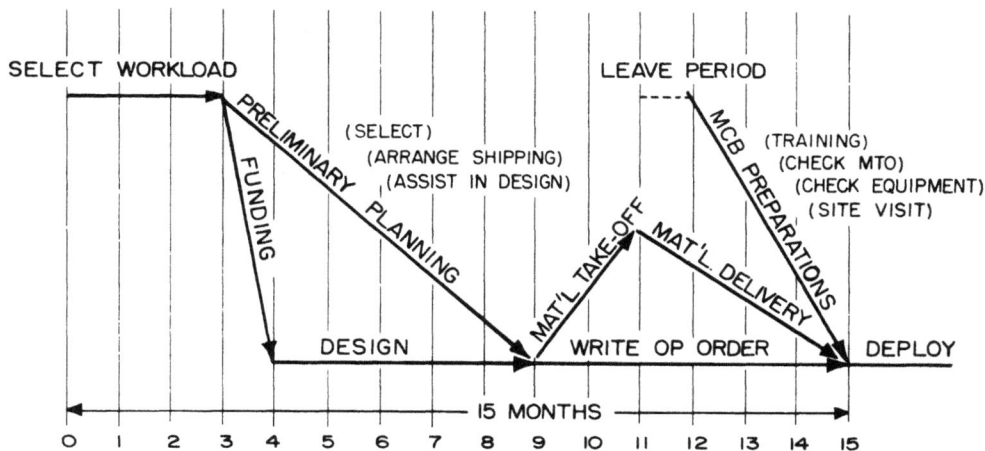

FIGURE 98
PEACETIME DEPLOYMENT PREPARATIONS

177

FIGURE 99
HOMEPORT NAVAL CONSTRUCTION REGIMENT

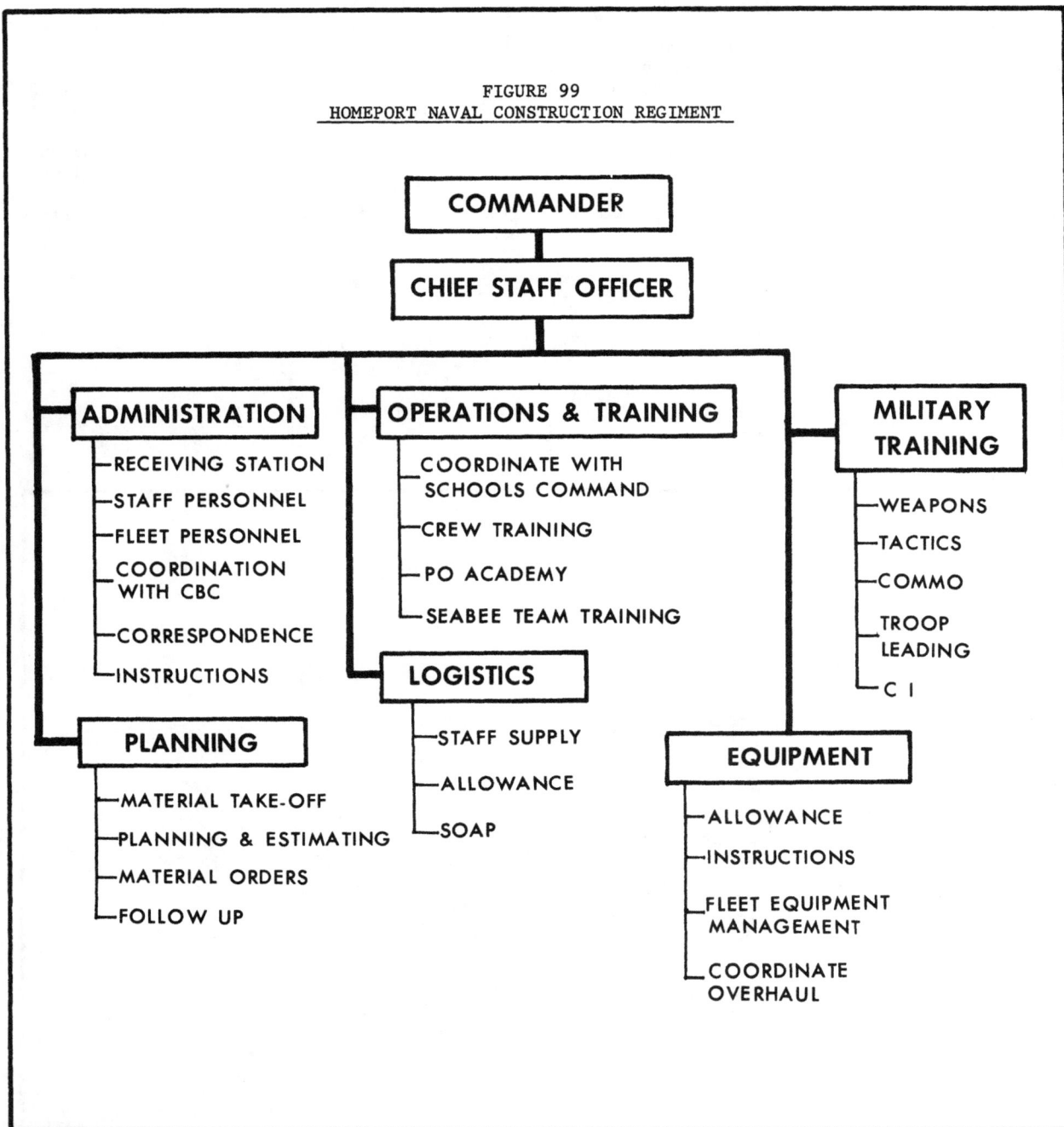

COMMANDER

CHIEF STAFF OFFICER

ADMINISTRATION
- RECEIVING STATION
- STAFF PERSONNEL
- FLEET PERSONNEL
- COORDINATION WITH CBC
- CORRESPONDENCE
- INSTRUCTIONS

PLANNING
- MATERIAL TAKE-OFF
- PLANNING & ESTIMATING
- MATERIAL ORDERS
- FOLLOW UP

OPERATIONS & TRAINING
- COORDINATE WITH SCHOOLS COMMAND
- CREW TRAINING
- PO ACADEMY
- SEABEE TEAM TRAINING

LOGISTICS
- STAFF SUPPLY
- ALLOWANCE
- SOAP

MILITARY TRAINING
- WEAPONS
- TACTICS
- COMMO
- TROOP LEADING
- C I

EQUIPMENT
- ALLOWANCE
- INSTRUCTIONS
- FLEET EQUIPMENT MANAGEMENT
- COORDINATE OVERHAUL

SECTION C - OPERATIONAL
BRIGADES AND REGIMENTS (NCB/NCR)

1. <u>Introduction</u>. Operational Naval Construction Brigades and Regiments are commissioned by CNO to control several Naval Mobile Construction Battalions operating in the same theatre. Other NCF units in the area may be placed under their control. The Regiments control one or more Battalions and the Brigades control one or more Regiments.

2. <u>Mission and Tasks</u>. The mission of these headquarters is to control the construction effort of all NCF Units assigned to their operational control. Some of the tasks in support of this mission include:

    a. Assignment and realignment of construction projects to provide a balanced workload for all subordinate units.

    b. Assignment of priorities and deadlines for completion of construction projects consistent with directives from higher headquarters.

    c. Monitor construction progress and quality.

    d. Monitor equipment requirements of assigned units, reassign equipment among units, provide equipment to units from headquarters pools and manage the SEABEE Equipment Program within the theatre.

    e. Provide logistics support and assistance to all units assigned, particularly construction material and Tactical Support Functional Components.

    f. Provide initial review of plans and specifications and perform planning, estimating, and material take-off in order to facilitate workload distribution and long lead procurement.

    g. Provide engineering and technical assistance to assigned units.

    h. Develop data and planning factors relating to capabilities of assigned unit.

3. <u>Staff Organization</u>. The staffs are organized similar to the MCB Executive Staff with the exception that the S-2 functions differ and the S-4, Logistics, and the S-5 (N-5) equipment, are actually doers rather than purely staff officers. The Staff Sections on the Brigade Staff are designated (N), and on a Regiment Staff (R), rather than (S).

    a. <u>Commander</u>. The Commander of an NCB is a Rear Admiral, CEC. The Regiment is commanded by a Captain (Commodore).

    b. <u>Chief of Staff</u>. The Chief of Staff coordinates the efforts of all executive and special staff officers. This position is properly titled Chief Staff Officer when the Commander is not a Flag Officer.

    c. <u>R-1 or N-1 - Administration</u>. The operational NCBs and NCRs do not have administrative control over assigned NCF Units. Therefore, this section is concerned only with headquarters administrative and personnel problems. The other important responsibilities are public affairs and communications.

    d. <u>R-2 or N-2 - Planning</u>. This Section is quite different from the S-2 of an MCB. The intelligence function remains, but training is a very minor function in the actual contingency situation. This section's main responsibility lies in the area of further construction operations. The construction backlog is analyzed as well as the capabilities of the assigned units. New jobs are estimated and preliminary material take-off is performed.

    e. <u>R-3 or N-3 - Operations</u>. The Staff Operations Officer's position is very similar to that of a Battalion S-3. He manages, directs and coordinates the construction program of all assigned units. Projects are assigned to the various units along with priorities. Progress is closely monitored, and unit reports studied. Unit workloads are balanced and equipment requirements studied.

    f. <u>R-4 or N-4 - Logistics</u>. This Section provides supplies to the staff, construction material and TSFCs to the Battalions, and performs budget and fiscal functions. The Headquarters Administrative Pool is also operated by this section.

    g. <u>R-5 or N-5 - Equipment</u>. This Section will be organized only in the senior SEABEE Headquarters in the area of operations. The equipment program for the area is managed as well as providing a pool of heavy construction equipment for augmentation of assigned battalions.

4. <u>Contingency Regiments</u>. Contingency Regiments are planning staffs located geographically with the Marine Expeditionary Force Planning Headquarters. The staff will prepare, with the force engineer, the engineer support plans to all MEF Contingency Plans. The staff will, of course, become the NCR Staff upon execution of any plan (See Figure 12, page 14). The NCR will have operational control of assigned MCBs under the Force Engineer Group Commander until chopped to the Navy Component Commander.

5. <u>Alert Battalions</u>. A certain number of MCBs are designated Alert Battalions for support of MEF Plans or Fleet Commander Plans. One or more Alert Battalions may well be located contiguous to the MEF and contingency regiment headquarters in a short reaction time posture. In this case the contingency regiment staff may be at full strength and exercise some degree of operational control of the Alert Battalions, particularly in the field of training.

6. <u>Command Relationships</u>. As pointed out in Chapter I, a SEABEE operation that begins as support of expeditionary forces can easily grow to a base development eff-rt. The two different operations require two different chains of operational control. The Naval Construction Regimental Staff and assigned Mobile Construction Battalions will go ashore as part

FIGURE 100
NAVAL CONSTRUCTION BRIGADE/REGIMENT

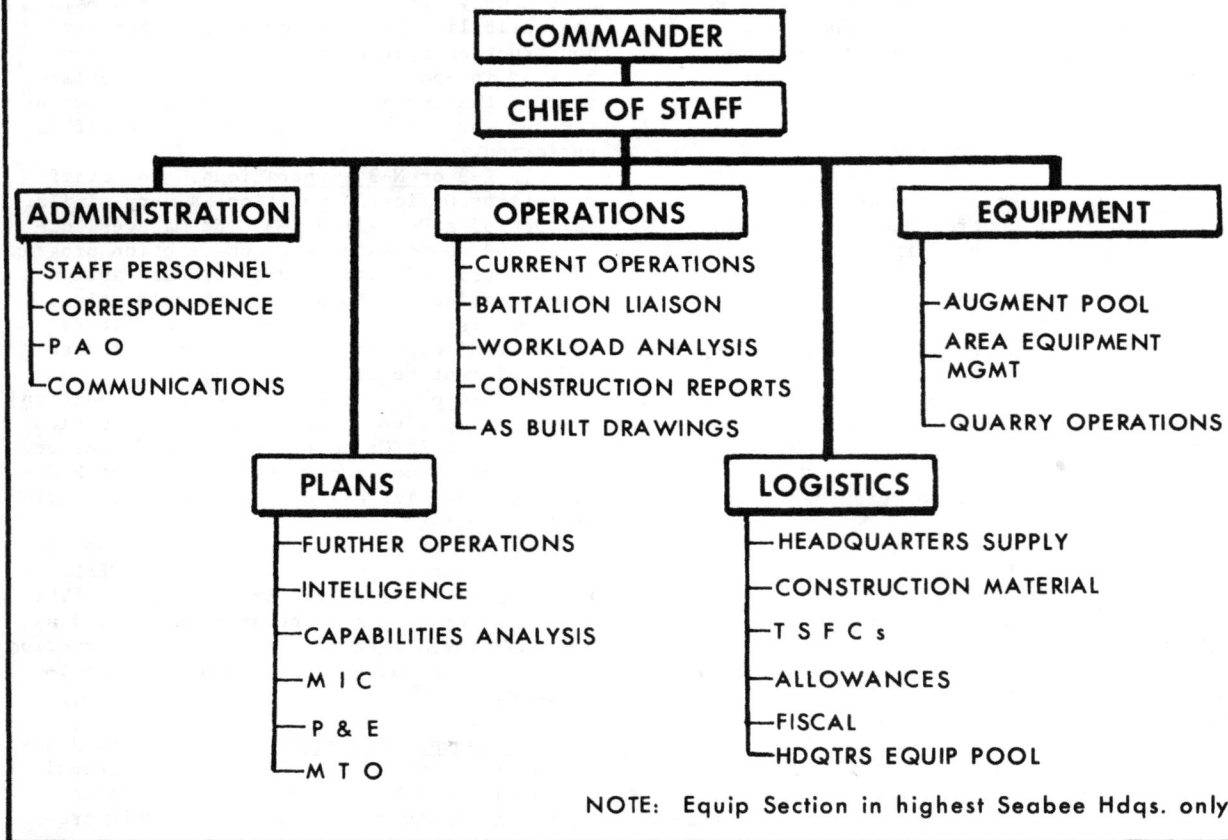

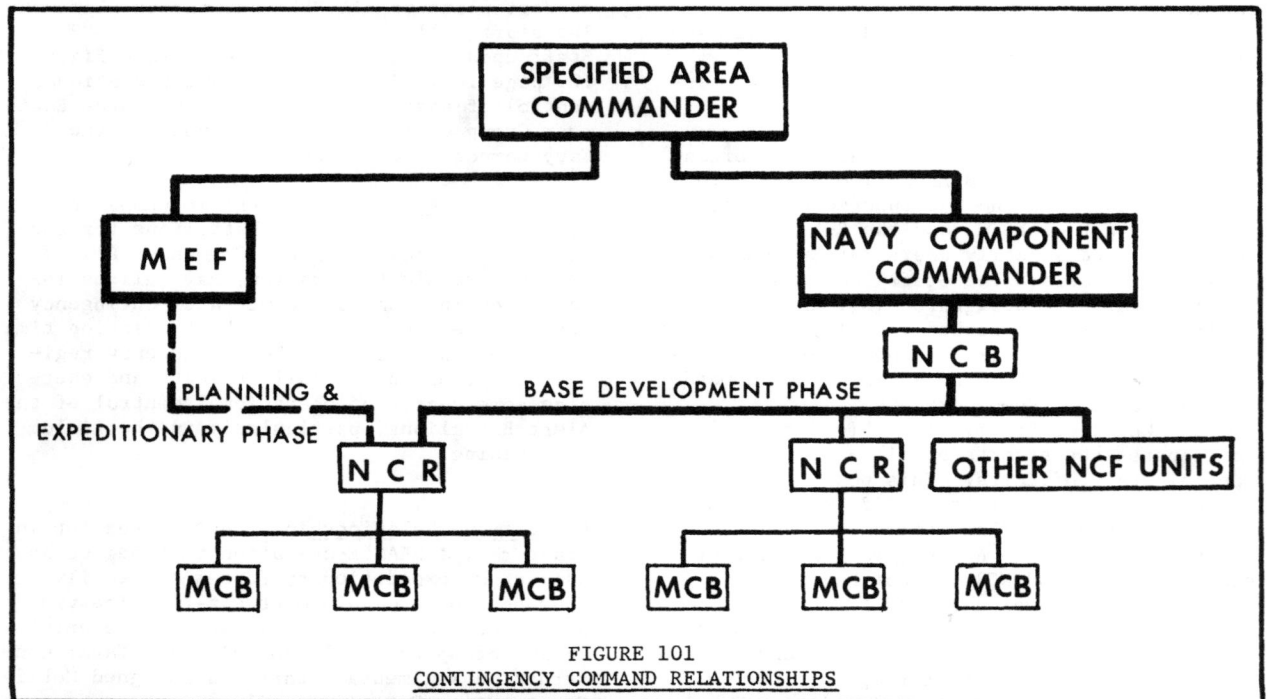

COMMANDER

CHIEF OF STAFF

ADMINISTRATION
- STAFF PERSONNEL
- CORRESPONDENCE
- P A O
- COMMUNICATIONS

OPERATIONS
- CURRENT OPERATIONS
- BATTALION LIAISON
- WORKLOAD ANALYSIS
- CONSTRUCTION REPORTS
- AS BUILT DRAWINGS

EQUIPMENT
- AUGMENT POOL
- AREA EQUIPMENT MGMT
- QUARRY OPERATIONS

PLANS
- FURTHER OPERATIONS
- INTELLIGENCE
- CAPABILITIES ANALYSIS
- M I C
- P & E
- M T O

LOGISTICS
- HEADQUARTERS SUPPLY
- CONSTRUCTION MATERIAL
- T S F C s
- ALLOWANCES
- FISCAL
- HDQTRS EQUIP POOL

NOTE: Equip Section in highest Seabee Hdqs. only

SPECIFIED AREA COMMANDER

M E F

NAVY COMPONENT COMMANDER

N C B

PLANNING &
EXPEDITIONARY PHASE

BASE DEVELOPMENT PHASE

N C R

N C R

OTHER NCF UNITS

MCB   **MCB**   **MCB**

MCB   **MCB**   **MCB**

FIGURE 101
CONTINGENCY COMMAND RELATIONSHIPS

of FEG under the operational control of the MEF. When the specified command commissions a Navy Component Commander (COMNAV for Area), the Naval Construction Force will be chopped to him. At the same time, additional MCBs, NCRs and probably a NCB will arrive for the base development phase.

SECTION D - CONSTRUCTION BATTALION CENTERS

1. General Support Rendered to MCBs. The mission of the CBCs is to support Naval Construction Forces and other fleet units deployed from or homeported at the CBC; to provide storage, preservation and shipping facilities for Advanced Base and Prepositioned War Reserve Stocks (PWRS); to perform engineering and technical services and such other tasks as may be assigned by the Commander of the Naval Facilities Engineering Command. Since the ACBs are not homeported at the Construction Battalion Centers, the support rendered to them is limited to engineering and technical services whereas the support of the MCBs is analogous to the support that tender ships give to their vessels. Furthermore, the mission to maintain the PWRS is actually indirect

support to the MCBs because these groupings become the project material and construction equipment needed by the NCF units in carrying out their mission to support expeditionary forces and advanced base development.

Complex equipment such as trucks, generators, pumps, etc., is maintained by the Construction Equipment Department while other equipment and supplies such as heaters, quonset huts, hand tools, etc., is maintained by the Supply Department. The PWRS must be kept in a high state of readiness and available for immediate outloading. The NAVFAC P-400, Construction Battalion Center Material Management Guide will give SEABEE officers an insight into the complex operation of providing material at the required readiness condition to construct logistic support facilities for the Navy and Marine forces in limited war emergency.

There are three Construction Battalion Centers located at Gulfport, Mississippi, Port Hueneme, California, and Davisville, Rhode Island. They offer support to assigned units of the active Naval Construction Forces, and the Commanding Officer of the CBC exercises area coordination control of homeported MCBs. Figure 102 shows the organization of the CBC,

FIGURE 102
ORGANIZATION OF THE CONSTRUCTION BATTALION CENTER, PORT HUENEME

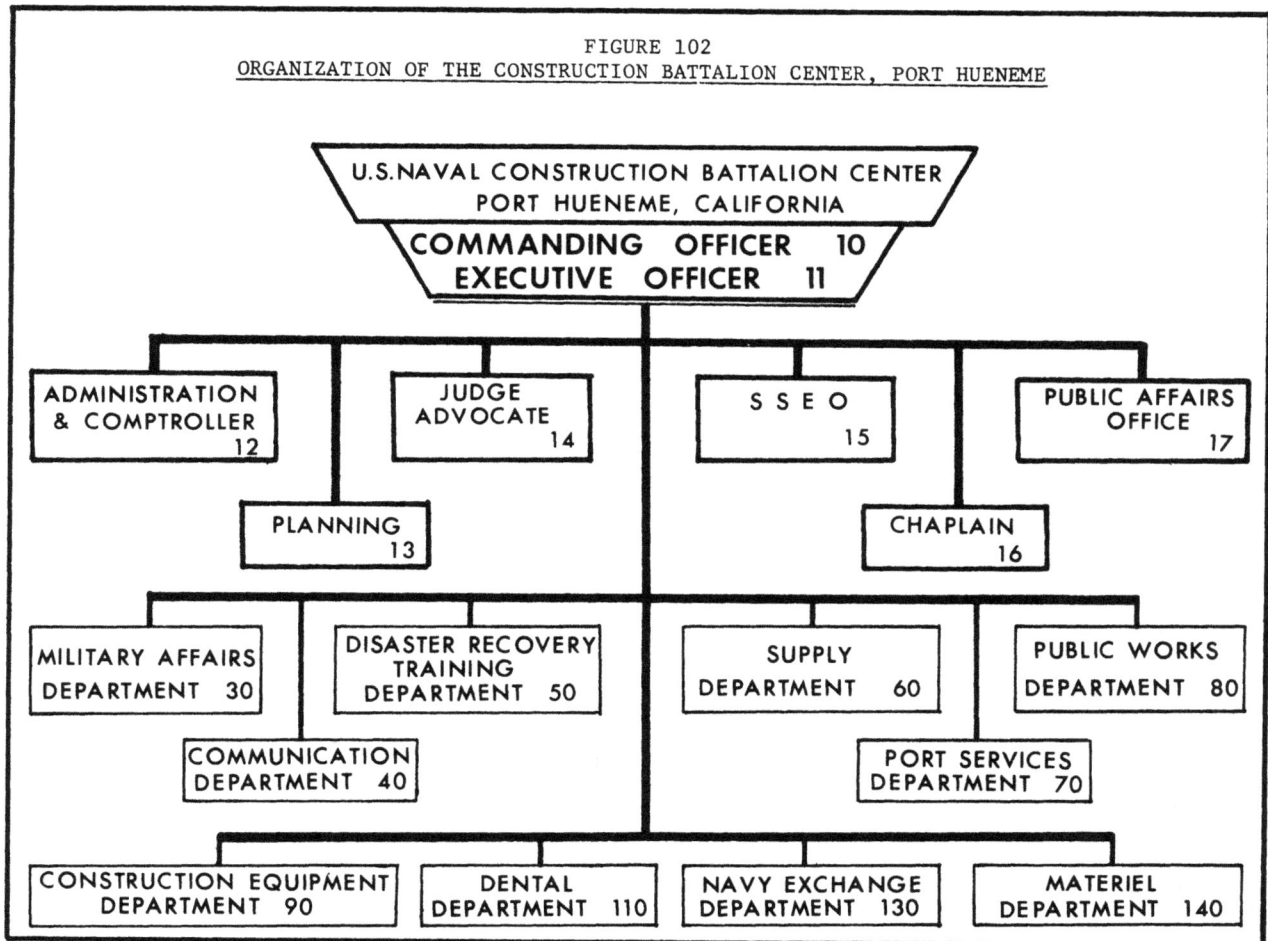

U.S. NAVAL CONSTRUCTION BATTALION CENTER
PORT HUENEME, CALIFORNIA
COMMANDING OFFICER 10
EXECUTIVE OFFICER 11

- ADMINISTRATION & COMPTROLLER 12
- JUDGE ADVOCATE 14
- S S E O 15
- PUBLIC AFFAIRS OFFICE 17
- PLANNING 13
- CHAPLAIN 16
- MILITARY AFFAIRS DEPARTMENT 30
- DISASTER RECOVERY TRAINING DEPARTMENT 50
- SUPPLY DEPARTMENT 60
- PUBLIC WORKS DEPARTMENT 80
- COMMUNICATION DEPARTMENT 40
- PORT SERVICES DEPARTMENT 70
- CONSTRUCTION EQUIPMENT DEPARTMENT 90
- DENTAL DEPARTMENT 110
- NAVY EXCHANGE DEPARTMENT 130
- MATERIEL DEPARTMENT 140

Port Hueneme. In general, the support provided to the MCBs falls into these three classes:

    a. General Logistics Support Including:
        (1) Office Spaces
        (2) Warehousing
        (3) Clothing and Small Stores
        (4) Fiscal Accounting
        (5) Port Terminal Services

    b. Military Personnel Services Including:
        (1) Barracks and Housing
        (2) Messing Facilities
        (3) Disbursing
        (4) Medical and Dental
        (5) Special Services and Navy Exchange

    c. Mission Support Including:
        (1) Training Facilities
        (2) Base Recovery Course
        (3) Procurement of Project Material
        (4) Requisition Follow-up
        (5) Fifth Echelon Equipment Maintenance
        (6) Prepositioned War Reserve Stock

2. The Construction Equipment Department. In support of the Construction Battalions the Construction Equipment Department has three direct functions namely:

    a. Overhaul and rehabilitation of complex equipment needed by the battalions to conduct routine training deployments. Automotive, construction, weight handling and service equipment that requires fifth echelon maintenance is brought back from the deployment sites and turned over to CED for repair.

    b. Maintenance of specified complex equipment in a ready-for-issue state. This equipment may be part of the basic allowance, part of the supplemental functional component pool or part of special allowances designed to support SEABEE operations.

    c. Technical support in providing repair parts to deployed battalions by assisting the CBC Supply Department to process requisitions. This assistance is in the form of identification of manufacturers part numbers, determination of the acceptability of substitutions and recommendation for sources of supply.

In addition to this direct supply to the battalions, CED provides indirect support to the NCF by performing the technical management, maintenance, repair and overhaul of the NAVFAC Prepositioned War Reserve complex equipment. This function, which is the bulk of the workload for CED, includes inspection of incoming equipment for completeness and conformance to contract specifications, preservation packing and marking of complex stock for storage or shipment, analysis of preservation methods and cost data and by inspection, maintenance and quality control of all preserved and stored equipment.

The PWRS complex equipment maintained by CED constitutes much of the NAVFAC contribution to various functional components. This equipment mmay be used as part of the equipment needed at an advanced base or as part of the equipment needed by a construction battalion to build the advanced base.

3. The Supply Department. The Supply Department provides homeported MCBs with the general mess, clothing and small stores and disposal services. It also processes requests for material and equipage and handles the water-borne movement of materials, supplies, tools and equipment needed at the deployment site. Project material, ordered by the battalion or the material-take-off team, is given to the Supply Department to procure from the Navy Supply System or local vendors. Delivery status information is provided to the battalion along with follow-up services to assure prompt receipt of materials. When received the Material Division of the Supply Department inspects, packs and forwards it to the proper destination Project material is delivered to the Marine Terminal Division for shipment overseas but tools or equipment are normally delivered to the battalion warehouses.

The Marine Terminal Division is responsible for all water-borne shipping functions. They receive, assemble and check cargo delivered dockside. As mentioned above, project material is usually sent directly from the Material Division. However, tools and other equipage are turned over to the battalion. The Marine Terminal Division provides transit storage facilities, plans the cargo assembly, coordinates space allocations, prepares cargo stowage plans, controls stevedoring operations and prepares ocean manifests and other transportation cargo movement documents.

The Supply Department has a special program section which develops and initiates procurement for the repair parts assemblies that accompany each battalion deploying overseas.

4. SEABEE Systems Engineering Office (SSEO). Establishment of the SEABEE Systems Engineering Office (SSEO) was approved by the Commander, Naval Facilities Engineering Command in July 1968. Its mission is to provide a systems analysis capability to the Naval Facilities Engineering Command; to develop, operate and maintain the SEABEE Tactically Installed Navy Generated Resources System (STINGER), and to perform engineering and technical services and such other tasks as may be assigned by higher authority. It is to accomplish this mission by:

    a. Defining, developing, coordinating, maintaining and supporting the STINGER System.

    b. Analyzing STINGER sub-systems in being and determining if they should be retained, improved or discontinued.

    c. Developing and implementing new sub-systems applicable to the STINGER System or as directed by NAVFAC.

    d. Developing and maintaining systems engineering expertise to provide this support to NAVFAC. This would include the evaluation of potential impact of new or existing mandatory systems and the development of implementing procedures if requested.

    e. Developing procedures and criteria to measure cost effectiveness and operational readiness capabilities of men, material, equipment, and data for the STINGER System.

    f. Developing and implementing directed portions of an integrated logistics support systems for the Naval Construction Forces and NAVFAC. This includes designing, engineering,

logistics support and procurement management for men, material, equipment and data necessary to operate the STINGER System or for other NAVFAC systems if directed.

    g. Supporting the DOD Standardization and Cataloging Programs.

    h. Accomplishing NBy contracting and effect procurement and provisioning as directed by NAVFAC.

    i. Other tasks as assigned by NAVFAC

    The SSEO is located at the Construction Battalion Center, Port Hueneme, California. Its organizational chart is indicated in Figure 103. The Director of the SSEO is a Department Head of CBC.

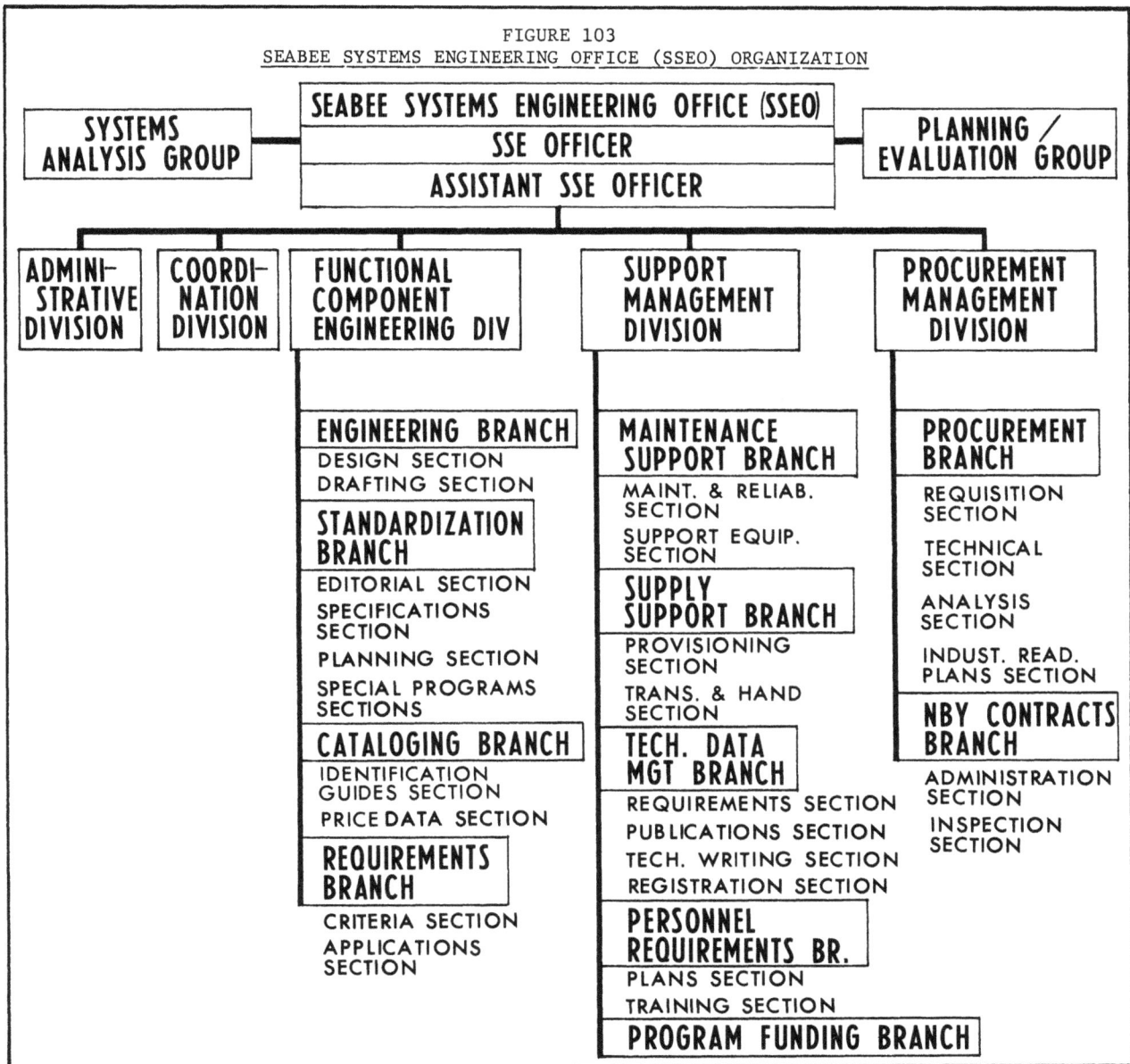

FIGURE 103
SEABEE SYSTEMS ENGINEERING OFFICE (SSEO) ORGANIZATION

DOCTRINE AND POLICY FOR MCBs

DEPARTMENT OF THE NAVY
OFFICE OF THE CHIEF OF NAVAL OPERATIONS
WASHINGTON, D.C. 20350

OPNAV 5450.46D
Op-401K
Ser 307P401
12 March 1962

OPNAV INSTRUCTION 5450.46D

From: Chief of Naval Operations
To: Distribution List

Subj: U.S. Naval Construction Forces; doctrine and policy concerning

1. <u>Purpose</u>. This Instruction promulgates basic doctrine and policies for the Naval Mobile Construction Battalions.

2. <u>Cancellation</u>. This Instruction cancels and supersedes OPNAV INSTRUCTION 5450.46C.

3. <u>Doctrine</u>. The existing active Naval Construction Force Units are integral commissioned units of the Naval operating forces. In wartime they will provide military and construction support to Naval, Marine and other forces in military operations and the construction and maintenance of Naval base facilities. In peacetime, the following policies shall apply:

   a. <u>Operational Readiness</u>
   (1) A minimum of two-thirds of all Naval Construction Force Units assigned to each fleet will be capable of being organizationally redeployed from their deployment sites within ten days.

      (a) Sixty days after return to homeports, such units will be capable of redeploying in ten days.

      (b) During the first forty days after return, such units will be capable of redeploying in thirty days.

      (c) During periods while enroute to or from deployment sites, such units will be capable of immediate diversion to emergency, contingency or mobilization assignments.

   b. <u>Materiel Readiness</u>
   (1) Functional components for outfitting of Naval Construction Force Units will be tailored to sustain, for ninety days without augmentation, operations planned or envisioned for contingencies, i.e., under emergency, sublimited war conditions, or in a limited or general war. The allowance will be predicated on emergency consumption rates and are to support construction operations for two ten-hour shifts per day seven days a week - a total of 1,800 construction hours

      (2) The P-25 functional component is the basic table of allowance for a Naval Mobile Construction Battalion. Other basic components will be developed and maintained as necessary for other Naval Construction Forces that may be organized.

      (3) The basic P-25 and other functional components required for counterinsurgency operations of active Naval Construction Force Units will be assigned to cognizant Fleet Commanders for reassignments. Fleet Commanders will be responsible for assuring that these components are maintained in a continuing state of readiness for mounting out for a contingency assignment within a period of ten days. These functional components may be utilized for peacetime assignments,

augmented as necessary to assure continuing readiness for contingency assignments. Such augmentation will be accomplished from pools maintained under the cognizance of Fleet Commanders or by the activities for which Naval Construction Force Units accomplish work. Portions of the functional components not required on peacetime deployments may be positioned in other areas ready for immediate movement to join with the battalion.

     (4) Supplemental functional components will be developed, maintained and assigned to the cognizant Fleet Commander for reassignment to assure the capability of Naval Construction Force Units to accomplish specific special construction assignments.

     (5) Satisfactory state of readiness will be attained in the following manner:

        (a) Equipment (including spare parts) in use with and maintained by the active Naval Construction Forces shall be ready to move in ten days or less if required by specific plans.

        (b) Preserved, packed and stored materials at prepositioned locations shall be ready to move in ten days or less if required by specific plans.

        (c) Stock levels of required items of packaged POL, ammunition, rations, special clothing, and other consumable supplies needed for contingency deployment of MCBs may be increased, as necessary, at individual stocking points to insure availability for issue within ten days.

    c. Peacetime Employment and Deployment
     (1) MCBs will normally be assigned construction projects contributing to training and readiness for fulfillment of contingency assignments. Peacetime Naval Construction Force Unit deployment and employment requires definitive planning and emphasis placed on mobility and readiness to meet emergency situations. Training and planning with fleet units, including Fleet Marine Forces will be accomplished whenever possible.

     (2) Foreign aid-type projects may be undertaken upon specific approval by the Chief of Naval Operations. Each individual project will be assessed in terms of over-all worth to the nation and the Navy considering all of the political and military factors involved. Normally, projects will not be approved without concurrence of the cognizant Fleet Commander in Chief unless overriding factors or the urgency of the requirement dictate otherwise.

     (3) Mobile Construction Battalions will be deployed to, and employed in, such areas as are necessary to insure adequate training under various environments to enable these forces to adequately meet emergency requirements that may develop in similar areas of the world. Deployment sites will be chosen wherever possible to insure proximity of forces to possible areas of contingency assignment.

     (4) Lack of funds shall not, in itself, be sufficient justification for the employment of military construction units.

     (5) Under conditions of emergency, disaster or catastrophe caused by enemy action or natural causes, Naval Construction Units may be utilized as directed by appropriate cognizant authority. This includes the furnishing of assistance to civilian agencies.

     (6) For special situations not covered by the foregoing specific requests may be submitted to the Chief of Naval Operations for consideration.

4. Responsibility

    a. The Chief of Naval Operations
     (1) Shall approve annual deployment plans.

     (2) Shall approve foreign aid-type projects.

    b. The Commanders in Chief:
     (1) Will determine existing deficiencies to implement contingency, limited and general war plans. Deficiencies, phased as to time and space, will be reported to the Chief of Naval

Operations with recommendations for appropriate action and requests for release of required components or materials.

(2) Will determine and coordinate transportation requirements to implement the policies set forth herein relative to both peacetime and emergency deployments.

(3) Will insure routine deployment schedules and assigned projects and training are in consonance with the policies set forth herein.

(4) Will determine requirements and develop procedures for resupply of the battalions under normal and contingency operations.

c. The Chief of Civil Engineers is the technical advisor to the Chief of Naval Operations on Naval Construction Forces.

d. The Technical Bureaus shall provide material and funding support.

e. The Shore Establishment will assist the Fleet Commanders to the fullest extent to insure that the operational and material readiness of the Naval Construction Force are maintained at the levels set forth herein.

C. O. TRIEBEL
By direction

DOCTRINE AND POLICY FOR ACBs

DEPARTMENT OF THE NAVY
OFFICE OF THE CHIEF OF NAVAL OPERATIONS
WASHINGTON, D.C. 20350

OPNAV 5440.62
Op-403E
Ser 3288P40
30 Oct 1961

OPNAV INSTRUCTION 5440.62

From: Chief of Naval Operations
To: Distribution List

Subj: Functions, deployment, and employment of the U.S. Naval Amphibious Construction Battalions

1. <u>Purpose</u>. This Instruction sets forth the basic policies with respect to the U.S. Naval Amphibious Construction Battalions.

2. <u>General</u>

   a. <u>Wartime</u>: The Amphibious Construction Battalion provide military and amphibious assault construction support to Naval, Marine and other Forces in military operations as directed.

   b. <u>Peacetime</u>

      (1) Under peacetime conditions, the primary consideration governing the assignment of fleet training exercises or construction projects to the Amphibious Construction Battalions shall be contribution to training and readiness for fulfillment of the wartime missions.

      (2) "Foreign Aid" type projects will be undertaken. Each of these projects shall be assessed on an individual project basis in terms of overall worth to the Nation and the Navy considering all of the political and military factors involved. These projects will be approved by the Chief of Naval Operations only with prior concurrence of the cognizant Fleet Commander in Chief unless overriding factors or the urgency of the requirement dictate otherwise.

      (3) Amphibious Construction Battalions will not be employed in competition with any form of civilian labor in the United States. Exceptions may be made when the nature of the work necessitates security which could not otherwise be assured, or where the isolated location of the work to be performed makes it impractical of contractor accomplishment.

      (4) Lack of funds shall not, in itself, be sufficient justification for the use of military construction units.

      (5) The training of Naval Reservists as organized units shall be accomplished in order that the full potential of these forces may be readily available when required in wartime or other emergency.

   c. <u>Disaster Recovery</u>. Under conditions of emergency, disaster, or catastrophe caused by enemy action or natural causes, Amphibious Construction Battalions, including reserve units, may be utilized as directed by cognizant authority including the furnishing of assistance to civilian agencies.

   d. <u>Special Situations</u>. For special situations not covered by the foregoing, specific requests may be submitted to the Chief of Naval Operations for consideration.

4. Assignment, Command and Deployment

    a. The Amphibious Construction Battalions are an integral part of the Naval Operating Forces to which assigned.

    b. Amphibious Construction Battalions will maintain organizational command integrity under any assignment.

    c. Deployment schedules for the Amphibious Construction Battalions and/or detachments thereof will be approved by the applicable Amphibious Force Commander.

5. Readiness. Peacetime Amphibious Construction Battalion deployment requires detailed and austere planning with primary emphasis continually placed upon mobility and operational readiness to meet all emergency and contingent operations involving amphibious assault operations.

6. Technical Advisor. The Chief of Civil Engineers is the technical advisor to the Chief of Naval Operations on Naval Construction Forces.

                                                 MILES H. HUBBARD
                                                 By direction

Annex C
THE FUNCTIONAL COMPONENT SYSTEM

1. **Background.** It has been said that strategy and tactics provide the scheme for the conduct of military operations and that logistics provides the means therefore. The ability to procure items required by a series of military commanders and distribute these items on a priority basis, exploiting a changing military situation, requires enormous effort on the part of logistics officers and tactical commanders at all levels of command. During the Pacific Campaign of World War II, materials ordered for one base were appropriated for the more immediate need of another that was nearer the front. At times it was impossible to know what part of how much of a grouping of materials had been planned for a project and which had been diverted from a project of lower priority. Therefore, early in the war, logistic planners attempted to standardize and catalog in related groups all the radios, typewriters, boats, men and nails from the numerous allowance lists. The problem was tackled from the point of tasks to be performed, and who and what was needed for each task. The modular or building block concept was developed, with each component designed to serve a specific function no matter where it was placed. Thus the Functional Component System as a tool of Naval logistics, evolved out of the experience with early advanced base planning and shipment in World War II. It is the result of an analysis of all foreseeable overseas requirements of men and materials and the organization of such requirements into compact and related units for assignment and shipment, individually or in multiples, to fill any sized operational requirement. The aggregate of all these units or functional components, large and small, in their relationship to each other, is the Functional Component System as developed by the Navy. All advanced base logistical planning is expressed in terms of functional components or their equivalents. The Functional Component System has become the quantitative expression of measurement of planning, procurement, assembly and shipping of material needed for war and other operational requirements. It is the decimal system of Naval logistics.

The assembly or availability of materials for a component of an advanced base is directly related to the determination of the requirements. Assembly or procurement of advanced base functional components is coordinated through each systems command headquarters in Washington, D.C. CNO deals directly with them

on all matters of assembly and continually advises them of all changes in requirements. Changes in requirements stem principally from changes in the strategic concept which may throw emphasis on one area, and impose additional requirements or different demands because of climatic conditions, terrain, type of area considered, and the like.

2. **Functional Components.** A functional component is a grouping of personnel and material designed to perform one of the specific tasks of an advanced base. A functional component contains the technical personnel and the technical equipment necessary for the performance of these tasks including as pertinent, workshops, housing, vehicles, boats, office equipment, tools and a 30 to 90-day initial supply of consumables. It is in effect, a bill of material prepared for a typical facility. It consists of a listing of line items and assemblies or kits as shown in Figure 104.

If a CEC officer is to accomplish practical work with functional components, he must know many of the details of the system and how it is organized. Components and assemblies are arranged into major groupings. The responsibility for developing, maintaining, and revising the initial outfitting list requirements for advanced base component material, and for personnel when included in a component, is assigned to the system command designated as dominant by the CNO. Two factors are used to determine which systems command is selected as dominant: When the technical function of a component is a matter of responsibility of a systems command, that command is designated as dominant; when the function of a component does not fall within the field of one of the systems commands, the command which furnishes the major portion of the material is designated as dominant. Other systems commands which contribute material are termed contributing commands. For example, the dominant systems command for an air base magazine functional component is the Naval Ordnance Systems Command, since it is responsible for ammunition handling. However, the Naval Facilities Engineering Command is a contributor since it supplies the quonset huts, magazines and automotive equipment, as is the Navy Supply Systems Command which contributes the bedding and special clothing needed, etc.

Each component is given a number and letter designation. Generally, the components are grouped by broad functions. There are

FIGURE 104
EXAMPLE OF A DETAILED LISTING OF A FUNCTIONAL COMPONENT

COMP OR ASSY	E C H CODE	STOCK NO./ASSEMBLY REFERENCE/DESCRIPTION	U N I T	QUANTITY REQUIRED	Z O N E	WEIGHT POUNDS	CUBIC FEET	COMP OR ASSY

Functional Component

H14E

H14E AVIATION TANK FARM BASIC   H14E
Y-D DRAWING 1028191

SECTION 1 STRUCTURES

1		1357 HUT ARCH RIB 20X48	304040	1		11276	432.5	
1		1687 HEAD CONSTRUCTION AREA	889734	1		3118	117.6	

SECTION 2 MECHANICAL

1		2223 CONNECT CRS 8N PIPE-4N COUP	1028191	9		10314	213.3	
1		2230 CONNECTION F/FIRE TRUCK FOAM	1028203	2		1056	13.2	
1		2262 TANK 10000BBL 6 AND 8N PIPE	1028196	5		509000	9218.0	
1		2274 PUMP STA 1PUMP 8N HEADR	1028198	4		36516	1405.2	
1		22 FILTER SEPARATOR ASSY JET FUEL	1028202	2		39264	1477.4	
1		22 FILTER HEATER/POWER	1028202	2	N	5508	348.0	
1		22 FUEL DISPENSING A/C-REFUEL	1028322	2		24184	2260.6	
2	2C	4320-264-4573 PUMP CENT 55GPM 50TDH	EA	1		230	14.0	
1	9C	4710-203-0183 PIPE GBE 6X20F	EA	450		67500	3060.0	
1	9C	4710-203-0188 PIPE GBE 8X20F	EA	450		94500	5040.0	
1	9C	4710-639-9441 PIPE BLK STL 3/4	FT	63		72	.5	
2	2C	4720-954-4209 HOSE ASSY LWT QCPLG 2X50F	EA	1		70	2.5	
2	2C	4720-991-8534 HOSE SUC LWT 2X10 QUIK C	EA	2		40	2.4	
2	2C	4730-087-9285 COUPLING HLF QUIK 2FX2M	EA	2		180	.2	
1	9C	4730-142-1591 COUPLING CLAMP 8N GRV PP	EA	525		14700	210.0	
1	9C	4730-187-7571 COUPLING PP STL 3-4NPT	EA	30		11	.1	
1	9C	4730-189-2617 UNION MI 3/4	EA	15		10	.1	
1	9C	4730-193-7080 ELBOW PP MI 90D 3/4	EA	42		18	.3	
2	9C	4730-196-2062 NIPPLE PP CLS 2X2	EA	2		1		
1	9C	4730-202-7215 CROSS PP GRV ENDS 6IPS	EA	1		47	.9	
1	9C	4730-202-9506 REDUCER PP GBE 8X6IPS	EA	14		294	5.9	
1	9C	4730-273-8056 ELBOW PP GBE 90D 8IPS	EA	9		540	18.0	
1	9C	4730-273-8178 COUPLING PP GRV-BVL 8IPS	EA	12		171	3.4	
1	9C	4730-273-8313 ELBOW PP GBE 45D 6IPS	EA	4		80	1.6	
1	9C	4730-273-8314 ELBOW PP GBE 45D 8IPS	EA	4		140	4.4	
1	9C	4730-273-8359 ELBOW PP GBE 90D 6IPS	EA	14		420	9.4	
1	9C	4730-278-2669 COUPLING PP GRV-BVL 6IPS	EA	12		132	2.6	
1	9C	4730-288-9514 COUPLING CLAMP GRV 6 1PS	EA	505		7878	101.0	

COMP OR ASSY	E C H CODE	STOCK NO./ASSEMBLY REFERENCE/DESCRIPTION	U N I T	QUANTITY REQUIRED	Z O N E	WEIGHT POUNDS	CUBIC FEET	COMP OR ASSY

(2262)

Assembly

ASSEMBLY NUMBER 2262 Y-D DRAWING NUMBER 1028196   2262
TANK 10000 BBL 6 AND 8 INCH PIPE

SECTION 1 STRUCTURES

	2C	5430-925-4571 TANK LIQUID STRGE 10000B	EA	1		85730	1560.0	

SECTION 2 MECHANICAL

		2230 CONNECTION F/FIRE TRUCK FOAM	1028203	1		528	6.6	
	9C	4710-203-0183 PIPE GBE 6X20F	EA	6		900	40.8	
	9C	4710-203-0188 PIPE GBE 8X20F	EA	6		1260	67.2	
	9C	4710-639-9441 PIPE BLK STL 3/4	FT	21		24	.2	
	9C	4730-142-1591 COUPLING CLAMP 8N GRV PP	EA	20		560	8.0	
	9C	4730-187-7571 COUPLING PP STL 3-4NPT	EA	8		3		
	9C	4730-189-2617 UNION MI 3/4	EA	4		3		
	9C	4730-193-7080 ELBOW PP MI 90D 3/4	EA	12		5	.1	
	9C	4730-202-7208 COUPLING PP 6N GRV-FLG	EA	2		64	1.0	
	9C	4730-202-7209 COUPLING PP 8N GRV-FLG	EA	2		90	2.4	
	9C	4730-202-9506 REDUCER PP GBE 8X6IPS	EA	1		21	.4	
	9C	4730-273-8056 ELBOW PP GBE 90D 8IPS	EA	3		180	6.0	

NOTE: These pages from the Detailed Catalog of Equipment and Material Requirements for Advanced Base Functional Components, NAVFAC P-103, list the items furnished by NAVFAC for a functional component (H14E) and for a typical functional-component assembly (2262).

FIGURE 105
TOOL KIT 7002 IS A TYPICAL ASSEMBLY

This assembly may be found in many functional components. It consists of the basic hand tools for a Builder.

over 340 components grouped in the following categories:

A - Administration		11
B - Harbor Control and Defense		33
C - Communications		41
D - Supply and Disbursing		53
E - Ship and Boat repair		18
F - Cargo Handling		3
G - Medical and Dental		34
H - Aviation		32
J - Ordnance		31
N - Camp and Welfare		33
P - Construction and Public Works		47
S - Special Groups		--

For example, the A components are designed to furnish the necessary personnel and equipment to establish and administer an advanced base. The Intelligence Office, Shore Patrol Company Headquarters, Terminal Post Office and Legal Offices are types of components included within the A category.

After the material has been procured, it is necessary to preserve, store, pack, mark and assemble it in the component form. The location of the assembly point of any particular component is determined by the amount and type of material involved, the type of preservation required, and the prospective time of employment and ultimate destination of the material. The material the Naval Facilities Engineering Command contributes to the components is a large percentage of all the material required for the base. Because of the size and bulk of this material, it is ordinarily handled and processed near the shipping point. The bills of material will provide the Commanding Officer of a base with a list of the material furnished, package numbers, and echeloning codes, so that he will know exactly what is packaged and how it can be located.

The components must be continually reviewed in order to ensure that they are compatible with current weapons systems and operational systems. Constant effort is applied to simplify each component and reduce its weight and volume by substitution of new material and methods. A description of each component is given in OPNAV INST. 4040.22, Table of Advanced Base Functional Components. This catalog is a one-volume publication issued by the Chief of Naval Operations. It is a general summary and reference, and gives a brief description of each functional component, its personnel complement, and material requirements in broad terms and the approximate weight and cube. The catalog is a ready reference for use by area commanders and their planning staffs. It is designed as a tool for high level planning and for that reason indicates, in broad terms only, the personnel and equipment in the components.

For procurement, specifications are further broken down in the Advanced Base Initial Outfitting List (Detailed), on EAM cards in the several controlling systems commands. For the Naval Facilities Engineering Command, this information is printed and distributed in NAVFAC P-103, the "Detailed Catalog of NAVFAC Equipment and Material Requirements for Advanced Base Functional Components."

A revised collection of the more commonly used advanced base drawings is issued by NAVFAC as a planning aid in NAVFAC P-140, "Advanced Base Drawings." The drawings themselves are schematics, designed to provide field personnel with sufficient information to guide them in laying out basic structures and equipment; the bills of material included, indicate the quantity of material required for the installation. In meeting specific field conditions, it will sometimes be necessary to deviate from these plans as drawn, but major deviations will call for adjustments in the bills of material.

The Army has also published technical manuals describing the functional components that they have developed. They are: Tables of Engineer Functional Components (TM-5-301), Drawings of Engineer Functional Components (TM-5-302), and Bills of Material of Engineer Functional Components (TM-5-303).

3. Tailoring Components. Functional Components are designed and accumulated to meet the needs of typical bases, therefore, many requisitions can adequately be filled by selecting the proper functional components without variations. However, modification of a functional component, or the addition of other elements, may be necessary. This means that some functional components may have to be tailored to meet the needs arising at a specific base. This tailoring involves deletions or additions that will change the character or the amount of the support provided. Planners should realize that because components have been developed in terms of typical needs, there may be individual situations where considerable tailoring will have to be done to fit a component to the needs of a particular base.

Because of the overall need for conserving national resources, it is vital that each base plan use for only the functional components necessary for the accomplishment of its mission The availability of trained manpower for immediate use is so limited that the planner cannot waste it in handling nonessential material or constructing nonessential facilities. In planning it is necessary to remember that the unit will be more mobile if the functional components have been tailored and the excess deleted.

Good judgment is the key to successful tailoring of the functional components. The following procedures have been found useful in the past in the matter of tailoring:

a. Outline the mission of the part of the facility being planned.

b. Break this mission into its logical component tasks or elements.

c. Check each of these tasks for its essential need to support the mission.

d. Review the site data covering climatic conditions, topography, operational needs, and the possible use of local labor and material, and existing facilities. This should include an analysis of the proximity of good anchorage, suitable rail and road access, ample water supply and appropriate raw materials such as rock and lumber.

e. After eliminating those functions which can be satisfied through the use of existing facilities and local resources, compare the remaining operational items with the assigned functional components.

f. Review alternate methods of construction.

g. After deciding on those components which are still needed, compare the capacity and capability of each component and the aggregate of the components, to support the anticipated base mission and population.

h. Add or subtract individual line items to adjust the components to the specific base.

i. Review the proposed grouping or dispersal of the components at their final locations to determine the type and extent of the utilities systems.

After the tentative list of necessary functional components is prepared, other items are added to the list. These are not normally contained in the functional components but may be needed to improve part of the facilities of the base or to handle special construction jobs An example of such additional equipment or materials that may be needed are cement, lumber, asphalt, steel and timber piling, pipelines and prefabricated bridges. More specific and detailed bills of material will be prepared for any special items, with the weight and cubic measurement listed. This is most important since the cargo planners will need to include this information at the time of shipment.

4. Echelon Shipment. Advanced base units and assemblies may be so large, running into many shiploads, that assembly and movement by echelon may be necessary. Care must be taken to

see that each echelon has only the personnel and material necessary for a particular stage of the installation. In this manner the people and the equipment at the base are not tied up by the work of handling and caring for items required much later in the development. Three principles to be followed in ordering units to be sent into an advanced base are:

a. Provisions should be made for early arrival of cargo handling battalions so that ships can be unloaded with a minimum of delay.

b. Arrival dates of other components should be as close to their operational date as possible.

c. All units reaching the advanced base site in the early period (during the first 90 days) should be self-sufficient both in equipment and the ability to build, modify and maintain their own initial housekeeping establishment. No unit should be separated from the equipment it needs to care for itself in the first few weeks at the base.

In general, an echelon code is assigned to all items listed in the detailed catalogs. These codes are: the construction phase (code 1), the fitting out phase (code 2), the operational phase requiring repair parts (code 3), and the maintenance phase (code 4). The assignment by the systems commands of an echelon code to an item does not predetermine the sequence of shipment but under this system only like coded items are packed together.

The construction schedule and echelon planning are so closely related that without coordination and careful planning the development of the base will be greatly handicapped. For example, construction materials must arrive before they are needed on the project but should not arrive before the facilities to off-load and store them have been established. The flow of material to the base is also dependent upon the availability of shipping space. On arrival at the base, the functional component material is off-loaded according to the prescribed unloading priority set up in the base development plan. Buildup ashore will then proceed in accordance with the supporting base construction plan as prepared by the commander of the construction forces, either as a part of the area commander's operational plan or as an independent construction OPORD.

5. Use of the System. The functional component system:

a. Enables planners to select preplanned packages when planning advanced base construction or expansion, thereby eliminating the need to determine every "nut and bolt" each time a base is being developed.

b. Simplifies accurate estimates of men and material requirements and provides con-trols on the quantities of material in the area of operation and in the supply pipeline.

c. Makes possible accurate forecasts of resource availability at all levels of command.

d. Provides check-off lists for proper scheduling of requirements.

e. Enables the Armed Forces to attain great strategic mobility economically by making it unnecessary to build and maintain (vulnerable) permanent bases in all potentially dangerous areas.

f. Assists procurement agencies to establish priorities and maintain optimum operating procedures.

Even though no two bases ever had exactly the same characteristics, base development was greatly expedited by the formulation of comprehensive and complete standards, with deviations anticipated and permissible for the specific bases. This theory of standards proved correct because revisions, corrections, additions, or deletions were easier than the preparation of independent lists for each individual base.

This method was not as wasteful of material, equipment, and personnel as it might seem. The requirements for particular assemblies and subassemblies at many bases regardless of type, are identical, and it is possible to design them for use almost anywhere. Fundamentally, under the Functional Component System, the problem is not to calculate all the needs of an advanced base, since the broad requirements have already been determined, but to find the differences and tailor available units to fit the local situation. It should be remembered that an advanced base will probably never be composed of functional components exactly as they are listed in the catalog. This system offers the speediest means for logistical planning, and eliminates the needless repetition of calculations which would be necessary if each base were designed separately.

6. System Review and Updating. One of the major contributions of the STINGER System will be review/updating of the NAVFAC dominant ABFCs. This will be accomplished by the SSEO. The SSEO will receive input from the field as to the acceptability from a user/customer basis of existing ABFCs. Further the SSEO will review the current engineering status of the art and incorporate the latest advances in technology under the ABFC System. The need for this type of review was illustrated drastically during the early phases of the Vietnam building. Medical technology as exemplified by hospital ABFCs was discovered to be 20-30 years behind the current state of the art.

Annex D
STANDARD FORMAT FOR OPERATION ORDERS

An operation plan (OPLAN) is generally designed for operation extending over a considerable period of time. It is a detailed statement of the course of action to be followed to accomplish a prescribed or anticipated mission. It is the formal statement issued by a commander to subordinate commanders outlining their area of responsibility. The OPLAN is prepared well in advance of anticipated operations and is based upon stated assumptions.

An operation order (OPORD) is a directive issued by a commander to his subordinates for the purpose of achieving coordinated execution of an operation and serves as an order to conduct the operation. Unless otherwise stated, it is effective from the date and time it is signed by the commander. An OPLAN becomes an OPORD when it is ordered executed by the commander.

An order may be complete or fragmentary. A complete order covers all essential aspects of the operation and includes the mission of all subordinate units. A fragmentary order consists of separate instructions to one or more units prescribing the part each is to play in the operation. Fragmentary orders are issued when speed in delivery and execution is imperative and are used often to direct operations as decisions are made or as situations develop. Orders may be written or oral. They may be transmitted by message, letter, signal or messenger. Oral orders are usually followed by written confirmation.

The basic format for an OPORD consists of a heading and a body containing the task organization and five basic paragraphs accompanied by the required annexes, appendices and tabs.

The heading contains the security classification, a statement concerning change from oral orders, copy number (hand-written), issuing headquarters, the place of issue, date and time of issue, file notation, title and serial number of the order, references (maps, charts, and photomaps), and the time zone to be used throughout the order. If used, the code name of the operation will follow, and be on the same line as the operation title and number.

The body contains the task organization and five main numbered paragraphs. The five paragraphs cover the following topics: situation, mission, execution, administration and logistics, and command and communications. SMEAC, a simple code word formed by utilizing the first letter of each topic is often used

as a memory aid. The main topics of an operation order must be covered whether the order is from a battalion commander, platoon commander, squad leader, or a fire team leader. Naturally, battalion operation orders are quite lengthy whereas a patrol leader's order is usually brief. A format of a patrol leader's order is shown in Figure 106. A patrol leader's order is usually oral, with the patrol members taking notes.

The task organization includes the task subdivisions or tactical components which comprise the command together with the names and grades of the commanders. Units in support are shown under the headquarters of the major unit which commands them, not under the headquarters that the unit supports. Attached units are shown under the headquarters of the unit to which attached.

Paragraph 1, Situation: This paragraph contains background information plus information regarding enemy forces, friendly forces and attachments and detachments. It does not include plans or instructions. In OPLANS the assumptions made by the commander in developing the plan are stated here.

Paragraph 2, Mission: A clear, concise statement in strong and aggressive language telling the unit exactly what it is to do and the details of time, place and purpose (if appropriate) are included. This paragraph is written in the sequence the mission is to be accomplished. Subparagraphs are never used in this paragraph.

Paragraph 3, Execution: This paragraph summarizes the concept of operation, describes the course of action to be followed by the entire unit, assigns specific tasks to each subordinate unit and in the last sub-paragraph provides detailed instructions for coordinating the actions of these subordinate units.

Paragraph 4, Administration and Logistics: This paragraph sets forth personnel, logistical and administrative instructions covering such details as evacuation, hospitalization and transportation. If the material to be covered is lengthy, some or all of it would be included in an administrative annex which would be referenced in this paragraph.

Paragraph 5, Command and Communications: This paragraph gives the necessary information regarding the unit's operational control and other facets of command such as command post location and relationship with other military and military construction organizations. Communications information and liaison instruc-

```
                          FIGURE 106
               FORMAT FOR A PATROL LEADER'S ORDER


  1.  SITUATION
      a.  Enemy Forces: Weather, terrain, identification, location, activity, strength.
      b.  Friendly Forces: Mission of next higher unit, location and planned actions of
          units on right and left, fire support available for patrol, mission and route
          of other patrols.
      c.  Attachments and Detachments.

  2.  MISSION - What the patrol is going to accomplish.

  3.  EXECUTION - (Subparagraph for each subordinate unit.)
      a.  Concept of operation.
      b.  Specific duties of elements, teams, and individuals.
      c.  Coordinating instructions.
          (1)  Time of departure and return.
          (2)  Formation and order of movement.
          (3)  Route and alternate route of return.
          (4)  Departure and reentry of friendly area(s).
          (5)  Rallying points and actions at rallying points.
          (6)  Actions on enemy contact.
          (7)  Actions at danger areas.
          (8)  Actions at objective.
          (9)  Rehearsals and inspections.
         (10)  Debriefing.

  4.  ADMINISTRATION AND LOGISTICS
      a.  Rations.
      b.  Uniforms and Equipment (state which members will carry and use).
      c.  Arms and ammunition.
      d.  Method of handling wounded and prisoners.

  5.  COMMAND AND COMMUNICATIONS
      a.  Signal.
          (1)  Signals to be used within the patrol.
          (2)  Communication with higher headquarters - radio call signs, primary and
               alternate frequencies, time to report and special code to be used.
          (3)  Challenge and password.
      b.  Command
          (1)  Chain of Command
          (2)  Location of patrol leader and assistant patrol leader in formation.
```

tions are also given in this paragraph if necessary.

Annexes to an OPORD amplify and supplement the basic order and identified by upper case letters. Annexes discuss broad topics generally too lengthy to be placed in the basic order. The format should be patterned after the basic operation order when possible but this is not a rigid requirement (See Figure 107). Some examples of annexes most commonly used by SEABEE units are:

Organization of Command Relationships
Concept of Operations
Intelligence
Training
Engineer Tasks
Communications
Logistics
Administration and Personnel

Safety
Medical and Dental
Disaster Control
Public Information

Appendices supplement annexes and are identified by Roman numerals. Appendices discuss portions of annexes which readily lend themselves to separate discussions and provide clarity or emphasis. For example, Allowance Equipment could be an appendix to the Logistics Annex. The appendices for each annex are always consecutively numbered beginning with Roman numeral I in each annex. Appendices need not be the same in each OPORD although this is desirable in most instances. Appendix page number are preceded by the annex letter and the appendix Roman numeral; for example, A-I-1, A-I-2, A-I-3. There is no specified format for appendices but they may also be arranged in the five paragraph format.

Tabs supplement appendices and will be identified by uppercase letters. Tabs discuss portions of appendices which readily lend themselves to separate discussion and improve clarity or provide emphasis. For example, allowance equipment to be returned to homeport for overhaul at the end of the deployment could be a tab to an appendix on Equipment in general.

Tabs are always alphabetically lettered beginning with the letter A in each appendix. Tabs need not be the same in each OPORD although this is also desirable. Tab page numbers are preceded by the annex letter, the appendix Roman numeral, and the tab letter; for example, A-I-A-1, A-I-A-2, A-I-A-3. Tabs may be provided in any format.

FIGURE 107
TYPICAL ANNEX TO A BATTALION OPERATION ORDER

(Classification)

--(Copy Number)--
--(Issuing Battalion)--
--(Place of Issue)--
--(Date/time of Issue)--

OPERATION ORDER
MCB-SIX OPORD 210-67

ANNEX D
ENGINEER TASKS

Ref: (a) ---
       (b) ---
Time Zone: ---

TASK ORGANIZATION. (Assignments of projects to elements of the battalion; size and composition of crews, attachments or detachments are shown here or in appendices.)

1. SITUATION. (Any items that have been omitted elsewhere and that are pertinent only to the construction tasks should be listed here.)

2. MISSION. (A one paragraph statement of the construction mission to be accomplished by the battalion in executing the operation.)

3. EXECUTION. (Includes instructions to specific company commanders and coordinating instructions that involve more than one project or support function.)

4. ADMINISTRATION AND LOGISTICS. (This includes supplies, facilities and services available; who is responsible for providing them; and how, when, and where they will be provided. The basic OPORD, separate orders or standard battalion operating procedures or instructions may be referenced here.)

5. COMMAND AND COMMUNICATIONS. (Instructions necessary for the exercise of control during the construction phase, location of command posts, etc., are shown here. The communications annex may be referenced.)

--(Signature)--
--(Name)--
--(Rank and Service)--
--(Command Title)--

APPENDICES (Typical examples)

   I - Road Plan
  II - Extracts from Base Development Plan
 III - Construction Priorities
  IV - Facilities for Fuel Storage
   V - Control of Project Material

(Classification)

D-1

199

Annex E
PERTINENT ARTICLES OF NAVAL REGULATIONS

A. THE COMMANDING OFFICER

Article 0701. Responsibility of the Commanding Officer.

1. The responsibility of the Commanding Officer for his command is absolute, except when, and to the extent, relieved therefrom by competent authority, or as provided otherwise in these regulations. The authority of the Commanding Officer is commensurate with his responsibility, subject to the limitations prescribed by law and these regulations. While he may, at his discretion, and when not contrary to law or regulations, delegate authority to his subordinates for the execution of details, such delegation of authority shall in no way relieve the Commanding Officer of his continued responsibility for the safety, well-being and efficiency of his entire command.

2. A Commanding Officer who departs from his orders or instructions, or takes official action which is not in accordance with such orders or instructions, does so upon his own responsibility and shall report immediately the circumstances to the office from whom the prior orders or instructions were received.

Article 0703. Relationships with Executive Officer.

The Commanding Officer shall keep the Executive Officer informed of his policies and normally shall issue all orders relative to the duties of the command through that officer.

Article 0704. Effectiveness for Service.

The Commanding Officer shall:

1. Exert every effort to maintain his command in a state of maximum effectiveness for war service consistent with the degree of readiness prescribed by proper authority.

2. Reporting to his appropriate senior any deficiency which appreciably lessens the effectiveness of the command.

3. Report, with his recommendations, to the bureau or office concerned whenever, in his opinion, his authorized allowances of personnel or material exceed or fall short of requirements.

Article 0708. Inspections, Musters, and Sighting of Personnel.

1. The Commanding Officer shall, when circumstances permit, hold an inspection of personnel and material of the command on such day of each week, except Sunday, as may be most expedient. The Commanding Officer shall insure that, consistent with their employment, the personnel under his command present at all times a neat, clean, and military appearance.

2. Aboard ship and, where appropriate, in other commands, the Commanding Officer shall require that, under ordinary circumstances, quarters for inspection are held daily. Officers not attached to divisions shall report their presence at quarters in person.

3. The Commanding Officer shall require a daily report of all persons confined, a statement of their offenses, and the dates of their confinement and release.

4. A muster of all persons attached to the command shall be held daily. Persons who have not been sighted by a responsible senior shall be reported as absent.

Article 0709. Welfare of Personnel.

The Commanding Officer shall:

1. Use all proper means to promote the morale, and to preserve the moral and spiritual well-being of the personnel under his command.

2. Endeavor to maintain a satisfactory state of health and physical fitness in the personnel under his command.

3. Afford an opportunity, with reasonable restrictions as to time and place, for the personnel under his command to make requests, reports, or statements to him, and shall insure that they understand the procedures for making such requests, reports or statements.

4. Insure that noteworthy performance of duty of personnel under his command receive timely and appropriate recognition and that suitable notations are entered in the official records of the individuals.

5. Insure that timely advancement in

201

ratings of enlisted personnel is effected in accordance with existing instructions.

Article 0710. Training and Education.

The Commanding Officer shall:

1. Endeavor to increase the specialized and general professional knowledge of the personnel under his command by the frequent conduct of drills, classes, and instruction, and by the utilization of appropriate fleet and service schools.

2. Encourage and provide assistance and facilities to the personnel under his command who seek to further their education in professional or other subjects.

3. Require those lieutenants (junior grade) and first lieutenants who have less than two years commissioned or warrant service, and all ensigns and second lieutenants:

   a. To comply with the provisions prescribed for their instruction by the Chief of Naval Personnel, the Commandant of the Marine Corps, or the Chiefs of other appropriate bureaus.

   b. To keep journals, to attend classes, and to receive appropriate practical instruction, as the Commanding Officer deems advisable.

4. Detail the officers referred to in paragraph 3 of this article to as many duties successively as practicable. This rotation of duties should be completed during the first two years of the officer's commissioned service. The Commanding Officer shall indicate on the fitness report of each such officer the duties to which he has been assigned, the total period of assignment, and the degree of qualification in such duties.

5. Designate a senior officer or officers to act as advisers to the officers referred to in paragraph 3 of this article. These senior officers shall assist such junior officers to a proper understanding of their responsibilities and duties, and shall endeavor to cultivate in them officer-like qualities, a sense of loyalty and honor, and an appreciation of naval customs and professional ethics.

Article 0712. Safety Precautions.

The Commanding Officer shall require that persons concerned are instructed and drilled in all applicable safety precautions and procedures, that these are complied with, and that applicable safety precautions, or extracts therefrom, are posted in appropriate places. In any instance where safety precautions have not been issued or are incomplete, he shall issue or augment such safety precau-

tions as he deems necessary, notifying when appropriate, higher authority concerned.

Article 0717. Economy Within the Command.

The Commanding Officer shall be responsible for economy within his command. To this end he shall require from his subordinates a rigid compliance with the regulations governing the receipt and expenditures of public money and stores, and the accounting therefor.

Article 0740. Code of Conduct for Members of The Armed Forces of The United States.

1. The Code of Conduct for Members of the Armed Forces of the United States shall be carefully explained to each enlisted person:

   a. Within six days of his initial enlistment.

   b. After completion of six month's active service, and

   c. Upon the occasion of each reenlistment.

2. Instruction in the Code of Conduct for Members of the Armed Forces of the United States shall be included in the training and educational program of the command.

3. A text of the Code of Conduct for Members of the Armed Forces of the United States shall be posted in a conspicuous place, or in conspicuous places, readily accessible to the personnel of the command.

Article 1244. Direct Communication with the Commanding Officer.

1. The right of any person in the naval service to communicate with the Commanding Officer at a proper time and place is not to be denied or restricted.

2. Officers who are senior to the Executive Officer have the right to communicate directly with the Commanding Officer.

3. A head of department, or of any other major subdivision of an activity, has the right to communicate directly with the Commanding Officer concerning any matter relating to his department or subdivision, but, in such cases, shall keep the Executive Officer informed.

B. THE EXECUTIVE OFFICER

Article 0801. Status, Authority and Responsibility.

1. The status, authority, and responsibility of the Executive Officer is derived from Title 10, United States Code, paragraph

5953 which is quoted herewith: "(a) The Secretary of the Navy may detail a line officer of the Navy as Executive Officer of a vessel or a naval station. When practicable, the officer so detailed shall be one who is next in rank to the Commanding Officer. (b) While executing the orders of the Commanding Officer, the Executive Officer takes precedence over all officers attached or assigned to the vessel or station, and his orders are considered as coming from the Commanding Officer. However, the Executive Officer has no independent authority by reason of his detail as such an officer. (c) Any officer in a staff corps who is attached or assigned to a vessel or naval station and who is senior to the Executive Officer of that vessel or station may communicate directly with the Commanding Officer."

2. The Executive Officer is the direct representative of the Commanding Officer. All orders issued by him as such representative shall have the same force and effect as though issued by the Commanding Officer, and shall be obeyed accordingly by all persons within the command. In the performance of his duties, he shall conform to and effectuate the policies and orders of the Commanding Officer and shall keep him informed of all significant matters pertaining to the command. The Executive Officer shall be primarily responsible, under the Commanding Officer, for the organization, performance of duty, and good order and discipline of the entire command. He shall recognize the right and duty of a head of department to confer directly with the Commanding Officer on matters specifically relating to his department.

3. An officer who is acting as Executive Officer during the temporary absence or disability of that officer shall have the same authority and responsibility as the Executive Officer, but he shall make no change in the existing organization, unless ordered to do so, and shall endeavor to have the routine and other affairs of the command carried on in the usual manner.

Article 0802. Assuming Command.

The Executive Officer shall be prepared to assume command should the need arise. During action, he shall be stationed where he can best aid the Commanding Officer and, if practicable, where he would probably escape the effects of a casualty disabling the Commanding Officer, and yet would be able to assume command promptly and effectively.

Article 0803. Specific Duties.

The Executive Officer, subject to the orders of the Commanding Officer and assisted by the appropriate subordinate shall:

1. Prepare and maintain the bills and

orders for the organization and administration of the command as a whole.

2. Assign the personnel to departments or other major subdivisions of the command.

3. Maintain the records of personnel, except those records assigned as the responsibility of some other officer.

4. Supervise and coordinate the work, exercises, training and education of the personnel of the command.

5. Supervise and coordinate the operational plans and schedules of the command as a whole.

6. Prepare and promulgate, as appropriate, a daily schedule of employment, and such other advance schedules as may serve to aid subordinates in planning their work.

7. Make frequent inspections, in company, when practicable with the subordinates concerned; and take such remedial action to correct defects as appears necessary.

8. Insure that all prescribed or necessary security measures and safety precautions are understood and strictly observed.

9. Maintain high morale within the command. The discipline, welfare, and privileges of the individuals of the command shall be a chief concern of the Executive Officer, and he shall, to the extent of his authority, insure that these and related matters are administered in a just and uniform manner.

10. Perform such other duties as may be assigned.

C. COMPANY COMMANDERS AND DEPARTMENT HEADS.

Article 0901. Status, Authority and Responsibility.

The head of a department of a command or other activity is the officer detailed as such by competent authority. He is the representative of the Commanding Officer in all matters that pertain to the department. All persons assigned to the department shall be subordinate to him and all orders issued by him shall be obeyed accordingly by them. In the performance of his duties as a head of department, he shall conform to the policies and comply with the orders of the Commanding Officer.

Article 0902. Direct Communication with Commanding Officer.

1. The head of a department shall confer directly with the Commanding Officer concerning any matters relating to his department whenever he believes such action to be neces-

sary for the good of his department or of the naval service.

2. He shall keep the Commanding Officer informed as to the general condition of all machinery and other installations of his department, and especially of any circumstance or condition which may adversely affect the safety or operation of the command. He shall inform the Commanding Officer of the need for and the progress of repairs other than those of a minor nature.

3. He shall not disable the machinery or equipment for which he is responsible, when such action may affect adversely the safety or operation of the command, without permission of the Commanding Officer.

4. The Executive Officer shall be kept appropriately informed of all matters described in this article.

Article 0903. Specific Duties.

The head of a department, subject to the orders of the Commanding Officer, shall:

1. Organize and train his department to insure readiness for battle.

2. Prepare and maintain the bills and orders for the organization and operation of the department.

3. Assign the personnel to stations and duties within the department.

4. Be responsible for the effectiveness of the department, and to this end he shall plan, direct, and supervise the work and training of personnel within the department.

5. Insure that all prescribed or necessary security measures and safety precautions are strictly observed by all persons within the department and by others who may be concerned with matters under his control. He shall insure that all applicable safety precautions are kept properly posted, in conspicuous and accessible places, and that the personnel concerned are frequently and thoroughly instructed and drilled in their observance.

6. Make frequent inspections of the personnel and material of the department, including the spaces assigned thereto, and take necessary actions to correct defects and deficiencies. In ships, and in other commands as directed by the Commanding Officer, each head of department, or his representative, *shall each day inspect and report the condition* of the department to the Executive Officer, who shall make a similar report to the Commanding Officer. These inspections and reports normally shall be made between 1800 and 2000.

7. Control the expenditures of funds alloted, and operate the department within the limit of such funds.

8. Insure economy in the use of public money and stores.

9. Be responsible for the proper operation, care, preservation, and maintenance of the equipment and other material assigned to the department, and for the submission of such data in connection with the accounting therefor, including periodic inventories of assigned material, as may be prescribed by competent authority.

10. Be responsible for the maintenance of records and the submission of reports of the department.

11. Be the custodian of the keys to all spaces and storerooms of the department, except such as are assigned by regulation to the custody of another officer. He may designate and authorize subordinates within the department, as necessary, to have duplicates of such keys.

12. Be responsible for the cleanliness and upkeep of the spaces assigned to the department except as prescribed by regulation or other competent authority.

13. Anticipate the personnel and material needs of the department, and submit timely requests to fulfill requirements.

14. Contribute to the coordination of effort of the entire command by appropriate cooperation with other heads of departments.

15. Perform such other duties as may be assigned.

D. PLATOON LEADERS

Article 1044. Responsibilities and Duties.

1. A division officer shall be responsible, under the head of his department, for the proper performance of the duties assigned to his division, and for the conduct and appearance of his subordinates, in accordance with regulations and the orders of the Commanding Officer and other superiors. He shall keep himself informed of the capabilities and needs of each of his subordinates and, within his authority, he shall take such action as may be necessary for the efficiency of his division and the welfare and morale of his subordinates. He shall train his subordinates in their own duties and in the duties to which they may succeed, and shall encourage them to qualify for advancement and to improve their education He shall suppress any improper language or unseemly noise or disturbance, and he shall re-

port to the Executive Officer all infractions of regulations, orders, and instructions which are deserving of disciplinary action.

2. He shall, by personal supervision and frequent inspection, insure that the spaces, equipment, and supplies assigned to his division are maintained in a satisfactory state of cleanliness and preservation. He shall report promptly to his head of department any repairs which may be required or other defects which need correction and which he is unable to effect.

3. He shall carefully instruct his subordinates in all applicable safety precautions and shall require their strict observance.

4. He shall maintain a corrected copy of the watch, quarter, and station bill and other bills and orders for his division and shall insure that pertinent parts thereof are kept posted where they will be accessible to his subordinates.

E. THE DUTY OFFICER

Article 1001. Definition of Watch Officers.

A watch officer, within the meaning of these regulations, is one regularly assigned to duty in charge of a watch or of a portion thereof.

Article 1003. Assignment of Watch Officers.

1. Subject to such restrictions as may be imposed by a senior in the chain of command, or by these regulations, a Commanding Officer may assign to duty in charge of a watch, or to stand a day's duty, any commissioned or warrant officer who is subject to his authority and who is, in the opinion of the Commanding Officer, qualified for such duty.

2. Marine officers below the grade of major may be assigned to duty as officers of the deck in port. Those Marine Officers on the junior watch list may stand junior officer watch at sea.

3. At times when the number of commissioned or warrant officers qualified for watch standing is reduced to an extent which may interfere with the proper operation of the command or may cause undue hardship, the Commanding Officer may assign to duty in charge of a watch, or to stand a day's duty, subject to such restrictions as may be imposed by a senior

in the chain of command, or by these regulations, any petty officer or non-commissioned officer who is subject to his authority and who is, in the opinion of the Commanding Officer, qualified for such duty.

Article 1005. General Duties of Watch Officers.

1. An officer in charge of a watch shall be responsible for the proper performance of all duties prescribed for his watch, and all persons on watch under him shall be subject to his orders.

2. He shall remain in charge and at his station until regularly relieved. He shall scrupulously obey all orders and regulations and shall require the same of all persons on watch under him. He shall instruct them as may be necessary in the performance of their duties, and shall insure that they are at their stations, attentive, alert, and ready for duty. He shall endeavor to foresee situations which may arise, and shall take such timely and remedial action as may be required.

3. At all times he shall present and conduct himself in a manner befitting his office. His orders shall be issued in the customary phraseology of the service.

4. He shall promptly inform the appropriate persons of matters pertaining to his watch what they should know for the proper performance of their duties.

5. Before relieving, he shall thoroughly acquaint himself with all matters which he should know for the proper performance of his duties while on watch. He may decline to relieve his predecessor should any circumstance or situation exist which, in his opinion, justifies such action by him, until he has reported the facts to and received orders from the Commanding Officer, or other competent authority.

6. Any officer standing a day's duty shall insure, by personal attendance and frequent inspections, that the duties prescribed by these regulations or by the Commanding Officer are properly performed.

Article 1213. Exchange of Duty.

No person in the naval service shall exchange an assigned duty with another without permission from his Commanding Officer or other superior.

PROCEDURE FOR CAPTAIN'S MAST

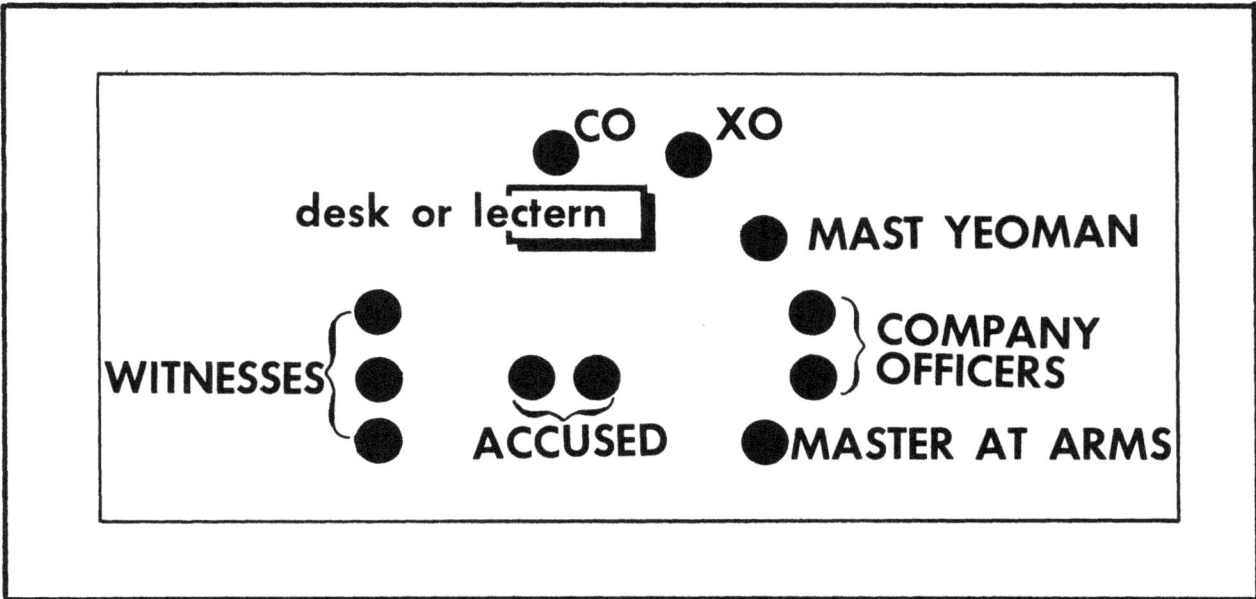

Captain's mast is normally held in a quiet place. The surroundings should provide privacy and reflect an atmosphere of court decorum. The proceedings are carried out with the solemnity that prevails during a courts-martial. The Commanding Officer stands behind a desk or lectern with head covered. The Executive Officer stands nearby, also with head covered. The accused stands with head bare. If there are witnesses present, they stand at right angles to the accused. If company commanders are present, they may also stand at right angles to the accused, on the opposite side of the room from the witnesses. The master-at-arms is usually near the door. A suggested mast layout is shown above. It is not necessary that all of the accused be in the room at the same time. An alternate procedure is to have the accused wait outside the mast room and enter one at a time as directed by the MAA.

The Commanding Officer has the charge sheet before him. Either the Commanding Officer or the Executive Officer has the service record of each accused person. Before inquiring into the facts in the case, the Commanding Officer warns the accused that they are not required to incriminate themselves. He may give this warning to the entire group before taking up any case or may warn each accused individually as his case is taken up. The following is a paraphrased warning that may be used:

Commanding Officer (addressing the accused as a group):

"All of you are accused of committing an offense under the Uniform Code of Military Justice. It is my purpose to inquire into the facts surrounding the offenses of which you stand accused.

Before any case is called, I want to inform and warn everyone here (including you witnesses) of your right to refuse to answer any question or make any statement which incriminates you or may tend to incriminate you.

I want to hear both sides of the story so that I may intelligently decide how to dispose of your cases. That does not mean, however, that you will or can be compelled to answer any questions or make any statement regarding the offense of which you are accused, or which you feel may involve you in another offense. You should thoroughly understand this right to remain silent because any testimony voluntarily given may be used against you if a trial by courts-martial develops.

Do you understand? Is there anyone who does not understand?"

If anyone indicates a lack of understanding, the Commanding Officer must explain in detail the meaning of his statements.

The Commanding Officer then continues as follows:

"This hearing will be conducted in the following manner: First, I will advise you of the nature of the offense of which you are accused.

Second, I will interrogate your accuser (if there be one) both to bring out the facts and to inform you further of the details of the charge against you.

Third, you will have an opportunity to explain or defend your conduct in your own words, and, if you desire, by calling witnesses in your behalf.

After I have heard all the evidence, I will decide whether your individual cases can be handled here at mast, or whether they must be referred to courts-martial for trial."

The Executive Officer then hands the Commanding Officer a report slip saying, "The first (or next) case is the case of ———————" (last name of accused). The Commanding Officer then reads the report slip, which alleges an offense. If there is an accuser, the Commanding Officer asks him to state his version of the facts in the case. Any witnesses are then asked to testify. The Commanding Officer may ask the accuser or witnesses any questions that appear pertinent.

The accused is then given an opportunity to state his version of the case. He is again warned that he is not required to make a statement, and that anything that he does say may be used against him in a later proceeding.

After the accused makes a statement (or declines to make a statement), the Commanding Officer then may ask the company commander of the accused if he would like to make a statement. This officer then has the opportunity to make any remarks about the accused that appear pertinent. These remarks may be favorable or unfavorable depending upon the man's work and his character, as judged by the company commander.

The Commanding Officer then decides and announces the disposition of the case. He may take any of the following steps:

(1) Dismiss the case.

(2) Dismiss the case with a warning to the accused.

(3) Impose nonjudicial punishment.

(4) Award punishment, but suspend its execution.

(5) Refer the case to an officer for further investigation.

(6) Announce that the case will be referred for trial by courts-martial.

In a case where non-judicial punishment is imposed, the accused must be informed that he has the right to appeal. The Commanding Officer phrases the notice of the right of appeal in words similar to these:

"If you consider this punishment unjust or out of proportion to this offense, you may appeal through the proper channels to my superior in command. If you wish to make such an appeal Mr._____, your company commander will assist you in preparing it. Any appeal must be in writing and must be made within a reasonable length of time. This appeal will be promptly forwarded by me, but, in the meantime, you may be required to undergo the punishment awarded."

The Commanding Officer then asks the accused if he understands. Appropriate entries are made in the unit punishment book. The Commanding Officer departs with the Executive Officer after the last case is heard. After the company commanders and witnesses are dismissed the chief master-at-arms takes charge of those persons to whom punishment is awarded.

Appeals from non-judicial punishment are rare. When prepared, they must state the reasons for regarding the punishment as unjust or disproportionate. The battalion Commanding Officer, when forwarding an appeal to his next superior, may include a copy of the case record. The reviewing authority may modify the punishment or set it aside but not increase it. He returns the papers through channels to the appellant (accused) with a statement of the disposition of the case.

An announcement is usually made in the Plan of the Day of the action taken on each case at Captain's Mast.

Annex G
SEABEE RATING STRUCTURE

There are seven SEABEE Ratings included in in the Construction Group (GROUP VIII). The path of advancement for a recruit is to become a Construction Recruit CR, then a Construction Apprentice CA, then a Construction Man CN. From CN he may "strike for" any of the ratings in the Construction Group. A Construction Man performs routine or non-skilled duties in construction battalion operations. If he is a graduate of a Class A school or if he has passed the examination for third-class petty officer but was not promoted because of quota limitations, he may be designated as a striker in the rating that he has studies. Each Constructionman (CN) is qualified to:

a. Select proper hand tools to perform command and elementary tasks in carpentry, plumbing, rigging, automotive repair, and electric wiring.

b. Read simple diagrams and sketches used in construction work.

c. Serve as chainman in a survey party, using chaining pins, tapes, and range poles.

d. Operate passenger vehicles and light trucks and perform 1st echelon maintenance on them including such tasks as adding gasoline, oil and water, change and repair tires, change lamps and lamp fuses.

e. Know the safety precautions for quarry and earthmoving operations, shop work, painting or preservation tasks and handling firearms and ammunition.

f. Have sufficient knowledge of mathematics to add, subtract, multiple and divide numbers, fractions and decimals.

g. Understand the battalion organization and mission and the duties of each of the Group VIII Ratings.

h. Serve as a member of a fire team demonstrating knowledge of basic formations and field cleaning of assigned weapons.

i. Serve as a member of the interior or perimeter guard.

The Manual of Qualifications for Advancement in Rating (NAVPERS 18068) contains a description of the general ratings and a list of the practical and knowledge factors to be learned and demonstrated in order to qualify for advancement. The duties of the general ratings and the Rating NEC Codes are described on the following pages.

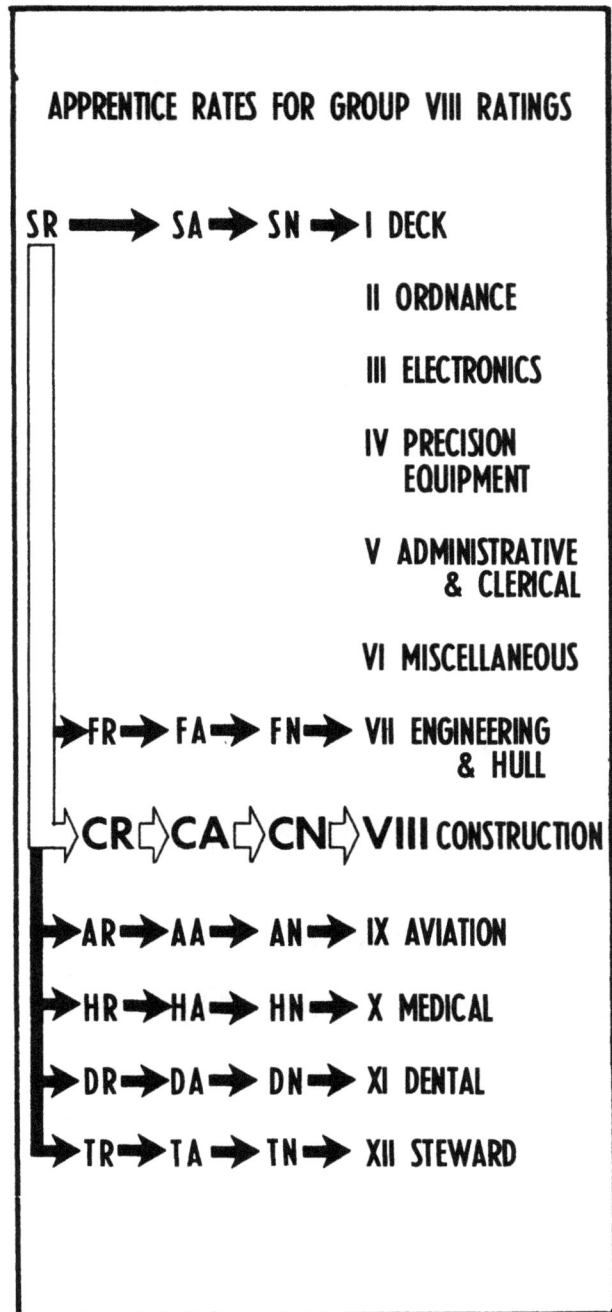

APPRENTICE RATES FOR GROUP VIII RATINGS

SR ➝ SA ➝ SN ➝ I DECK

II ORDNANCE

III ELECTRONICS

IV PRECISION EQUIPMENT

V ADMINISTRATIVE & CLERICAL

VI MISCELLANEOUS

FR ➝ FA ➝ FN ➝ VII ENGINEERING & HULL

CR ➝ CA ➝ CN ➝ VIII CONSTRUCTION

AR ➝ AA ➝ AN ➝ IX AVIATION

HR ➝ HA ➝ HN ➝ X MEDICAL

DR ➝ DA ➝ DN ➝ XI DENTAL

TR ➝ TA ➝ TN ➝ XII STEWARD

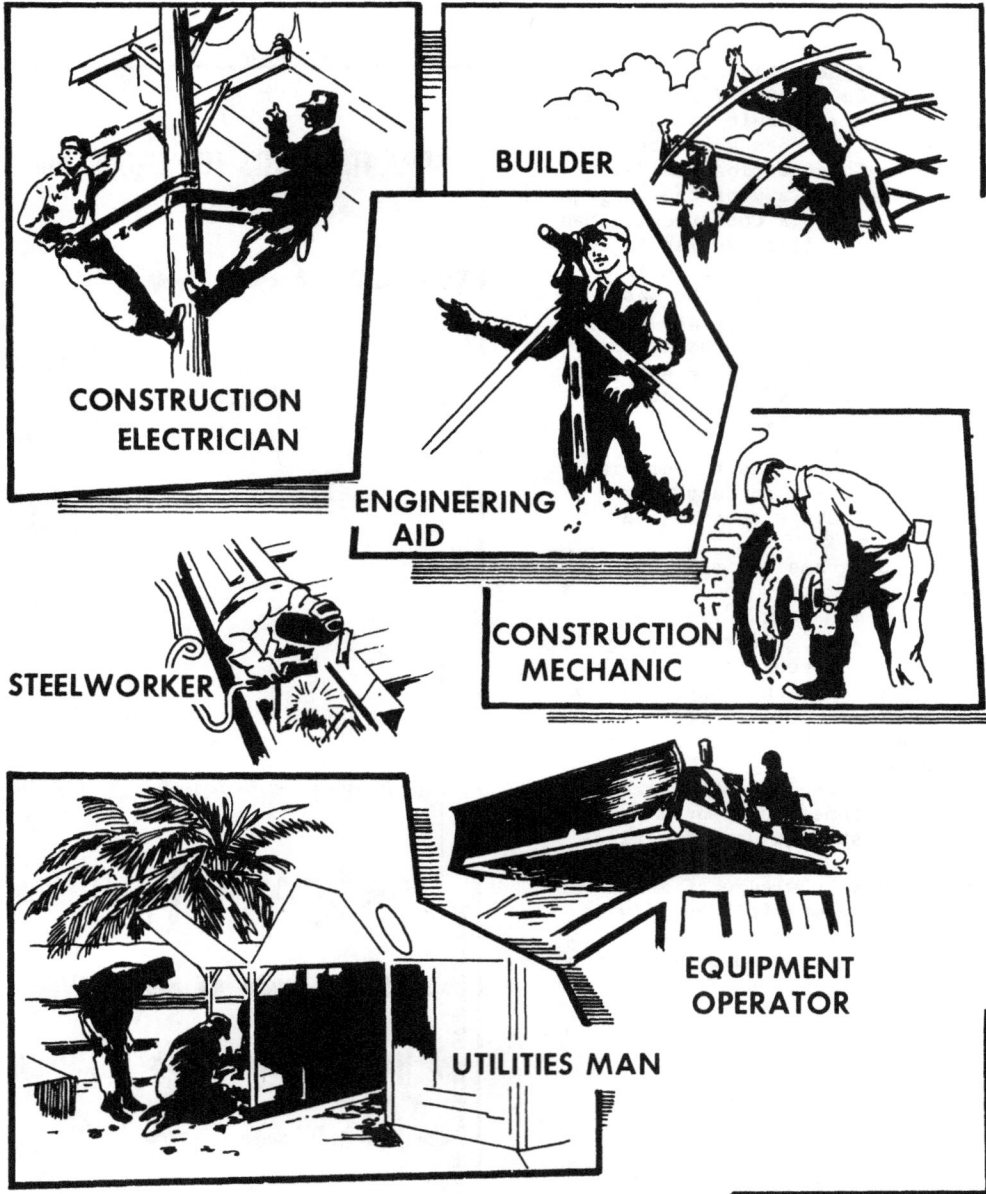

FIGURE 108
SEABEE RATINGS

CONSTRUCTION
ELECTRICIAN

BUILDER

ENGINEERING
AID

STEELWORKER

CONSTRUCTION
MECHANIC

UTILITIES MAN

EQUIPMENT
OPERATOR

BUILDER (BU)

1. CAREER PATTERNS.

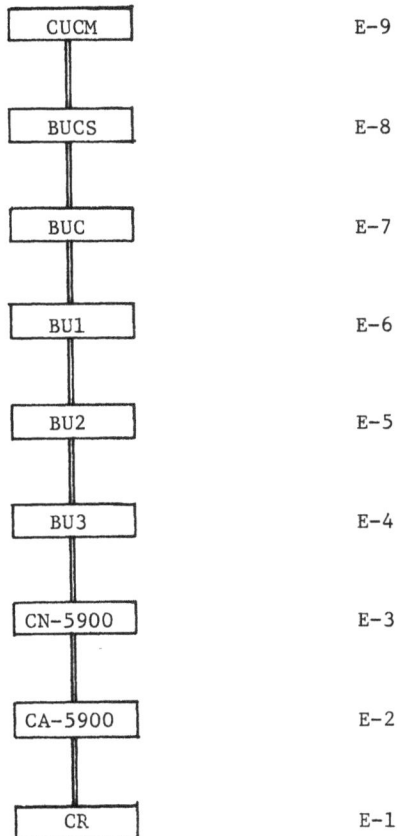

CUCM	E-9
BUCS	E-8
BUC	E-7
BU1	E-6
BU2	E-5
BU3	E-4
CN-5900	E-3
CA-5900	E-2
CR	E-1

2. GENERAL RATING (PO3 through CPO).
Builders plan, *supervise, and perform tasks re-*quired for construction, maintenance, and repair of wooden, concrete, and masonry structures, concrete pavement waterfront, and underwater structures; initiate procurement and direct storage of building materials; form and direct efforts of crews to perform rough and finish carpentry; erect and repair waterfront structures, wooden and concrete bridges and trestles; fabricate and erect forms; mix, place, and finish concrete; and lay or set masonry; paint and varnish new and refinished surfaces; and operate concrete batch plants.

3. RATING NEC CODES.

BU-5901 Concrete Worker (Source: BU, E-5 and above): Supervises the mixing, placing and finishing of concrete for foundations, slabs, walls and underwater structures: Operates concrete pipe and concrete block manufacturing and concrete batching plants; operates concrete paving machines; concrete spreaders and finishing machines used in placing concrete for roads and runways; supervises the batching and transportation of dry mix materials.

BU-5902 Mason and Tile Setter (Source: BU, E-4 and above): Mixes mortar, and lays brick, concrete block, and rubble masonry: Prepares surface, determines quantities, types and shapes, and sets ceramic tile; mixes and applies plaster and stucco.

BU-5904 Millworker (Source: BU, E-4 and above): Operates woodworking machines and handtools in the manufacturing and repair of furniture, cabinets, fabrication of doors, window sash, stairway members, molding, trim and picture framing.

BU-5906 Heavy Construction Worker (Source: BU, E-5 and above): Performs all duties related to the construction of timber trestles and towers, bridges, piers, wharfs, sheet pile and concrete bulkhead systems, cofferdams, seawalls, jetties, and breakwaters: Directs the operation of skid, barge and crane mounted pile driving rigs.

BU-5908 Carpenter Shop Maintenanceman (Source: BU, E-4 and above): Installs and maintains carpenter shop and saw mill machinery: Sharpens, repairs, sets, adjusts, installs, and removes all types of saws, blades, knives, and chisels used in carpentry; files, sets, gums, and joins all circular and band saws; files and grinds knives or blades for carpenter shop or lumber mill machines; cuts and splices band saw blades; maintains and adjusts grinders, gummers, and other shop maintenance tools and machines.

4. SENIOR AND MASTER CHIEF PETTY OFFICERS.

Senior Chief Builder (BUC). Chief Builders (BUC), eligible to participate in the E-8 examination for Senior Chief Builder (BUCS), should be knowledgeable of both the knowledge factors and the knowledge aspects of the practical factors currently required of the SW3, SW2, EA3 and EA2.

Master Chief Constructionman (CUCM). Senior Chief Builders (BUCS), eligible to participate in the E-9 examination for Master Chief Constructionman (CUCM) should be knowledgeable of both the knowledge factors and the knowledge aspects of the practical factors currently required of the SW1, SWC, EA1, and EAC.

CONSTRUCTION ELECTRICIAN (CE)

1. CAREER PATTERNS.

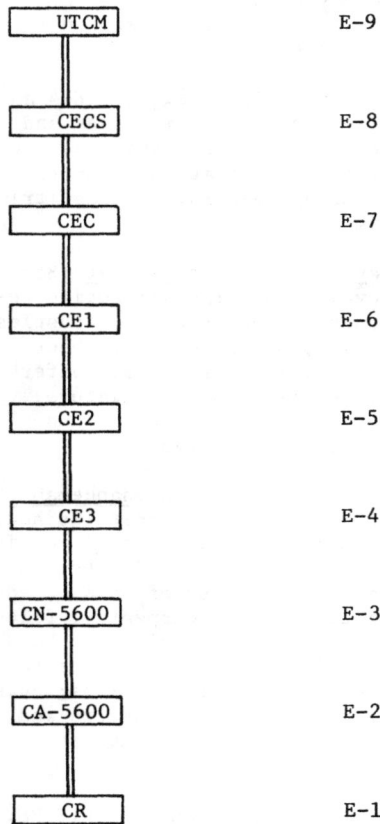

Rate	Grade
UTCM	E-9
CECS	E-8
CEC	E-7
CE1	E-6
CE2	E-5
CE3	E-4
CN-5600	E-3
CA-5600	E-2
CR	E-1

2. GENERAL RATING (PO3 through CPO).
Construction Electricians plan, supervise, and perform tasks required to install, operate, service, and overhaul electric generating and distribution systems and wire communication systems; control activities of individuals and crews who string, install, and repair interior, overhead, and underground wires and cables, and attach and service units, such as transformers, switchboards, motors, and controllers; schedule and evaluate installation and operational routines.

3. RATING NEC CODES.

CE-5632 Shore-Based Power Plant Operator/Maintenanceman (Source: CE and EM, E-5 and above): Operates, services, and repairs electrical systems of shore-based power plants.

CE-5642 Central Office Exchange Repairman (Source: CE, E-5 and above): Installs, inspects, maintains, and repairs automatic PBX or PAX telephone exchange equipment, including switchboards, dial selection mechanisms, and telephones: Performs troubleshooting operations; performs necessary adjustments and replacements; cleans and adjusts contacts and adjusts contact clearance; test circuits using spring tension gauges, voltmeters, and ohmeters, performs daily or periodic ground or insulation tests on switchboards.

CE-5644 Electrical Cable Splicer (Source: CE, E-5 and above): Splices multiple conductor communication and electric-power transmission systems.

4. SENIOR AND MASTER CHIEF PETTY OFFICERS.

Senior Chief Construction Electrician (CECS): Chief Construction Electricians (CEC), eligible to participate in the E-8 examination for Senior Chief Construction Electrician (CECS) should be knowledgeable of both the knowledge factors and the knowledge aspects of the practical factors currently required of the UT3, and UT2.

Master Chief Utilitiesman (UTCM). Senior Chief Construction Electricians (CECS) eligible to participate in the E-9 examination for Master Chief Utilitiesman (UTCM), should be knowledgeable of both the knowledge factors and the knowledge aspects of the practical factors currently required of the UT1 and UTC.

CONSTRUCTION MECHANIC (CM)

1. CAREER PATTERNS:

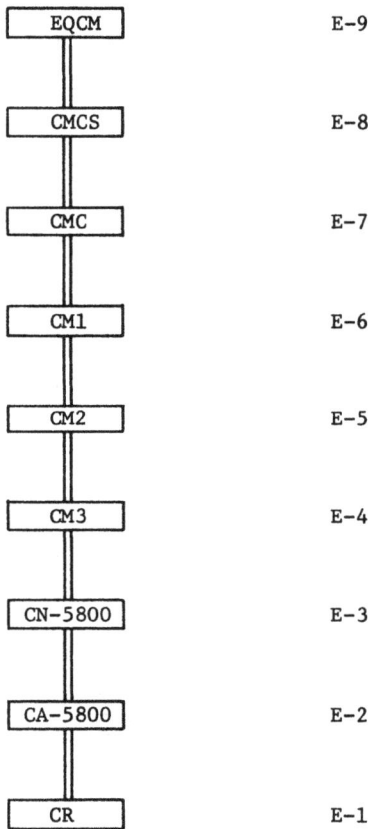

EQCM	E-9
CMCS	E-8
CMC	E-7
CM1	E-6
CM2	E-5
CM3	E-4
CN-5800	E-3
CA-5800	E-2
CR	E-1

2. GENERAL RATING (PO3 through CPO).
Construction Mechanics schedule, oversee, and perform tasks involved in maintenance, repair, and overhaul of automotive, materials- handling, and construction equipment; assign and supervise activities of assistants who locate, analyze, and correct malfunctions in equipment, and issue repair parts; maintain records, prepare requisitions and reports; train assistants in repair procedures and techniques.

3. RATING NEC CODES.

CM-5802 Automotive Electrician (Source: CM, E-4 and above): Services, adjusts, analyzes, and repairs automotive, materials, handling, and construction equipment electrical systems.

CM-5804 Stationary Diesel Engine Mechanic (Source: CM and EN, E-5 and above): Maintains repairs, adjusts, and overhauls stationary diesel engines over 600 horsepower.

CM-5801 Automatic Transmission/Hydraulic Systems Mechanic (Source: CM, E-4 and above): Troubleshoots, dismantles, repairs, and reassembles all types of automatic transmission and torque converters; analyzes and corrects malfunctioning hydraulic pumps, controls and relief valves, rams and cylinders; repairs or replaces hydraulic lines and hoses.

4. SENIOR AND MASTER CHIEF PETTY OFFICERS

Senior Chief Construction Mechanic (CMCS) Chief Construction Mechanics (CMC), eligible to participate in the E-8 examination for Senior Chief Construction Mechanic (CMCS), should be knowledgeable of both the knowledge factors and the knowledge aspects of the practical factors currently required of the EO3 and EO2.

Master Chief Equipmentman (EQCM). Senior Chief Construction Mechanics (CMCS), eligible to participate in the E-9 examination for Master Chief Equipmentman (EQCM), should be knowledgeable of both the knowledge factors and the knowledge aspects of the practical factors currently required of the EO1 and EOC

ENGINEERING AID (EA)

1. CAREER PATTERNS:

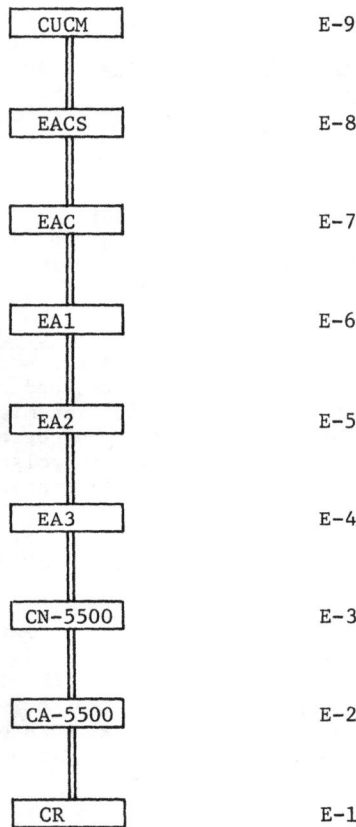

CUCM	E-9
EACS	E-8
EAC	E-7
EA1	E-6
EA2	E-5
EA3	E-4
CN-5500	E-3
CA-5500	E-2
CR	E-1

2. GENERAL RATING (PO3 through CPO).
Engineering Aids plan, supervise, and perform tasks required in construction surveying, construction drafting, planning and estimating, and quality control; prepare progress reports, time records, construction schedules, and material and labor estimates; establish and operate a basic quality control system for testing soils, concrete, and construction materials; prepare, edit and reproduce construction drawings; and make and control surveys, performing such tasks as running and closing traverses, staking out for excavations, and obtaining and converting field notes into topographic maps.

3. RATING NEC CODES

EA-5502 Quality Controlman (Source: EA, E-5 and above): Performs tests necessary to design soil stabilization in the laboratory and field: Performs laboratory tests of wet mechanical analysis, Atterberg Limits, specific gravity, soundness of aggregates, Marshall Test, and bituminous materials identification; performs field tests on soils for density using radiac equipment and moisture by the calcium carbide method, concrete for cement-ratio and strength analysis using cylinders and beams, and bituminous materials identification; write soil and materials reports of proposed construction sites and routes with regard to types of soils, soil profiles, drainage requirements, and usable construction materials.

EA-5515 Construction Planner and Estimator Specialist (Source: EA, BU, CE, SW, and UT, E-5 and above): Plans and estimates material, manpower, and equipment requirements for various construction jobs: Performs scheduling, procurement, production control and management reporting of construction projects.

4. SENIOR AND MASTER CHIEF PETTY OFFICERS

Senior Chief Engineering Aid (EACS). Chief Engineering Aids (EAC), eligible to participate in the E-8 examination for Senior Chief Engineering Aid (EACS), should be knowledgeable of both the knowledge factors and the knowledge aspects of the practical factors currently required of the BU3, BU2, SW3 and SW2.

Master Chief Constructionman (CUCM). Senior Chief Engineering Aids (EACS), eligible to participate in the E-9 examination for Master Chief Constructionman (CUCM), should be knowledgeable of both the knowledge factors and the knowledge aspects of the practical factors currently required of the BU1, BUC, SW1 and SWC.

EQUIPMENT OPERATOR (EO)

1. CAREER PATTERNS.

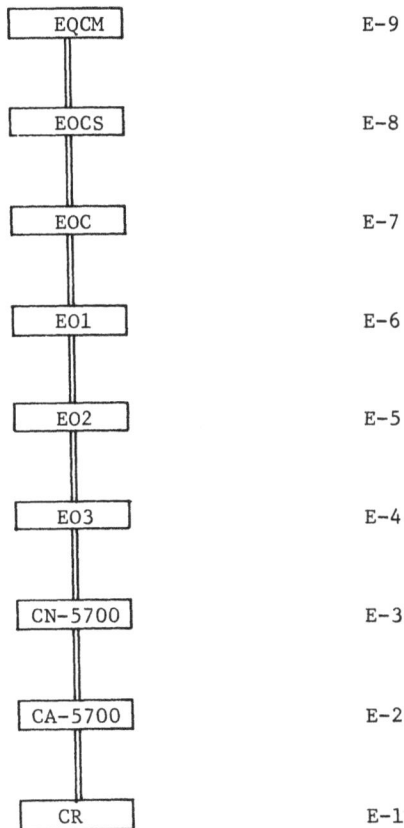

EQCM	E-9
EOCS	E-8
EOC	E-7
EO1	E-6
EO2	E-5
EO3	E-4
CN-5700	E-3
CA-5700	E-2
CR	E-1

2. GENERAL RATING (PO3 through CPO). Equipment Operators plan, supervise, and perform tasks involving deployment and operation of automotive, materials-handling, weight-lifting, and construction equipment; direct and coordinate efforts of individuals and crews in execution of construction, earthmoving, roadbuilding, asphalt paving, and quarrying assignments; maintain records and reports on mobile and stationary equipment. Organize and supervise automotive and construction equipment pools.

3. RATING NEC CODES.

EO-5707 Rotary and Percussion Water Well Driller (Source: EO, E-5 and above): Sets up and operates well drilling machines (rotary and precussion) to drill water wells: Hoists tubular casing and bolts necessary connections; manipulates levers to control drill and drive casing; threads additional casings and drives casings into well as hole deepens; removes samples of subterrain.

EO-5708 Blaster (Source: EO, E-5 and above): Places and detonates explosives to clear sites, excavation, or obtain raw materials for rock crushers: Transfers explosives from magazine to blasting area; exercises specified safety precautions; bores holes, notes soil formation, determines amount of explosives required; explodes charge by fuse or electrically; oversees stowage of explosives in magazine; maintains records of explosives expended and in stock.

EO-5709 Rock Crusher Operator (Source: EO, E-5 and above): Sets up, operates, and performs operator's lubrication on jaw and roll crushers and screening plants used in the production of aggregate.

EO-5711 Grade Foreman (Source: EO, E-6 and above): Supervises personnel and controls equipment engaged in excavating, filling, and grading areas to specifications: Prepares and works from mass diagrams and construction schedules; computes equipment and material estimates from drawings, sketches, and specifications; uses leveling instruments to set and verify grade stakes; supervises soil stabilization operations.

EO-5712 Asphalt Paver (Source: EO, E-5 and above): Sets up, operates, and maintains bituminous pavement plants: Performs all duties in asphalt paving; performs prescribed operator's lubrication and service on asphalt paver, asphalt distributor, and self-propelled road rollers.

4. SENIOR AND MASTER CHIEF PETTY OFFICERS.

Senior Chief Equipment Operator (EOCS). Chief Equipment Operators (EOC), eligible to participate in the E-8 examination for Senior Chief Equipment Operator (EOCS), should be knowledgeable of both the knowledge factors and the knowledge aspects of the practical factors currently required of the CM3 and CM2.

Master Chief Equipmentman (EQCM). Senior Chief Equipment Operators (EOCS), eligible to participate in the E-9 examination for Master Chief Equipmentman (EQCM), should be knowledgeable of both the knowledge factors and the knowledge aspects of the practical factors currently required of the CM1 and CMC.

STEELWORKER (SW)

1. CAREER PATTERN.

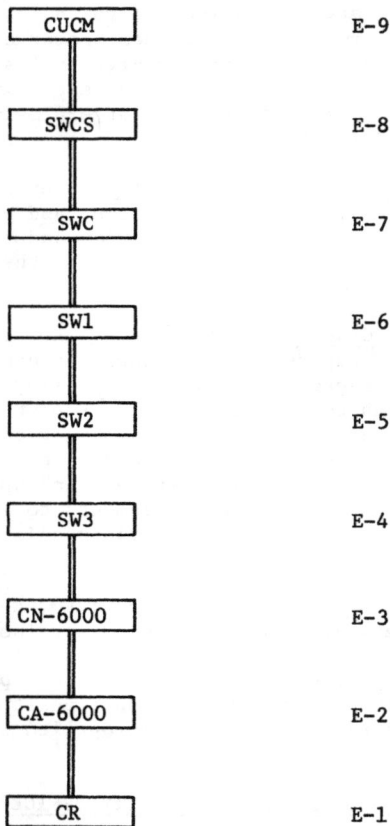

CUCM	E-9
SWCS	E-8
SWC	E-7
SW1	E-6
SW2	E-5
SW3	E-4
CN-6000	E-3
CA-6000	E-2
CR	E-1

2. GENERAL RATING (PO3 through CPO).
Steelworkers plan, supervise, and perform tasks directly related to fabrication and erection of pre-engineered structures; control job site deployment of materials and equipment; and direct and coordinate the composiition, training, and efforts of crews who fabricate, assemble, erect, position, and join structural members and fabricated sections.

3. RATING NEC CODES.

SW-6012 Certified Welder, 1st level
(Source: SW, E-4 and above): Performs manual shielded metal-arc welding on carbon steel of any thickness in all positions with the type and size of electrode specified. All welding must be done in accordance with specifications listed for Test No. 1 in Military Qualifications Test for Welders. MIL-STD-00248B (SHIPS). Must requalify annually.

SW-6013 Certified Welder, 2nd level
(Source: SW, E-5 and above): Meets requirements of SW-6012 and performs manual shielded metal-arc welding on carbon steel pipe up to 3/4 inch thick in all positions, with the type and size of electrode specified. Performs manual gas welding on carbon steel up to 1/4 inch thick in all positions, with the type and size rod specified. All welding must be done in accordance with specifications for Tests No. 2A and 3 in Military Qualifications Test for Welders, MIL-STD-00248B (SHIPS).

SW-6014 Certified Welder, 3rd level
(Source: E-5 and above): Meets requirements of SW-6013 and performs semi-automatic, gas (MIG) welding on aluminum up to 3/4 inch thick in all positions, with the type and size welding wire specified. All welding must be done in accordance with specifications for Test No. 1, Military Qualifications Test for Welders, MIL-STD-00248B (SHIPS).

SW-6017 Sheetmetal Worker (Source: SW, E-4 and above): Lay-out, cut, prepare joints, bend and fabricate complex sheet metal, aluminum, and copper shapes: Performs soft-soldering employing torch and soldering copper; operates treadle and power driven squaring shears, ring and circle shears, finger and cornice brakes, roll forming, crimping, and beading machines; operate spot welding machines and rivet guns.

SW-6018 Blacksmith and Foundryman (Source: SW, E-5 and above): Conditions molding sands for small simple castings, melts and casts aluminum or copper base alloys, lead and babbit; forges, heat-treats, casehardens, and forge-welds various types of metal; conducts hardness tests; heats, draws, shapes, upsets, and twists various metals into required forms; operates electric furnaces, automatic ovens, and coal, gas, or oil-fired forges and furnaces; manufactures simple hand tools from bar stock; tempers, dresses, grinds, and sharpens hand and machine tools using hand or power equipment.

SW-6021 Safety Inspector (Source: All Group VIII Ratings, E-6 and above): Organizes and supervises the operation of the safety department: Investigates accidents and analyzes accidents and problem areas and recommends methods to decrease frequency and/or eliminate accidents; collects data to ascertain accident trends; inspects project sites, grounds, buildings, and machinery to isolate hazards to life, health, and equipment; conducts safety education campaigns by preparing and/or distributing literature, posters, charts, and displays; organizes and directs safety committee; directs placement of traffic control signs and devices.

4. SENIOR AND MASTER CHIEF PETTY OFFICERS.

Senior Chief Steelworker (SWCS). Chief Steelworkers (SWC), eligible to participate in the E-8 examination for Senior Chief Steelworker (SWCS), should be knowledgeable of both the knowledge factors and the knowledge as-pects of the practical factors currently re-quired of the BU3, BU2, EA3 and EA2.

Master Chief Constructionman (CUCM). Senior Chief Steelworkers (SWCS), eligible to participate in the E-9 examination for Master Chief Constructionman (CUCM), should be know-ledgeable of both the knowledge factors and the knowledge aspects of the practical factors currently required of the BU1, BUC, EA1, and EAC.

UTILITIESMAN (UT)

1. CAREER PATTERNS.

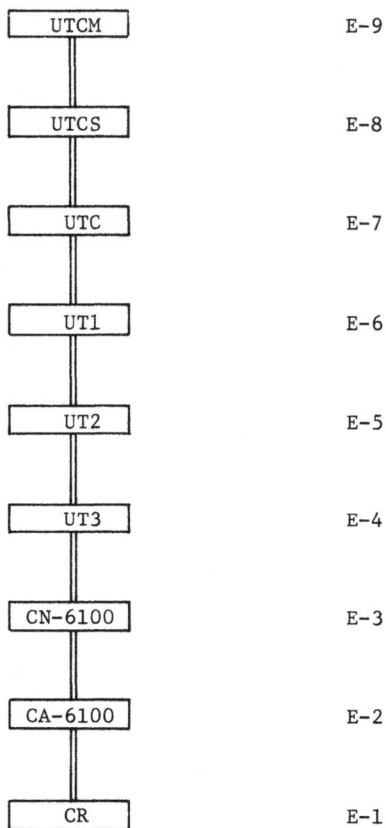

UTCM	E-9
UTCS	E-8
UTC	E-7
UT1	E-6
UT2	E-5
UT3	E-4
CN-6100	E-3
CA-6100	E-2
CR	E-1

2. GENERAL RATING (PO3 through CPO).

Utilitiesmen, plan, supervise, and perform tasks involved in installation, maintenance, and repair of plumbing, heating, steam, fuel and water distribution and treatment systems, air-conditioning and refrigeration equipment, and sewage disposal facilities, as prescribed by drawings and specifications. Schedule and evaluate installation and operational tasks, prepare records and reports.

3. RATING NEC CODES.

UT-6102 Shore-Based Boiler Operator/Repairman (Source: UT, E-4 and above): Operates, services, and repairs high and low pressure stationary steam boilers and ancillary equipment including pumps and blowers; conducts boiler, feed water, and fuel oil tests; performs operational tests; replaces parts including tubes, gages, casings, and gaskets; cleans fire and water sides; maintains operational logs.

UT-6104 Shore-Based Refrigeration and Air-Conditioning Mechanic (Source: UT, E-5 and above): Installs, operates, maintains, and repairs refrigeration, air conditioning, water cooling equipment, cube and flake ice machines and block ice manufacturing plants.

UT-6117 Petroleum Tank Farm Operator (Source: UT, E-5 and above): Operates and maintains storage and transfer equipment for petroleum products at storage terminals and tank farms: Reserves and distributes petroleum by connecting available tanks; operates pump engines and valves to transfer petroleum; reads meters and gauges to determine completion of petroleum transfer; verifies amount and type of petroleum in storage; operates fire fighting equipment, personnel protective and safety equipment common to petroleum storage facilities.

UT-6121 Water Treatment and Sanitation Systems Operator (Source: UT, E-5 and above): Installs, operates, and maintains water purification, fluoridation, distillation equipment, water wells, and well equipment: Operates and maintains sewage disposal plants.

4. SENIOR AND MASTER CHIEF PETTY OFFICERS.

Senior Chief Utilitiesman (UTCS). Chief Utilitiesmen (UTC), eligible to participate in the E-8 examination for Senior Chief Utilitiesman (UTCS), should be knowledgeable of both the knowledge factors and the knowledge aspects of the practical factors currently required of CE3 and CE2.

Master Chief Utilitiesman (UTCM). Senior Chief Utilitiesmen (UTCS), eligible to participate in the E-9 examination for Master Chief Utilitiesman (UTCM), should be knowledgeable of both the knowledge factors and the knowledge aspects of the practical factors currently required of the CE1 and CEC.

MCB ALLOWANCE

1. Development. As was mentioned in Annex C, a functional component is a grouping of personnel and material to perform one of the specific tasks of an advanced base. But the actual construction of these advanced bases is one such task, and the functional component designed to perform this task was the Naval Construction Battalion, designated a P-1 functional component. This component was developed during World War II and consisted of 33 officers and 1082 enlisted men and the necessary camp equipment and tools. Since this NCB was considered too large to support highly mobile fleet operations, it has been replaced in the Table of Advanced Base Functional Components, OPNAV INST. 4040.22 by a smaller unit designated as a P-25.

The P-25 functional component, which consists of 21 officers and 563 enlisted men, had about one-half the equipage of the P-1. It was structured to provide 90 day's support for limited advanced base construction operations running two ten-hour shifts per day, seven days per week under emergency wartime conditions. This P-25 functional component became the allowance for the Mobile Construction Battalion and NAVFAC published detailed information pertaining to outfitting an MCB in the publication "The Personnel, Equipment and Material Allowance for P-25 Mobile Construction Battalion." It is used as the basic planning guide for outfitting mobilized reserve MCBs.

However, the commissioned MCBs were required not only to build advanced bases but also to provide engineer support to expeditionary forces (See Chapter I). It was decided that the P-25 functional component did not completely provide for this support. Therefore, for the active duty MCBs, another allowance list was drawn up using the P-25 functional component as a guide but adding or subtracting items where necessary. This new listing is referred to as the MCB Allowance. Some of the concepts considered in tailoring the P-25 functional component were these:

    a. Increased responsiveness to contingency support of Fleet Marine Forces.

    b. Compatibility of equipment with the forces supported to make equipment repair easier.

    c. The ability to embark the allowance in amphibious shipping and land across the beach in a tactical situation.

    d. The capability of performing projects of equal magnitude at two separate locations.

    e. Decreased comforts, increased defensive capability and increased mobility.

    f. The ability to move the allowance overland and the possibility of airborne movements.

    g. The need for supplementing the allowance to allow the battalion to accomplish larger single projects.

Therefore, the MCB Allowance is a tailored edition of the P-25 functional component that provides the tools and equipment for general construction support at a beachhead.

Although it was envisioned that most detachments would be outfitted from the equipment and tools in the parent battalion allowance, from time-to-time special allowances must be developed. One such example is the Basic Outfitting Lists for the SEABEE Teams. This list consists of two parts. The first part consists of the basic outfitting for SEABEE Teams operating in a construction or disaster area. All material and equipment in this category is completely air transportable and is designed to require minimal tailoring for any foreseeable SEABEE Team employment. The second part consists of equipment which follows the air-lifted team by "sealift", and is designed to meet the specific requirements of the assigned tasks. While designed as a sealift allowance, all items included in this category can be broken down into air-transportable components in emergency situations.

2. Categories. The MCB Allowance has three general categories:

    a. Basic Allowance. This allowance was developed to enable the battalion to carry out the operational requirements of OPNAV INST. 5450.46D. It is designed to support construction operations for 90 days under emergency conditions before resupply is required. Since the battalion will be expected to build any number of different facilities in any climatic condition, judicious selection of items is necessary to prevent the allowance from reaching gigantic proportions. A compromise must be reached, balancing bulk against capability. The MCB Allowance represents the best selection for providing general construction capability.

    b. Supplemental Functional Components: When an assigned project requires tools or equipment in excess of the battalion's capability, the allowance is supplemented. These additional tools and equipment are grouped into functional components designed for one specific task. For example, a component for quarry operations would include a power shovel, several dumptrucks, air drills and a rock crushing plant along with the necessary hand tools and repair parts. These components are maintained by the COMCBLANT/COMCBPAC headquarters and are issued to the battalion when required.

FIGURE 109
BOXES MUST BE DESIGNED TO SERVE A DOUBLE PURPOSE

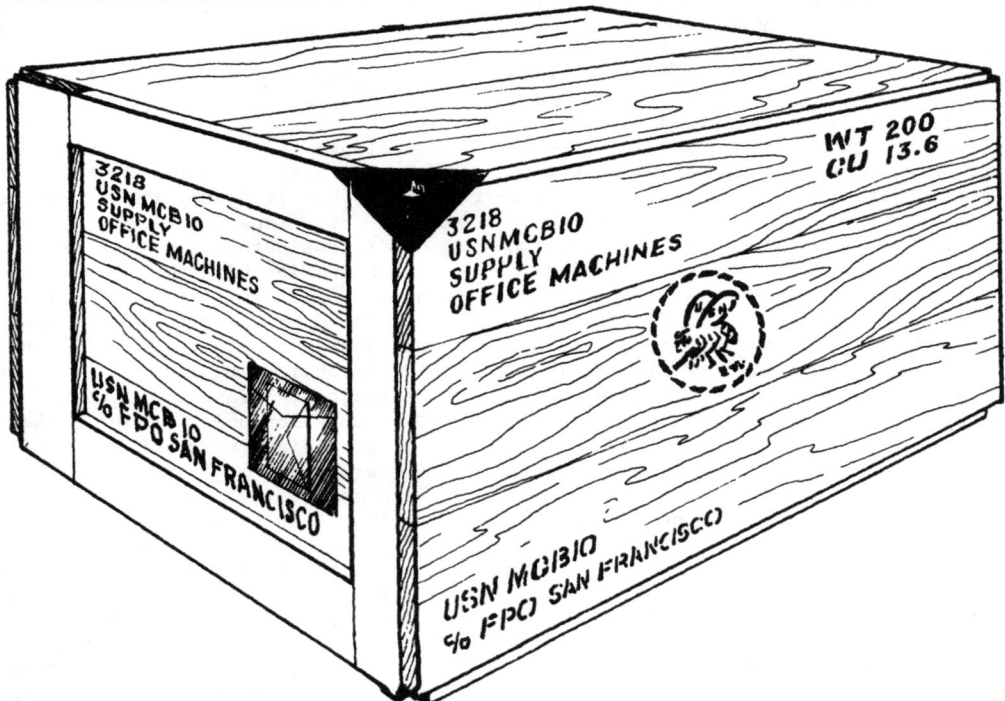

FIGURE 110
"BOX CONTROL" IS THE KEY TO BATTALION MOBILITY

c. _Augmenting Allowance_. Since all items not needed under emergency conditions have been tailored out of the allowance, several problems arise because of the nature of MCB peacetime operations. The allowance is continually in use on overseas construction projects, therefore, normal wear and tear and occasional breakdowns erode the capability of the battalion to perform emergency assignments. To offset this action, allowance items are increased by a certain percentage. This "over allowance", called augmenting allowance, protects the mount-out level and permits the battalion to carry out its normal on-the-job training. Another difficulty is that of austerity. Living in tents and eating from mess kits, etc., over a long period of time becomes a morale problem and it is costly. Therefore, some additional items, commonly referred to as "garrison gear," are added as augmenting allowance to provide a substitute for field equipment. These items are for peacetime use only and would be left behind when the MCB embarked for a contingency assignment.

In summary, it is therefore convenient to think of the MCB allowance for the active forces in terms of a basic allowance that is augmented for training purposes and that can be supplemented to increase MCB capability for a specific operation.

It is important to note that this allowance does NOT contain construction materials to provide facilities. These materials must be supplied to the battalion by the using activity either in the form of project material or additional functional components. These construction consumables generate a shipping requirement in addition to the MCB Allowance and must be included when ships are allocated.

Transportation of the entire basic allowance to and from normal peacetime deployment sites is not economical. Therefore, many large items are turned over to the custody of the relieving battalion and the control is maintained by COMCBLANT/COMCBPAC. The allowance is divided into two parts. Part I is a listing of those items which remain in the custody of each MCB wherever it is deployed. This is called "Organic Equipment and Material". Part II is a listing of the items which are prepositioned and is called "Equipment and Supplies Controlled and Supplied as Required by COMCBLANT/COMCBPAC." A listing of the sections in each part is given below:

PART I - _Organic Equipment and Material_

    A - Kit Assemblies
    B - Central Tool Room
    C - Kit Consumables
    D - NBC Warfare
    E - Damage and Safety Control
    F - General Equipment and Supplies
    G - Barber Equipment and Supplies

    H - Photographic Equipment and Supplies
    I - Administrative Equipment & Supplies
    J - Medical and Dental Equipment and
        Supplies
    K - Special Clothing and Bedding
    L - Communication Equipment
    M - Weapons
    N - Infantry Equipment
    O - Radiac Equipment
    P - Chaplain Equipment and Supplies

PART II - _Equipment and Supplies Controlled_
          and Supplied as Required by
          COMCBLANT/COMCBPAC

    A - Kit Assemblies
    B - Tent Camp
    C - General Equipment and Supplies
    D - Messing and Galley Equipment
    E - Cold Weather Clothing
    F - Provisions
    G - POL
    H - Automotive and Construction Equipment
    I - Supplemental Functional Components

3. _Packaging_. Since all allowance items must be packed and ready for embarkation within ten days or less, the allowance must be packaged and marked in a standard manner to facilitate emergency mount-out. All tool kits, ordnance gear, medical supplies, disaster control equipage, etc., are packed into standard sized boxes so that these boxes can easily be banded on pallets or loaded in metal shipping containers (CONNEX BOXES) without wasted space. The standard box is also built in one-half and one-quarter size in order to ensure that no single box exceeds two-hundred and fifty pounds in weight. This is done so that the boxes can be moved by hand by two men when equipment is not available. The insides of the boxes are individually configured to accommodate whatever is being carried. They also may be constructed with loose or fixed dividers which may later serve as shelves. The boxes may also have hinged lids or other features (See Figure 109). Each custodial officer orders the boxes he needs, and the "B" company commander constructs them and paints them olive green in his shops.

Every box is marked to indicate the general contents, to show the priority of need ashore and to permit grouping of like items on pallets The marking procedures are detailed in COMCB-LANT/COMCBPAC instructions. Generally, a color code is painted on diagonally opposite corners to indicate the general contents and the priority of need. In addition, the printed information, includes the box identification number, the name and address of the battalion, the weight and cube and the custodial staff officer (See Figure 110). Finally, a packing list containing a detailed inventory of the contents of the box is placed in a waterproof envelop attached to the outside of the box.

Annex I
STINGER AND THE NAVY/DOD PROGRAMMING SYSTEM

1. <u>Purpose</u>. The purpose of this annex is to describe the Navy Programming System - why it exists, what it does for the SEABEES, what input SEABEES have to it and the mechanics of its operations.

2. <u>NAVFAC System Responsibility</u>. NAVFACs direct responsibility in the system is through the Chief of Civil Engineer's role as technical advisor to CNO for SEABEES; practically speaking, this means in most cases short of a formal memorandum from the Chief of Civil Engineers to CNO, NAVFAC Code 06 acts as the staff for the OPNAV SEABEE Program Element Sponsors. A massive amount of the paperwork is required to validate changes to basic programming documents. Much of this paperwork is done at NAVFAC and passed on to the program element sponsor who guides it through the tortuous OPNAV/DOD path to approval and implementation. Before describing this decision-making process in detail, it is necessary to begin with an overview of the DOD/Navy Programming Systems concept.

3. <u>Programming</u>. Programming has been given a quite glamorous connotation in recent years, but it is certainly not a new concept. Programming is as old as the Navy itself - it is merely a <u>planned method or system of providing guidance and control for logical decision-making concerning all aspects of the Navy</u>. The complexities of the DOD have made it necessary to formalize, systematize and computerize this total system.

The current DOD/Navy System involves analytical studies, planning, determination of objectives, program development, budget analysis, and upon implementation, the control of resources - these areas have all been organized under a rational set of rules which when properly applied provide for decisions which determines force levels, weapons systems and support programs of the defense establishment.

It is a simple fact, and an obvious one, that this decision-making structure is only as good as quality of input. As shall be discussed, most of the basic program decision determinants affecting SEABEES are fed to OPNAV by NAVFAC. Thus NAVFAC input must be good to achieve good output. Steps are being taken to insure the best quality of input. First, NAVFAC is defining <u>our</u> system, the 26 year old SEABEE system. The vehicle for doing this, STINGER, is a <u>new</u> acronym describing an old system - NAVFAC is, as the rest of NAVY/DOD is, describing the total system on a rational ba-

sis to enable orderly decision-making. The second step, which makes the SEABEE system and the entire DOD system work, is a free flow of information between Washington and the field with regards to the systems approach and the input needed to make it work.

4. <u>Major Program Documents</u>. The DOD/Navy Programming System is based upon a series of major documents that vary in content from offering broad strategic conjecture to nailing down budget requirements for the next five years. If SEABEES do not make input to this cycle, they do not get either funds or authorized end strength.

Figure 111 depicts several of the basic programming documents that form the heart of the system. The JLRSS, Joint Long Range Strategic Study, and the JSOP, Joint Strategic Objectives Plan, has Navy parallels which are the 10 year qualitative NLRG - Navy Long Range Guidance and the 5 year NMRG, the Navy Mid-Range Guidance. Both the NLRG and the NMRG are annexes to the basic Navy Strategic Study (NSS) NSS and its annexes give qualitative guidance and have a quantitative parallel in the MRO, Mid-Range Objectives. The MRO gives force structure goals for the 11th fiscal year subsequent to approval. It is used for input to the JSOP, and on a level of more interest to SEABEES, initial guidance for the determination of Program Objectives. Other documents of this type of interest are the NSP, Navy Support Plan with its analysis of logistics capability for the next 8 years and the NCP, Navy Capability Plan, which provides guidance for mobilization, organization, training, and equipping of forces for the current fiscal year.

This is a brief summary of the interrelationship of the Navy Planning and Programming System to the DOD System. The rest of this Annex will be directed toward the TWO cornerstones of the system - the <u>FYDP</u>, Five Year Defense Plan, and the <u>PO</u>, Program Objectives. The DNFYP, Department of the Navy Five Year Program, is the Navy portion of the FYDP.

5. <u>The FYDP</u>. The FYDP is the summation of all DOD Components' approved programs. Unless SEABEE dollar needs are articulated in the latest FYDP, it is basically impossible to ask for money in the budget review cycle. There are, however, some emergency in nature abbreviated FYDP change procedures. The FYDP provides continuity and long range implications visible out to eight years for force levels and five in terms of specific resource levels. As the

PREP. or COORD. by	PRINCIPAL CONTRIBUTORS	DOCUMENTS	UPDATE MONTH	FISCAL YEARS
OP-60	OPS-01,03,04,05,07, 090,92 E 95 CNM, ONR E LABS	NAVY STRATEGIC STUDY	BASIC E ANNEXES 1 JAN	
JCS/OP-60	OP-07,93,92	JOINT LONG RANGE STRATEGIC STUDY	15 MAY	
OP-93	OPS-01,03,04,05,06, 07,90,91,93,92,94, E 95	NAVY MID-RANGE OBJECTIVES	1 JULY (FORCES)	
JCS/OP-60	DCNO'S,090,95	JOINT STRATEGIC OBJECTIVES PLAN	VOLUMN I 1 JULY VOLUMN II 1 APR	
OP-90	OP-09B,01,05,04, 05,06,07,92,93,94, 95 CNM, ONR, NAVCOMPT MARCORPS	DEPT OF NAVY PROGRAM OBJECTIVES	30 NOV E 1 MAR	
OSD/090	DCNO'S,090,95,94	DRAFT PRESIDENTIAL MEMORANDUM	JULY-OCT	
OSD/OP-90	DCNO/ACNO'S MARCORPS, NAVCOMPT OND, CNM, MSTS, 09B, ETC.	FIVE YEAR DEFENSE PROGRAM	CONTINUALLY UPDATED	
OP-40	MARCORPS, USCG, MSTS, OKM, OCMM OPS-098,90,92,94,01,03, 04,05,06,07 CINCFLTS SEAFRXS E DIST CMDTS	NAVY SUPPORT PLAN	CONTINUALLY UPDATED	
JCS/OP-60	DCNO'S	JOINT STRATEGIC CAPABILITIES PLAN	31 DEC	
OP-60	OPS 31,33,34,40 76,92,94,95,09B2 OCEANAV, 09BT, 09M	NAVY CAPABILITIES PLAN	1 JUN	

foundation of the DOD Programming System, it relates resources (inputs) to programs (outputs). The FYDP structure provides a method of aggregating forces, money, and manpower in various combinations within the three major building blocks; major programs, elements, and resource categories.

Major Programs. The ten major programs aggregate the entire defense posture into broad functional classifications of similar military missions. They are:

Program I — Strategic Forces
*Program II — General Purpose Forces
Program III — Intelligence & Communication
Program IV — Airlift and Sealift
**Program V — Reserve & Guard Forces
Program VI — Research & Development
***Program VII — General Supply & Maintenance
****Program VIII — Training and Medical and other General Personnel Activities
Program IX — Administration and Associated Activities
Program O — Support of other Nations

*The active NCF units are all grouped under Program II
**The Reserve NCF units are all in Program V
***The CBCs and NAVFAC are in Program VII
****CECOS and the NAVSCONS are in Program VIII

Program Elements. Major programs are sub-divided into smaller building blocks known as program elements. The program element is the smallest unit of military output controlled at the DOD level. It is defined as "an integrated combination of men, equipment and facilities which together constitute an identifiable military capability or support activity." It appears in the FYDP on data sheets containing information on the mission and related tasks, composition and major equipment, approved force levels for eight years, Total Obligational Authority (TOA) and manpower for five years. The TOA is subdivided into three categories: R&E, Investment Costs (OPN, MCON) and Operating Costs (MPN, O&MN). These are aligned with EOBs (Expense Operating Budgets) and RMS (Resources Management System) expense elements as of 1 July 1968. Each program element dealing with Navy programs has a sponsor in OPNAV. NCF units and support units are parts of several elements, but do not constitute an element by themselves.

Changes to the FYDP are made only upon approval of a PCR (Program Change Request). PCRs are initiated by the Program Element Sponsor upon approval of a Navy PO (Program Objective), these topics will be discussed in more detail further on in the Annex.

6. Program Objectives. The Department of the Navy Program Objectives (PO) are a Department-level programming document, similar in organization (i.e., the Program - Program Element relationship) to the Five Year Defense Program - FYDP. The two documents differ in one major respect: The FYDP addresses SECDEF approved programs, while the PO lists SECNAV approved Program Objectives. The PO represents the level of forces, manpower, and resources to which SECNAV supports the objectives established by CNO and Commandant, Marine Corps, in the JSOP. The PO includes all Departments of the Navy programs. With the assistance of Program Sponsors, the Director, General Planning and Programming Division (OP-90) develops and prepares the PO for the Navy. NAVFAC assists the SEABEE Program Sponsors in PO preparation. The PO is reviewed by CNO and is approved by SECNAV.

Prior to SECNAV approval the POs are designated as TPOs, Tentative Program Objectives; NAVFAC staffs the preparation of the TPO for the OPNAV Program Element sponsor.

These POs have a very practical relationship to what SEABEES do. The TPO (Tentative PO) submission time occurs in January of each year. Not until TPOs are approved as SECNAV POs can a PCR be submitted; i.e., no attempt can be made to change the FYDP until the PO is approved - and there is no basis for increased money requests, budget base increases, until the FYDP is changed.

The Program Objectives (PO) projects SECNAV approved resource levels, procurements, R&D, and supporting programs for five years, commencing two fiscals after the fiscal year in

which the PO is approved, and the force level objectives for an additional three years. (parallel to the FYDP structure). The PO serves as a control document for the Department of the Navy program planning and as a guide to PCR preparation. The PO must provide programs which are technologically, and fiscally feasible. The approved PO, provides the program planning guidance for budget submission

Program Objectives may be changed by submission of POCPs, Program Objective Change Proposals, initiated by the Element Sponsor, and in the case of the NCF, staffed extensively by NAVFAC.

7. Review/Change/Approval Cycle of the FYDP. In introducing this topic it is desirable to first examine the broad change procedures (somewhat parallel to the POs) which can effect the FYDP. This is the MFI/DPM/PCR cycle. During the early part of the calendar year when the JCS identifies MFI, Major Force Issues (i.e., questions, the resolution of which may cause major force level changes, (i.e., members of divisions, etc.). The tentative SECDEF decision on these are stated in IDPMS, Initial Draft Presidential Memoranda. After publication of these, 30 days are allowed for comment and PCR preparation which price out the IDPM. The final SECDEF decision comes in a FDPM or Final DPM.

PCRs are also, and primarily as far as SEABEES are concerned, the administrative tool in the DOD Programming System used to request changes to the FYDP. A PCR is used to request a change to forces, Total Obligational Authority (TOA), or personnel assigned to individual program elements in the FYDP. In addition, reclamas to SECDEF PCR decisions are submitted in the PCR format.

If a PCR is approved it is reflected as a change to force level or TOA in a FYDP element. This does not provide any extra dollars. It only provides SECDEF approval to request more dollars in the upcoming budget submission.

The Navy Department Program Information Center (NDPIC) is the Navy processing agent for PCRs. The PIC assists Program Sponsors in determining requirements for and in the preparation of PCRs, coordinates initial review, and obtains CNO, CMC, and SECNAV approval in conjunction with the Office of the Director, Navy Program Planning and the SECNAV Office of Program Appraisal (OPA).

Figure 112 depicts the PCR cycle and its relation to the JSOP, MFIs, and DPMs/DGMs. PCRs submitted in response to these documents are commonly called "DPM oriented." PCRs which are not related to the JSOP, MFIs, or DPMs/DGMs are referred to as "Non-DPM" or support PCRs; these last are of primary interest to the NCF. Non-DPM oriented PCRs may originate in any office which determines a requirement for a change to the FYDP. In the SEABEE case, this is usually at NAVFAC, based upon the NAVFAC estimate of field requirements. The originating office submits the requirement

FIGURE 112
PCR CYCLE

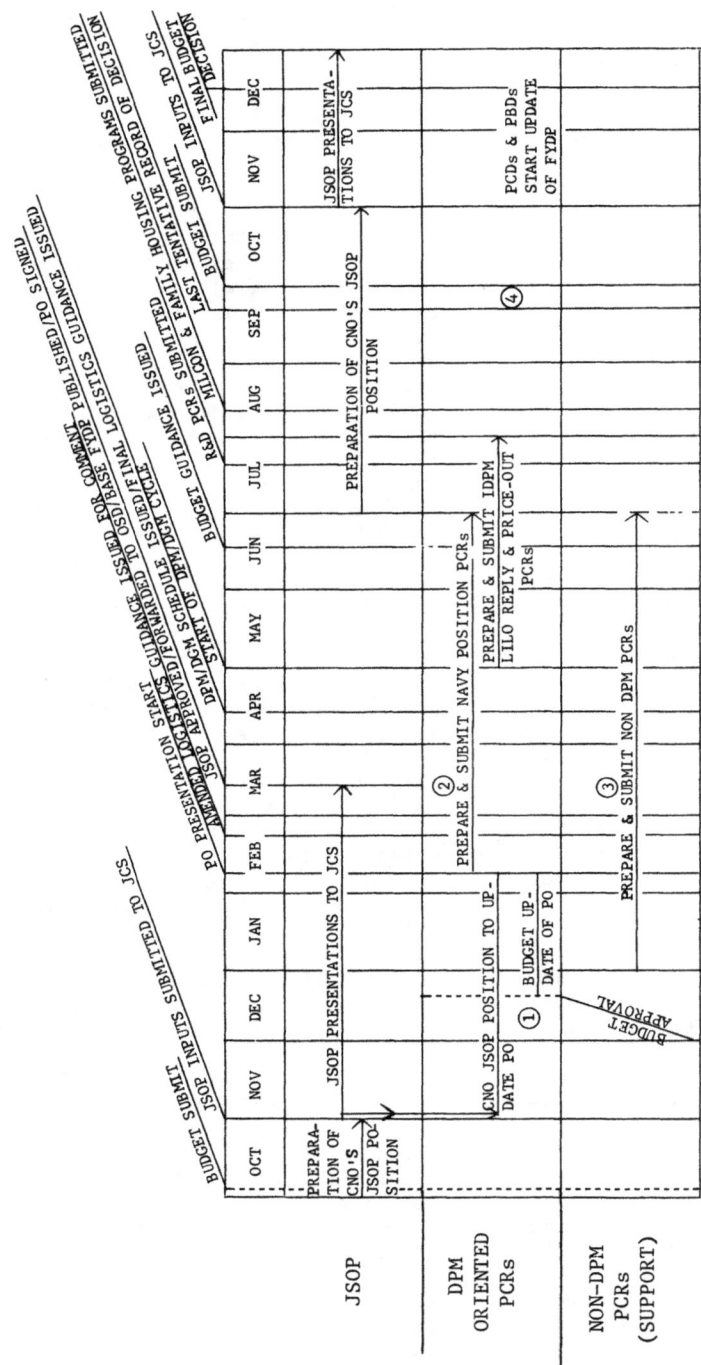

	OCT	NOV	DEC	JAN	FEB	MAR	APR	MAY	JUN	JUL	AUG	SEP	OCT	NOV	DEC
JSOP	PREPARA- TION OF CNO'S JSOP PO- SITION	JSOP PRESENTATIONS TO JCS								PREPARATION OF CNO'S JSOP POSITION				JSOP PRESENTA- TIONS TO JCS	
DPM ORIENTED PCRs		CNO JSOP POSITION TO UP- DATE PO	① BUDGET UP- DATE OF PO		② PREPARE & SUBMIT NAVY POSITION PCRs			PREPARE & SUBMIT IDPM LILO REPLY & PRICE-OUT PCRs							
NON-DPM PCRs (SUPPORT)			BUDGET APPROVAL		③ PREPARE & SUBMIT NON DPM PCRs							④		PCDs & PBDs START UPDATE OF FYDP	

Top labels (diagonal): BUDGET SUBMIT / JSOP INPUTS SUBMITTED TO JCS / PO PRESENTATION START / AMENDED JSOP APPROVD/FORWARDED / LOGISTICS GUIDANCE ISSUED TO OSD/BASE FYDP / SCHEDULE ISSUED/DPM/DGM CYCLE / START OF DPM/DGM CYCLE / PUBLISHED/PO SIGNED / FINAL LOGISTICS GUIDANCE ISSUED / DPM/DGM SCHEDULE ISSUED/FINAL CYCLE / BUDGET GUIDANCE ISSUED / R&D MILCON & FAMILY HOUSING PROGRAMS SUBMITTED / BUDGET PCRs SUBMITTED / LAST TENTATIVE RECORD OF DECISION / BUDGET SUBMIT INPUTS TO JCS / JSOP INPUTS FINAL BUDGET DECISION

① CNO'S JSOP POSITION AND BUDGET DECISIONS ARE USED TO UPDATE PO BY 15 FEBRUARY.

② DPM ORIENTED PCRs SUBMITTED TO PRESENT SECNAV'S POSITION ON ISSUES SHOULD BE ORIGINATED BY THE PROGRAM SPONSORS AS EARLY IN THE CYCLE AS POSSIBLE, GENERALLY WHEN THE CNO POSITION HAS BEEN DETERMINED FOR THE JSOP AND CONCURRENCE OBTAINED FROM SECNAV DURING THE PO PRESENTATIONS.

③ NON-DPM ORIENTED PCRs SHOULD BE SUBMITTED EARLY ENOUGH IN THE CALENDAR YEAR TO OBTAIN SECNAV'S APPROVAL PRIOR TO 1 JULY (SEE PARA 910).

④ BUDGET SUBMIT ON 30 SEPTEMBER WILL BE PREPARED IN ACCORDANCE WITH THE BASE PROGRAM (MARCH FYDP) AS MODIFIED BY PROGRAM CHANGE DECISIONS AND BY DPMs OR DGMs, ISSUED THRU AUGUST 30, 1968.

to the appropriate Program Sponsor who determines if the requirements meets the criteria for a PCR. The Program Sponsor then determines whether the new requirement can be included in a PCR already under consideration or a separate submission is required. If a separate submission is required, the Program Sponsor then prepares and submits the PCR with NAVFACs staffing, in the SEABEE case, in accordance with instructions contained in the Navy Programming Manual. The great majority of the Non-DPM oriented PCRs are identified during the presentation of the PO so that the PO, when approved, accurately reflects the Navy's requirements. These PCRs should be submitted as soon after approval of the PO as possible. PCRs may be submitted before approval of the PO if the requirement is urgent. In any case, they should be submitted early enough in the calendar year to obtain SECNAV approval prior to 1 July. This allows time for review and issuance of a decision by SEC-DEF prior to submission of the annual budget. PCRs can be submitted at any time during the yearly cycle if the requirement is urgent and so stated by SECDEF.

PCRs must be submitted whenever proposed changes meet any of the conditions stated below:

Force Changes. Any change to the force tables in the OSD published Five Year Defense Program.

Manpower Authorizations. Any change or aggregation of changes in the authorization stated in the FYDP which, if approved, would increase the end year strength by 100 or more military and/or civilian authorizations. Changes of less than 100 must be absorbed within current authorized end strength.

Issues. Proposed changes to the FYDP specifically addressed by the IDPM/IDGM.

Functional Transfers. Any transfers which involves an increase or decrease of a DOD Component's program TOA as stated in the FYDP.

Fact-of-Life Changes. Any uncontrollable change, such as, but not confined to: production slippage, operational accidents, or combat attritions; which cause force changes, manpower changes consistent with the above, or TOA changes from the Five Year Defense Program.

Total Obligational Authority (TOA). Any increase for a cost category of a Department of Defense Component in any fiscal year unless exception has been authorized (i.e., in the annual budget estimate memorandum or instruction).

Procurement Changes. Any change in procurement programs, in other than the current fiscal year, involving additional ships, additional aircraft, new items of equipment, additional missiles, or the introduction of new procurement items.

Reprogramming Action. A PCR must be submitted as the transmittal document for any Reprogramming Action (DD 1415) submitted in accordance with DOD Instruction 7250.10. "Implementation of Reprogramming of Appropriated Funds," 5 March 1963, for current or prior fiscal years. However, if a Program Change Decision has been issued which approved the Reprogramming Action for purposes of updating the FYDP, the requirement for a PCR is waived.

Confirmation Changes. Changes to the Five Year Defense Program which result from Secretary of Defense decisions expressed by other than PCDs or PBDs and made without benefit of prior PCR action, but which meet any of the criteria stated in the Paragraphs above.

After determination that a PCR is required the Action Officer designated by the Program Sponsor prepares or directs the preparation of the PCR.

It normally should include detailed data ("Backup Material") as tabs or appendices as required to support the change requested. The intent is to provide quantitative and qualitative rationale and justification and to identify and relate the PCR to other documentation. Content, length, and form of backup Material are not specified by directive since they depend upon the needs of the individual program addressed.

The Backup Material should normally contain:

a. Relationships to previous, current, or pending decisions, requests, or other closely related programming or budgeting documents.

b. Rationale for resources requested, including an economic analysis where applicable.

c. Cost-factors, planning factors, and calculations used in preparing the request.

d. Specific identification of sources of data.

e. Viable alternatives and the impact of each.

f. Consideration of cost-effectiveness implications of identified alternatives.

This backup information and the basic change rationale in the case of the SEABEE System is staffed by NAVFAC Code 06. Reports from the field are the basis of both the qualitative and quantative information.

The PCRs, after submission by the OPNAV Sponsor, are subjected to rigorous review throughout the Navy. If SECNAV approves the PCR, it is forwarded to the Office of the Secretary of Defense (OSD). The SECDEF decision is the final decision and it takes the form of a PCD, Program Change Decision and PBD, Program Budget Decision. When a PCD is published by SECDEF, it is returned to the Program Sponsor who may accept it or reject it. If he rejects it a point paper describing impact on the program element caused by the PCD is attached to the document and intent to reclama (in PCR format) is made known. The PCD is the final decision and is implemented upon publication and included in the FYDP. The reclama may result in a new decision, but while it is circulating the PCD is put into effect. The Program Budget Decision is the SECDEF action on the Navy budget submission.

8. Relationship of Programming and Budgeting. Throughout this Annex mention of the budget cycle has been all but ignored. It is and is not separate from the programming cycle. It is intimately related in that the POs and FYDP provide the budget base for all annual budget submissions, and the timing of the PO/FYDP approval effects our budget submission. The programming system is also the vehicle for obtaining budget base increases through submissions of PCRs which increase the TOA of a FYDP Program element; so the PCR becomes the budget system management tool. However, the programming cycle is separate from the budget approval and execution cycle which is primarily managed by separate offices in the Fleets, Syscoms, FRO, CNM, CNO, NAVCOMPT and SECNAV.

Also, all but passing mention of (RMS), the Resource Management System, has been avoided. One of the basic goals of the RMS Accounting System as established on 1 July 1968 is to align the programming, budgeting and accounting systems. Operating costs in the field will now be 100% tied into the specific FYDP program elements. This will give much more valid costing data for the preparation of PCRs for FYDP updating in the future.

9. Summary. Let us briefly summarize, as a broad spectrum of information has been presented in the Annex.

On the upper level of planning, the system is represented by a series of top level documents that when digested provide the impetus for major policy decisions.

On the working level, the FYDP is the major programming document. Basically, if a requirement is not articulated in the FYDP, there will not be any money flowing down the pipeline. If the dollars are inadequate in the FYDP, a PCR must be submitted to change them - but to do this there must be an approved PO from SECNAV. And to make NAVFAC support to the field timely, all of the documents - TPO, PO, POCP, PCR, PCD must be approved according to schedule or the changes won't be reflected in the current budget submittal. To properly staff these TPOs, PCRs, etc., good timely information from the field must regularly be forwarded to NAVFAC. It is evident, even from this quick overview of the system, the amount of responsibility for program guidance associated with the element sponsor's desk. This then explains the need for NAVFAC to become intimately involved in this cycle. The involvement may be formal with the Chief of Civil Engineers assuming his hat as Technical Advisor to CNO for SEABEES and submitting proposals directly to CNO via the Chain of Command or it may be informal involvement which means Code 06 staffing of PCRs and other point papers for the NCF Program Element Sponsors.

FIGURE 113
PROGRAMMING CYCLE

THE CYCLE INCLUDES:

    TPO - Tentative Program Objectives

    PO   - Program Objectives

    POCP - Program Objectives Change Proposals

    PCR - Program Change Request

    PCD - Program Change Decisions

    FYDP - Five Year Defense Program

    FYDP - DOES NOT EQUAL DOLLARS

Approval of POs and PCRs and changes to the FYDP constitute a license to request dollars in forthcoming budget submissions, not approved dollars!!

Annex J
SEABEE COMMAND

SECTION A - INTRODUCTION

This Annex is presented for use of all officers and senior petty officers of the Naval Construction Force Units. Sections B, C, and D are taken from the pamphlet, "SEABEE Command" prepared for prospective Commanding Officers by COMCBLANT, Captain G. A. Busbee, CEC, USN, and his Chief Staff Officer, Commaner A. C. Perkins, CEC, USN. The officers and petty officers of the unit can better serve the Commanding Officer if they understand the detailed responsibilities, problems, and essence of Command so well presented in this pamphlet.

SECTION B - THE COMMANDING OFFICER

1. <u>Introduction</u>. The mantle of SEABEE command falls on many Civil Engineer Corps officers. This sought after honor is widely regarded as a career high point -- and properly so. However, it sometimes comes to an officer with no previous SEABEE experience. While the Prospective Commanding Officer today may represent the best in technical competence with broad managerial and administrative experience, he has probably had very limited military command involvement. How can such an individual successfully meet the unique challenge of command? The answer is that most successful commanders are made -- not born. This learning process takes time and is inefficient at best. The following are offered from the experience of many to assit the PCO in reaching the plateau of the command learning curve as early as possible.

2. <u>Personality</u>. This is the crucial factor bearing on success and the hardest to alter. The more fortunate Commanding Officer will have a personality which insures his success in dealing with people. The less fortunate, who comprise the majority, must do some very objective self analysis and resolve to exploit psychology as a command tool.

3. <u>Knowledge</u>. Most of the early command errors of the CO stem from simple ignorance. His Academic ability would serve him well if he could but find what to study. Unfortunately, no well defined curriculum exists, and in its absence the school of failure has been the usual teacher. Her instruction follows the same pattern for each CO. Initially he places full trust in his subordinates and their assurance that things are under control. Full of complacency, and lacking knowledge, he does not check things for himself. Next comes the shock of the first time his command is tried in the balance and found wanting. The CO is shocked, angry, chagrined, and invariably wiser. He is about to start becoming an effec tive commander. From now on, three new and vital factors are evident in his exercise of command, <u>First</u>, he forces himself to acquire personal knowledge. <u>Second</u>, he no longer blindly trusts his subordinates and their judgments. <u>Third</u>, he checks things for himself.

The dangers and hazard to good organization and morale in the above are obvious. The CO must exercise skill in using these factors. However, his final realization that he needs knowledge and his determined action to acquire it are probably the most significant command steps he has taken.

4. <u>Image</u> of a unit may be compared to service reputation of an individual. It is intangible but most important.

The unit and its image directly reflect the Commanding Officer and his success as a commander. Few rules can be given on image except that like a good name or service reputation, it must be assiduously cultivated and jealously guarded. It cannot be a contrived or phony thing since in the long run it is reflected in solid measurable accomplishment.

Send only your very best men and officers to Seabee teams, detachments or other outside assignments where they will represent you. Not only will your image flourish with such successful representations, but you can also concentrate on developing the second level of talent thus forced into positions of responsibility within your main body.

The image that the next superior in the chain of command desires is "responsiveness." This implies not only unquestioning and cheerful willingness to comply -- but also competent, accurate, and fully effective action in that compliance.

5. <u>Motivation of Men</u>. The materials and equipment of the SEABEES are ordinary and available. The traditional strength of the SEABEES is their ability to accomplish an uncommon end from common resources. The key to such achievement lies in individual and group motivation -- and the Commanding Officer is central to this issue. Morale is his only collateral duty.

The Commanding Officer, as both the leader and father figure, must concern himself first and always with SEABEES -- and their motivation. He must continuously evaluate his command in terms of their basic needs of sense of belonging, recognition, and security. His actions, and his chain of command's actions, must always reflect the full realization that men are the ultimate tools of accomplishment.

Men respond best when they are kept informed and told the reason why. They appreciate being regarded as individuals with names and identities. They are proud to be part of a unit with high standards and a good reputation. They seek both a challenge and an example to follow. Each officer and man looks to his Commanding Officer to provide these things.

6. <u>Command Goals</u> must be defined as explicitly as possible. Progress toward these goals should be appraised regularly.

Command goals should be published. As a minimum they should include:

a. Increased readiness posture through training, planning, and timely preventive action.

b. Increased professionalism in command, administration, logistics, and operations.

c. Improvement, retention, and conservation of manpower.

d. Simplification of procedures and elimination of non-essentials.

7. <u>Chain of Command</u>. The Commanding Officer frequently vacillates until experience forces him across the Rubicon and into a total commitment to the principle of the chain of command. All other concepts and practices must be subordinate to this one. Authority and responsibility, orders and response, information and feedback flow through the chain of command -- the combined skeleton and nervous system of the command. Every action or evolution must be accomplished in terms of this single organizational structure. To depart from this is confusion and disaster.

The chain of command is the military organization. Permit no other organizations to exist. Think, plan, and act in terms of the chain of command. Stretch it, develop it, and insist upon it.

Use it as the principal communications channel both up and down. Some day it may be all you have. At quarters pass out the word to the company commanders. Insist that it flow directly then and there from company commanders to platoon commanders to squad leader to fire team leader to the man. Pencils and notebooks must, without exception, be part of the uniform of the chain of command. Test effectiveness by sampling the word passed as understood by the man in the rear rank.

Vary the routine and see how your chain reacts. Without notice keep the officers or chiefs or both from quarters and see what happens. Test your recall procedures. Trim your Master-at-Arms force and function to the small staff appendage to the Executive Officer that it should be. Put all your authority responsibility, and action in the chain of command. In short, get them involved completely and permanently.

8. <u>Leadership</u> is the essence of command. Personal example is its vital ingredient. Many techniques will demonstrate and support it.

Orders, once issued, must be enforced. Nothing is more damaging to military effectiveness than a climate in which orders are not enforced. A man should never have to decide whether you really mean what you have said or signed.

Most commands have unnecessary or even undesirable orders in effect - and largely ignored. Promulgate only a bare minimum of orders. Don't attempt to codify every possibility. Rely rather on the good judgment of your chain of command to control your unit. But once you have spoken or set your signature to an order, insist that it be enforced 100 percent.

A characteristic of an effective command is meticulous attention to detail. Small things make the difference. Create a climate wherein pride in a job well done motivates each man to do his best willingly -- and details will be taken care of.

9. The Executive Officer is your alter ego. You can afford no slippage or misunderstanding between you.

His position as second in command must be obvious and supported by the facts. While SEABEE operations may properly tend to be controlled directly by the Commanding Officer, be sure the Executive Officer participates to the maximum and is always kept fully informed.

Insist that he run the staff in an aggressive and forthright manner. Give him adequate opportunity to act for you. Don't be reluctant to go on leave.

10. <u>The Wardroom</u> is reflected *throughout the*

command -- and it in turn is a mirror of the command. The social relationships and camaraderie in the wardroom are the bases for meaningful cooperation and mutual support among the officers. The SEABEE CO exercises direct and personal control of the wardroom and its standards. His true nature is exposed more candidly and completely here than elsewhere.

The wise Commanding Officer not only manages the wardroom skillfully, but also uses it subtly as a training medium for his officers. Junior officer retention is often more influenced by the wardroom than other more obvious factors. Social graces, proper dress, and after dinner speaking ability often need developmental attention. Cliques must not be permitted to develop. One of the least understood principles is that a guest of a member is a guest of the mess, and should be scrupulously treated as such. Some officers must be made aware of their obligation to take their turn as host, both to the mess and to visitors to the Command. Formal occasions, such as mess nights or mess parties, afford a unique opportunity for social enrichment of mess members, wives, and friends.

The wardroom physical characteristics, furnishings, table service, and appointments should be the best available in the circumstances. However, the goals, attitudes, and conduct of the wardroom can always be of the highest order.

11. The Chief Petty Officer has long been recognized for his unique ability to get things accomplished. All too frequently his full potential is not utilized to the command's benefit. The CO must insure that his standards are understood in the CPO mess. He must insist upon and support the Chief Petty Officers in policing themselves. Insure that the traditional respect and privilege are afforded the Chiefs. The CO should be an occasional, friendly, well spoken, and genuinely sought after guest in the CPO mess.

Insure that no officer, who might tend to become a favorite of the Chiefs, becomes a too frequent visitor to the CPO mess. Be aware of and approve the entertainment and other practices of the CPO mess.

Talk to the Chiefs as a group periodically. Their participation in conferences is frequently beneficial. Insure that Senior and Master Chiefs are given appropriate responsibility, privilege, and recognition apart from the Chiefs.

12. The Squad Leader is a pivotal individual and has one of the hardest jobs in the unit. He has "eyeball" control of what is done on a twenty-four hour a day basis. His is the most common on-the-scene decision on on every action taken. He approves or disapproves. He corrects or fails to correct. He teaches or fails to teach. He checks or fails to check. He enforces or fails to enforce. In a very real sense, he is the ultimate agent in carrying out

your policies, enforcing your orders, and translating your will into action! He is very important and demands your continual scrutiny.

More significant failures occur at this level than any other. The squad leader is frequently ill equipped by experience, training, or knowledge to be a squad leader. Afford him every opportunity to learn and improve, but keep up a relentless pressure until he meets your standards. Take proper administrative action to reduce in rate those who can't or won't measure up. "Shape up or ship out" takes plenty of resolution in its execution.

The squad leader's responsibility for his squad is total. His involvement is personal and direct. Normally, the squad leader should live with his men. Appropriate privilege for squad leaders is sometimes difficult to establish, but it must be provided. Everything applicable to the squad leader applies commensurately to the fire team leader as well.

13. Preventive Action is the Commanding Officer's best tool if he uses it wisely. The CO should be away from his desk from 25 to 75 percent of the time. He should see and be seen. He should spot check and sample. This is communications at its best and a real test of leadership ability.

The first requirement of the CO is knowledge. He must know the area he intends to probe. He must know the questions to ask, the lids to be lifted, and the most probable deficiencies to be found. This requires preparation, perception, and psychology. The CO must not violate the chain of command in these informal inspections. When the CO shows up, the chain should automatically act to produce the senior men as rapidly as they can be obtained. The CO should give his informal comments on the scene, and reserve his severe criticisms for appropriate injection into the chain of command in his office later.

The CO must insure that he spot checks his entire command, and not just those functions in which he personally is particularly competent. He can best obtain a feel for the general condition by sampling multiple specifics in each area. He must be completely random in selecting areas to test. He strives for a climate in which every member of his command will be expecting the CO to be around checking the critical indicators, asking the right questions, and demonstrating an awareness of the importance of the area. This is a healthy condition and leads to a good tone of the command -- particularly when the Commanding Officer can leave some appropriate words of praise for a job well done.

Encourage your chain of command to develop the habit of preventive action -- spot checks and testing. Inspection check off lists and procedures should be in continuous use -- not just once yearly. Hold surprise drills without advance notice to anyone. Check the scene at midnight, just before reveille and on weekends. Require your OOD to make checks and log results for your review. Show up occasionally at the equipment yard when the equipment is lit off in the morning. Invite outside experts to come in and inspect you in specialty areas.

The trick is to do the above without creating the impression that your men and officers cannot be trusted to carry out their duties -- or don't know their business.

It deserves your full energy and ingenuity.

14. <u>Communications</u> is the lifeblood of progress and smooth operations. Foul ups invariably occur because someone didn't receive the proper information.

While the chain of command is your principal channel of communication, there is still room for plenty of horizontal communication. Just be sure that which is proper and essential flows through the chain.

Stress the communications feature of the chain of command. Reduce the plan of the day to a minimum level -- and never read it aloud at quarters. The plan of the day is a suitable channel for background information such as local customs and shopping information.

The Commanding Officer should speak to the unit regularly. This is one of the few occasions on which a loudspeaker system should be used. Hold officers call frequently and include Chiefs as appropriate. The Commanding Officer should give policy lectures on a regular basis. Conclude each officer's call or conference with a programmed short instruction period. The instructor should use charts and come to a definite conclusion.

Prepare at least two outstanding command presentations -- one slanted for civilian groups, the other for the rest of the Navy. Become a skilled briefer. Know the "Presentations Paralysis" series completely. Prepare a tailored briefing for each important visitor. Stretch and develop your officers in this vital area.

15. <u>Military Knowledge</u> can be a Civil Engineer Corps Officer's Achilles' heel. Chances are that he comes to the SEABEES far from proficient.

Fortunately this knowledge can be acquired if pursued early and vigorously. Search out the texts, attend every class, and ask questions. The burden of knowledge acquisition is clearly on you.

It is more than coincidence that military readiness and the armory are the most frequently low marked items on inspections. Hard work is the only way to become a truly competent military professional.

Probably you and your command are worse off than you realize. You personally must become the master of military basics and insure that every man knows them as well. Then you must move on to become as expert as possible in all military fields. Fire the course.

Know communications and tactics. Be the high scorer on any military test.

Require and encourage enrollment in correspondence courses such as those offered by the Marine Corps Institute. Make military proficiency a sought after status symbol at all levels.

16. <u>Training</u> is the primary SEABEE peacetime function. The best preparation for our wartime mission is today's training -- particularly of career personnel.

Unfortunately, this has not always been recognized in the past when pressure to put work in place frequently relegated training to a low priority. Production pressures dictated that the man skilled in a certain area be used in that area exclusively. This led to further narrowing of skills and produced some of the worst petty officer skill problems of today. Our objective is the well rounded, broad skilled petty officer.

Too often training is viewed as something the homeport regiment does for you, or deployment on-the-job training is considered sufficient. Neither assumption is valid. The technical competency and flexibility which have made the SEABEES famous are maintained largely through formal in-house training efforts within the command. Every officer, chief, and most petty officers should be in fact instructors.

Training will be done only if it is required, planned, programmed, supervised, and appraised by the CO. Training is accomplished by the chain of command. A comprehensive instruction must define the entire program, including technical, military, leadership, and all other components. Training priorities must be established. The best instructors for each subject must be found, trained, and given definite assignments. Lesson plans must be submitted and approved -- by the CO in some cases. Dry runs and critiques are necessary. Training must be ready to go on a moment's notice. Reports of training accomplished must flow immediately back to the training office and show on the records and status boards.

The ten minute training session in military drill, or leadership discussions at the platoon or squad level are excellent and can be tied in with the morning muster routine. Frequent short sessions are more effective than equivalent time in longer sessions

The keys to training are simply:

    a. Define the requirement.
    b. Maintain an up-to-date comprehensive and specific instruction assigning responsibilities.
    c. Get the chain of command involved.
    d. Evaluate lesson plans critically.
    e. Take advantage of every training opportunity.
    f. Evaluate training effectiveness and

knowledge retention.

    g. Keep up-to-date records and appraise results.

    h. Revise the training instructions.

17. Safety demands that the Commanding Officer be totally unyielding in his insistence upon full compliance with the highest standards. Nothing permanently scars a Commanding Officer's conscience like the memory of safety failures and accompanying human misery priced in terms of life, limb, and sight. Tragedy always seems to reach out for the young who have yet to reach their prime.

    Safety is a function of the chain of command. It is largely a matter of attitude. Everyone must be a safety inspector. Training must stress safety and its techniques.

    One of the biggest enemies of safety is the willingness of your men to take a chance in the interest of speeding up the job. Positive control is required to prevent such a misfortune. Train planners and estimators to note probable safety hazards on the drawings and in the specifications. Include and highlight safety equipment needed in the bills of material at time of take-off.

    Off the job accidents are the most frequent and serious. More SEABEES are drowned and killed in vehicle accidents than ever died on the job. Good example and chain of command action, particularly at the squad leader level, can create a climate of safety which will persist around the clock seven days a week.

    The Safety Chief is a widely misunderstood job. Unbelieveable as it seems, he has upon occasion been a low capability individual cast in the roll of "cop" with the rest of the unit as "robbers." He should be one of the top COPs in the unit. He functions on the personal staff of the Commanding Officer/Executive Officer or Operations Officer. He is a safety consultant to the chain of command. His effectiveness reflects the support and attitude of the Commanding Officer toward safety.

18. Non-Judicial Punishment is a challenge to any Commanding Officer. It can be one of his most effective leadership devices, or it can prove to be a source of difficulty and dissatisfaction. Knowledge, thorough case preparation, and human wisdom are required.

    Every mast action represents a failure -- for the man, his leadership, and the Navy. if your mast rate is high, first check any patterns which might lead you to find that you have some "chicken" orders. Men frequently react to such unjust situations by high violation rates. If you are satisfied with your orders, perhaps your chain of command is not providing the leadership it should. An ounce of squad leader prevention is worth ten pounds of NJP or court-martial cure.

    Require thorough and accurate pre-mast procedures. Go over the facts in advance with particular emphasis on legal aspects, the accused's record, and the views of his superiors

and their recommendations as to punishment and mitigation. Outline questions or areas you want brought out at mast. Make a tentative conclusion on range of punishments appropriate.

    Mast should be the most formal of ceremonies. Make sure it is precise, controlled, and well rehearsed. Be punctual, but deliberate. Have the accused's supervisors in attendance. Offer the accused the fullest opportunity to defend himself. It is frequently very impressive to read the table of maximum punishments for the offense listed in the Manual for Courts Martial.

    Remember that nowhere in our American system are the rights and properties of the individual so completely in the hands of another as at mast. There must be a reasonable doubt if you impose punishment. Better the guilty go free than the innocent be punished.

    Before assigning punishment to the accused, obtain in depth the views of his superiors in the chain of command. Reflect these views very openly in your final judgment. Your chain of command will flourish.

    Use the suspension and remission features of NJP to full advantage. Detention of pay is often as effective as forfeiture. Be sure the man understands these features.

    Correctional custody is very effective when done in your own facilities and under your supervision. Selection of the counselor is the key.

    Insist upon a follow up appraisal at definite points fixed at time of punishment. The chain of command must be charged with salvaging the individual and appraising his progress, It is a satisfying privilege to restore a man's rate or forfeiture upon positive evidence of change and the chain of command's recommendation. This is leadership in action.

19. Career Counseling and Retention effectiveness are right in the Commanding Officer's hands.

    Again, the secret of success is to get the chain of command interested and involved. The seeds of retention are sown every day in every operation. Like safety it is a matter of attitude and example.

    Send maximum members of officers and men to career counselor school. Carefully select an outstanding Chief Petty Officer who is highly motivated as your unit career counselor. His function is to be a consultant to the chain of command and put the finishing touches on the groundwork laid there. Give him a well appoined and comfortable office as an interview location. Publish a written program and adhere to it.

    Request waivers on qualifications for schools and assignments for outstanding individuals who are considering reenlistment. BUPERS will usually grant the exception if the CO presents a strong enough case for the man. The man is particularly impressed at this personal attention. Use a message if time is short.

Evaluate your retention climate. What is the composite influence of his shipmates upon the individual about to make a reenlistment decision? Use every means to create a favorable image for the man who has high potential and seeks a Navy career. Prohibit "short timer" chains and calendars. Make reenlistment a happy, respected, well publicized, and sought after event.

Personally receive each new first class petty officer and above at an appointed time in your office. Shake his hand and tell him what you expect of him. Talk to the other new arrivals in appropriate groups. Establish a specific routine which reflects great care in introducing new men to Navy life. Their first impressions are crucial to their motivation.

Exit interviews can be very revealing.

20. Public Affairs is generally not done as well as it should be. The Commanding Officer must control this program directly.

Be sure your PAO is motivated and psychologically suited for the billet. He should take the correspondence course and avail himself of any other training or apprenticeship available.

Have a written program for PAO. Develop standard procedures so that his staff can look for the unusual on an exception basis.

Most photographs are poor; train your photographers to take charge and get the picture you want.

Seek speaking engagements for yourself and your qualified officers and men. The demand is there.

Learn the local ground rules and use them to your advantage. Tie in with the local media representatives as well as the senior military PAO representative. Don't step on toes or violate local taboos.

Insure that your chain of command is PAO conscious and provides the necessary input and support.

21. Contingency Planning is a chronically weak area. This function has no pay off today, demands top talent, and is easily deferred.

Issue a definite written program of requirements in priority order. Assign talent and resources. Supervise the execution on a regular basis. Appraise results and update.

Be familiar with all plans. Have plans briefed and translated into specifics for your command. Embarkation data should always be up to date.

Insure that your personal staff, the chain of command, and every individual knows all he should of these specific contingency requirements. Think and stress mobility. Don't put down any roots. Exercise your staff and hold command post and mount out drills regularly.

22. Medical and Dental Officers frequently become isolated from the mainstream of events.

They should be included in battalion functions and briefings and be expected to perform as any other officer. Their programs should be controlled and evaluated just like any other function in the command.

These officers are frequently inexpert in administrative and supply matters which invariably leads to difficulties. Medical and dental supply status must be monitored.

Health and sanitation problems can remain undetected for a long time. Require scheduled medical sanitation and health inspections of all sites. Monitor the sick call rates and daily binnacle lists. Know the status of medical records, immunization and annual medical examination programs. The Medical Officer frequently works very well with the Chaplain in solving personal problems.

The dental status of your unit should be monitored thru analysis of up-to-date dental classification totals for your unit. Be aware of your dental prosthetics needs and procurement channels. Insure that the preventive dentistry programs and a climate of good dental hygiene are well established. Remember that the amount of time the Dental Officer spends in actual dentistry procedures is the best measure of dental progress.

Inspect the Medical and Dental spaces and equipment frequently. Know the sick call and appointment procedures. Are the preparations for detachments, disasters, or other emergencies complete? Insure that medical and dental support personnel are adequately trained and advanced in rate.

23. The Chaplain is an important officer. He must be very close to the Commanding Officer and yet be completely available to the entire command.

Support the religious program by your personal example and participation and encourage the chain of command to do likewise.

Keep the Chaplain's collateral duties to a minimum. He should not normally be the Public Affairs Officer or handle monies other than chapel funds. He needs adequate clerical and administrative assistance.

The Chaplain should cooperate with civilian church activities and project the Navy -- SEABEE image in a most favorable manner. The Chaplain in cooperation with the Medical and Dental Officers can be effectively used in the Civic Action Programs.

It is essential that the Chaplain gain the full respect and confidence of the men as a clergyman. He can also function as a command sounding board, spot trends or trouble areas, and should be frequently consulted by the CO. However, he must not supplant nor short-circuit the chain of command and its responsibilities.

Designate and train lay leaders at every level to insure total religious coverage of all denominations.

24. Supply traps many Commanding Officers --

and forewarned in this area is forearmed. A basic rule of thumb is that at any time the sum of the inventory plus the on order requisitions should equal the allowance -- available funds permitting.

Your Supply Officer must know what you expect. Be sure your intent is clear. Read the past inspection reports and evaluate yourself now.

Supply discipline is a function of the chain of command. Insist upon tight issue and inventory procedures. Be ruthless in eliminating waste and punishing negligent loss. Control special clothing and pilferable consumables with air tight procedures. Have a specific program for personally checking accountable personnel as part of your preventive action.

Supply must reflect specific mount-out or contingency plans adequately. Review status of outstanding requisitions. Monitor your financial status personally. Establish a budget board to advise you on how your funds should be utilized. Regular tool kit inspections are a must.

Do your warehouses and storage yards reflect your image?

25. <u>Heavy Equipment</u> can easily become a major problem area. Many a Commanding Officer has been down this primrose path.

Command attention is the most important ingredient for a successful equipment program. The COs interest should be evident to all. Firm NJP attention to negligence is appropriate. The COs weekly formal inspection of a class of equipment is a must.

Spare parts logistics are critical. This area must be continuously managed and monitored. Are allowances up to date and obsolete items screened out? Are parts ordered immediately at the proper priority? What follow up machinery operates? Are inventory and usage records perfect?

Maintenance procedures can easily be compromised. Are standing instructions enforced? Are required technical publications on hand? Are required checks and procedures followed completely? Who really determines what maintenance work is done? Who checks the quality of maintenance?

Poor operator discipline will deadline equipment and lead to accidents. Are pre-start checks scrupulously enforced? Are vehicles fueled after use? Are they left clean? Are speed limits enforced? Is safety equipment intact? What is the safety climate? Any evidence of "hot rodding"? Are sand and mud discipline enforced? Are all operators properly licensed? Is equipment field serviced properly? Any evidence of overloading?

Is your equipment conditions such that you will be proud to transfer it to the relieving command?

26. <u>Visitors</u> are important. They are fre-quently VIPs, and they always carry away large doses of your image.

Genuine hospitality and thoughtful consideration of your guests' interests are the watchwords. Build a command reputation for superlative reception of visitors.

The following checklist will assist in planning for visitors:

a. First impressions last.
b. Actions speak louder than words.
c. Details make the difference.
d. Have a definite plan and schedule prepared.
e. Send (or give) a written schedule and local information (uniform) earliest.
f. Assign continuous counterpart coverage and meet them at the plane.
g. Are visitor quarters the best available? Clean, towels, reading lights, bug bombs, comfortable?
h. Offer a shower, shave, haircut, laundry, and pills first.
i. Always a briefing -- big or little -- tailored to the visitor.
j. Give the briefing yourself -- or have your expert do it -- rapidly.
k. Offer a tour and introduce your officers and men.
l. Visitors are tired. Quit business early, and after a rest, start the social phase.
m. Charge guests for mandatory items only, and let the food and refreshment be the best available.
n. Introduce everyone. Counterparts are hosts. No cliques.
o. Retire early and insure that visitors are called in the morning.
p. Brief visitors on pertinent message traffic.
q. Visitors want to visit the Exchange and see local color.
r. Take pictures. Stage them to insure quality.
s. Quit while you are ahead and hold a brief departure conference.
t. Put them on the plane.
u. Provide feedback to your command. They are interested.
v. Write "nice to have had you aboard" letters immediately. Enclose pictures.

27. <u>General</u> items for the Commanding Officer's consideration:

a. Your uniform, shoes, and accouterments should be the best looking in your command. Remember that you are continuously being inspected.

b. Schedule formal inspections regularly Zone and material inspections should be thorough and detailed. Insist upon complete feedback and then reinspect on short notice. Take the Medical Officer with you on personnel inspections. Look for obesity, poor personal hygiene or chewed nails. Say something to each man. The Commanding Officer can really get the feel of his command at an inspection.

c. Military courtesy, like all forms of courtesy, is contagious. Be sure you have an epidemic going in your command continuously.

d. Never allow a subordinate to offer the excuse "I told them to do it ---."

e. To waste a man's time is an insult. Don't permit anyone to waste your SEABEES' time needlessly. Have the action planned and ready. Eliminate "hurry up and wait."

f. Impress the new man with your reporting-in-procedures. Provide transportation, a name list with the OOD, his name tag on a clean bunk, a hot meal, a guide and anything else to demonstrate that your command is interested in him and concerned with his well being.

g. Open the offices and support shops when the men can best use them.

h. Stamp out "little big shots." The wise striker behind the counter who tells the sweaty man just in from the job to come back tomorrow does more damage than all your career counselors can overcome.

i. SEABEES require a challenge. Be sure your men are given every opportunity to stretch and develop.

j. Delegate authority to the action level. Expect honest mistakes -- they are well worth the efficiency and self development of your chain of command that delegation brings.

k. Be free with your letters of commendation and meritorious masts.

l. Insure all hands understand and use request mast as often as appropriate.

m. Insure that no company level or informal punishments exist. Deprivation of liberty is your prerogative as a non-judical punishment and must never be infringed upon by others.

n. Cultivate the wives and mothers of your men. The distaff side exerts strong influence on your men's attitude and their level of satisfaction in your command.

o. Have frequent but well controlled all hands (dependents and guests) parties.

p. Promulgate a written special services/recreation program so that it starts to function the day you reach your deployment site.

q. Require individual weapons inspection at least once a week -- daily if conditions warrant. Every officer must be a weapons inspection expert.

r. Stress physical fitness with a specific written program. Encourage inter-company competitions.

s. A fat SEABEE is a potential liability. Promote thinness with diet foods and medical encouragement.

t. Commanding Officer/Executive Officer inspect some part of barracks and grounds daily.

u. Insure that the non-Group VIII personnel are given their share of formal training and advancement attention.

v. Assigning officer roommates can be a useful management tool in preventing potential problems. For example, assigning the Operations and Supply Officers as roommates insures full communications between these two important individuals.

w. Be sure fitness reports are given the serious thought and consideration they deserve. Avoid low marks just because you see more of the officer and observe more of his weaknesses than you might see in the usual shore station relationship.

x. The quality of the turnover to your relieving unit is the best test of your unit's image and your personal service reputation.

y. Consider fully your superiors' reaction to your proposed course of action.

z. Don't release that slightly critical message or complaint letter until you have slept on it.

aa. If in doubt, send it only to your boss.

bb. Sign commendatory and derogatory correspondence personally.

cc. If a Public Address System is to be used, insure that it is completely checked out and operating the way you desire. Be able to adjust it yourself and have a man on the controls.

dd. Check the NAVPERS form 1080 status each month.

ee. At the first indication that you may have a problem individual (alcoholism, incompetence, etc.), put him solidly on notice. Take the necessary documentary and disciplinary steps to support your later action action should he not improve. Too often the CO's failure to "put it on the record" makes later command action undefensible.

ff. Check customs regulations very carefully on every move. Don't delegate here.

gg. Casualty notice procedures are rarely done right. Prepare briefs of pertinent references, sample guides, and distribution lists in advance.

hh. Insure that your legal work is done properly and promptly. The key is to follow all of the manuals and have a legal course graduate on board.

ii. Nearly all commands have some flaws in their classified material control structure. There is no substitute for your detailed comprehensive review of all requirements. You alone must insure that your set up meets every requirement and is exactly what you personally approve.

jj. Answer messages and correspondence as rapidly and completely as possible. Use interim replies when appropriate.

kk. Check the armory condition and procedures in detail. Require a rigid condition inspection when weapons are issued or turned in.

ll. Personal weapons tend to show up in barracks before and after deployments. tragedy too frequently brings them to light.

mm. Most accidental discharges occur

while weapons are being cleaned or returned after an alert.

28. Epilogue. SEABEE command offers problems which will tax your reserve. In retrospect, you will not be satisfied with all of the decisions you are going to make. At times you will feel lonely or frustrated. Fortunately these occasions usually turn out to be inspirational starting points for further achievements.

You and your command are on your own. Regardless of circumstances, guidance, and support -- you ultimately must depend upon your own resources. The system about you is not perfect -- don't rely on it unduly or blame it for your problems.

Your allotted time to command is short, but the command becomes your mirror very rapidly. Take time each day to relax and think. Never underestimate the value of a smile. You will come to know the rare privilege and career highpoint of SEABEE command.

SECTION C - JUNIOR OFFICER GUIDELINES

1. You don't have to be a genius to be a successful officer. Assuming a reasonable IQ and a stable personality, all you need are proper attitude, adaptability, a sense of responsibility, and a willingness to work.

   a. Attitude. Without a positive attitude you will fail completely. With the drive stemming from an inner motivation to achieve, you will almost always succeed.

   b. Adaptability. One of the naval officer's greatest assets. Each task and tour and situation is different. If you can rapidly and accurately determine what is appropriate and not appropriate to the moment, you have a good basis for action.

   c. Sense of Responsibility. A must for success. You alone, not the system nor others about you, are answerable for your men and their actions.

   d. Willingness to Work. There is no substitute for individual effort. Committees don't win ball games. You must find the problem, accept it as yours, and focus your best efforts on it until it is resolved.

2. Priority. Be objective and discriminating. You cannot do your job properly unless you can pick out the important items requiring your immediate action.

3. Time Values. There is never "enough time," but there is almost always "sufficient time." Learn the difference.

4. Initiative. Be a self-starter. No one will ever be able to give you complete directions or detail the path you should follow. You are your own best personal programmer, evaluator, and critic.

5. Command Loyalty. Keep the interest of the command in view. Evaluate every proposed action in terms of its value to the overall organization -- not just a portion of it.

6. Give Your Opinion. Give your honest opinion on any proposed action. But once the decision is made, even if contrary to your recommendation, you must support it wholeheartedly. Feel free to recommend a change at any time if the supporting facts and conditions have changed.

7. Command Knowledge. Keep the Executive Officer and Commanding Officer informed. They would rather hear about their problems from you than be caught cold by outsiders.

8. Staff Work. Learn and practice completed staff work. Present the reasonable alternatives when seeking a decision, together with your recommendation as to which alternative is best. "Completed staff work" does not mean the elimination of all but one alternative before you submit it for approval.

9. Letter Writing. One of the most serious and widespread officer deficiencies is the lack of ability to write good letters. Pride yourself on developing an expertise so that your letters are signed out the first time. Force yourself to become an expert at dictation.

10. Briefings. Your professional progress will be increasingly measured in terms of your skill as a briefer. This vital ability can be acquired through practice and continuing study of the "Presentation Paralysis" series, NAVEXOS P-2328-3. Keep it simple and always have a target individual.

11. Public Speaking. A vital personal professional skill. Practice everytime you talk to two or more persons. Remember your voice sounds different to others than it does to you. Use a tape recorder. Above all -- don't be dull, indistinct, or inconclusive.

12. Move the Paper. Too much time is lost in too many baskets. If you can't act immediately on every item, shuffle through your basket several times a day to sift out the most important items. Carry your own papers if convenient.

13. Communicate. Communications is the lifeblood of management, the basis for operation, and the vehicle for command. Cut across organizational lines, but keep the chain of command informed.

14. No Excuses. Don't blame "them." Never permit yourself or your subordinates to explain a failure by saying "I told him to do it." Remember that nothing succeeds like success.

15. People. The real tools of your trade are people. Learn to use them wisely and well.

16. Be Courteous. It costs nothing -- but gains much. You can't afford not to be courteous and considerate of everyone. Be particularly courteous to those you wish to stimulate to more courtesy. It's infectious. Losing one's temper frequently is a sign of emotional immaturity. Repetitious or vulgar profanity flows easiest from small minds.

17. Be Rewarding. Be quick to reward both excellent and poor performance as appropriate. This will encourage the one while discouraging the other.

18. Military Excellence. Pay particular attention to military procedures. Go the extra step. Be outstanding -- the 1 in 100 mentioned on the Fitness Report. We are first and foremost Naval Officers and can never accept a second level of military smartness. You are constantly being evaluated by everyone, and the sum total of your "service reputation" is never static.

19. Personal Appearance. More important than most realize. Be superlative in personal grooming details and cleanliness. Even if your best friends won't tell you, they are offended. Shoeshines and fingernails are prime indicators.

20. Physical Fitness. Most people fall by the wayside because of physical failure, not brain failure. Are you really fit?

21. Professional Growth. Develop yourself in all the many aspects of our profession. Read as much as you can. Think ahead to other jobs in other locations, and remember that you will be a senior officer much sooner than you think. Be ready for whatever challenges and opportunities the future may hold.

22. Officer and Gentleman. This concept embodies moral courage, integrity, patriotism, and personal honor. No officer can lead well without understanding and reflecting these concepts in depth. "The word of an officer" is truly his bond.

SECTION D - SEABEE PETTY OFFICER GUIDELINES

1. You are a vital link in the chain of command, the organizational structure of the SEA-BEES. All of your authority and responsibility fall within this structure.

2. As a Petty Officer, you have the most important and the hardest job in the SEABEES. You directly control the work and training. You set the example and enforce orders and discipline.

3. As a Petty Officer, you are primarily a leader. It's not enough that you be a craftsman. You must know and teach and control your men. If they get in trouble or fail to develop themselves, it's your fault.

4. You must take appropriate action when a junior does something wrong or fails to do something that he should. If you do nothing, your failure is greater than his.

5. Your "service reputation" is being built up or broken down by everything you do or fail to do -- 24 hours a day. Guard your reputation jealously.

6. Safety is your responsibility. Nothing is more important. Know safety, think safety, and enforce safety. It's either safe -- or it isn't safe.

7. You are responsible to and for your men. Families have entrusted the welfare of their sons and husbands to you. You mold their outlooks and attitudes. Act as though you are going to meet these parents and families and explain to them what you have done with their boys.

8. As a Petty Officer, you have authority as well as responsibility. Use your men advantageously -- they each cost Uncle Sam many dollars each day.

9. Use your initiative. There is a better way to do everything. Find it. Put it into effect if you can, or send it up the line as a suggestion.

10. Communicate. As a supervisor you should be talking to everyone who is influenced by your job or who can help you do it. Most "foul ups" result from lack of communications. Cut across organizational lines as appropriate but keep the chain of command informed. You are the one who must pass the word to your men and be sure that they understand it. Insure that they know what the mission is, their part of it, and why it's important.

11. Be your own best critic. Don't wait for someone else to find mistakes on inspections. You are in the best position to find and fix. In doing so you build up everyone's confidence in you as a leader.

12. Small details make the difference. This is "pride" in its best sense.

13. Correct mistakes now, on the spot -- you owe it to your men, or they will never learn.

Never let an error pass by without your notice and action.

14. Be proud to be inspected. Don't offer excuses or blame others. Tell what you are doing about it.

15. Practice and demand military courtesy.

16. Wear your uniform with pride and be sure your men do likewise.

17. SEABEES are MOBILE. Be ready to mount out at anytime. Be sure each man knows his part.

18. If you have to bring someone to mast, do it. But remember that each mast or court-martial is failure -- for the man, the SEABEES, and the Navy. One of your biggest responsibilities is to spot the problem and correct it before it reaches the report slip stage.

19. Develop yourself; prepare for the future. Learn letter writing, Navy management, and administration. Read widely. Develop interrate skills. Think and act in terms of a job supervisor -- not just a tradesman. Learn planning and estimating. Prepare to advance -- and take your men with you.

20. The strength of the service is its personnel. The Navy is based upon respect and loyalty, both up and down. As a Petty Officer, you have a special trust and confidence placed in you. Be worthy of it.

21. You have the hardest job in the SEABEES. You set the tone and spirit. The success or failure of your unit largely depends upon you and how you carry out your responsibilities and use your opportunities.

SECTION D - GUIDELINES FOR ALL

1. Know first that you are needed in the SEABEES and are a welcome and integral part of them.

2. Our mission is important and much will be expected of you. We will accept nothing less than the very best you have to offer.

3. Words like duty, honor, country, and patriotism are very much alive and filled with meaning for SEABEES. Work, sweat, and effort are what give them substance.

4. There is only one organization in the SEABEES, the Military Organization. Normally, you will train, work, and fight, if necessary, in the same fire team and squad. Know your duties completely and where and how you fit into the organization.

5. SEABEES are MOBILE. Always be ready to

mount out and know what is expected of you. Keep your gear in tip-top shape.

6. It costs the taxpayers many dollars a day to support each SEABEE. Make sure that you give you full effort in return. Our equipment is worth many millions of dollars, so use it carefully.

7. Learn and practice military courtesy.

8. Safety always comes FIRST. Constant attention, knowledge, and common sense prevent accidents. Never take a chance. You can't afford to lose!

9. This is not an easy or particularly comfortable way of life. Help improve living conditions and make life as pleasant and rewarding as possible. Don't be a liability to your shipmates but keep yourself and your place clean.

10. Stay out of trouble. A mast or court-martial is a failure for you, your squad, the SEABEES -- everyone.

11. Make a good record and build a service reputation for excellence. It's important no matter where you go -- in the service or out.

12. Advance in rating.

13. Use your initiative. Find a better way.

14. If it's worth doing -- do it well. Take pride in everything you do.

15. Small details DO make the difference.

16. Build good habits now as a foundation for future success.

17. Get a high school diploma at least (or equivalent). Without it your progress will be limited and becomes increasingly difficult.

18. Save your money. Put at least 10% and as much more as you can into something permanent by means of an allotment.

19. Live within your means. If you get into debt, you become a military problem and may be discharged as unreliable.

20. Alcohol ruins more lives than you might think. No one ever started out deliberately to ruin his life, but it's easy to do.

21. Don't get involved with or married to anyone without long and realistic thought. Haste and the wrong woman have ruined many good men permanently.

22. Take an active part in the religious services of your choice.

23. Don't do anything you wouldn't do at home. Do those things that you would be proud to have your parents, friends and family know about.

24. You are the real strength of the SEABEES. Your abilities, dedication and willingness are our greatest assets. Improve yourself each day.

SECTION E - SELECTED LEADERSHIP REFERENCES FOR NAVAL CONSTRUCTION BATTALION OFFICERS

1. Roskill; The Art of Leadership, Archon Books, 1965

2. Hays and Thomas; Taking Command, Stackpole Books, 1967

3. Marshall; The Officer as a Leader, Stackpole Books, 1966

4. Gardner; Excellence, Harper Bros., 1961

5. Hitch; Decision Making For Defense, University of California Press, 1965

6. Eclles; Military Concepts and Philosophy, Rutgers University Press, 1965

7. Bush; Modern Arms and Free Men, Simon and Schuster, 1949

8. Finer; The Man on Horseback, Praeger, 1962

9. Millis; Arms and the State, Twentieth Century Fund, 1958

10. Fiedler; A Theory of Leadership Effectiveness, McGraw-Hill, 1967

11. Zaleznik; Human Dilemmas of Leadership, Harper-Row, 1966

12. McGregor, The Human Side of Enterprise, McGraw-Hill, 1960

13. McGregor; Leadership and Motivation, M.I.T. Press, 1966

14. Gellerman; Motivation and Productivity, American Management Association, 1963

15. Gellerman; Management by Motivation, A.M.A., 1968

16. Kelly; Lost Soldiers, M.I.T. Press, 1965

Annex K
THE PERSONNEL READINESS CAPABILITY PROGRAM (PRCP)

1. <u>Introduction</u>. The Personnel Readiness Capability Program (PRCP) is an integrated STINGER subsystem developed to provide detailed personnel skills information to all levels of Naval Construction Force commands. It deals with the type of information usually found only in a platoon commanders notebook. Developed by the Commander, Construction Battalions, U.S. Atlantic Fleet for use within the Atlantic MCBs, the system is being implemented through the Naval Construction Forces, both active and reserve, and shore activities having a significant number of SEABEES assigned. By late 1969, it is expected that 80% of the SEABEE community will under PRCP. Through the use of PCRP, commands at all levels will have knowledge of each SEABEE's skills and traits and the previously used rough estimates or laborious inventories of a command's capabilities will be eliminated. At all levels, the information will provide for better command and planning in matters of readiness, capability, training and logistical support.

2. <u>Criteria</u>. In order to be effective, workable, and incorporate the diversity of SEABEE units, employments, and command and support chains, the PRCP System had to meet several criteria:

    a. <u>Flexibility</u>: To meet changing requirements.
    b. <u>Simplicity</u>: To minimize the skill level required to operate the system.
    c. <u>Speed</u>: To provide instant response to command requirements.
    d. <u>Sophistication</u>: To benefit by use of modern computer techniques.
    e. <u>Accuracy</u>: To provide valid information in the time frame required.
    f. <u>Adaptability</u>: To insure availability of information at remote as well as Headquarters levels.

3. <u>Information</u>. The PRCP System tasks each field command to initially gather and continuously update the raw data on skills of each member of the command. This data is in a format most useful to the field commander, who has the capability to rapidly sort and print desired information. All of the equipment used at the unit level is portable and not dependent upon machine processing. Once collected or updated, the field command also transmits the information to the Headquarters level with minimum effort.

4. <u>Keysort Card</u>. An 8"x10½" coded card, called a McBee "Keysort" card is filled out and maintained for each man in the unit at the unit level. Information is encoded on the Keysort card (Figure 114) by punching slots in the edges. The card, when punched contains detailed information including:

    a. <u>Personnel Data</u>: Name, rate, serial number, company, anticipated loss date and reason for loss, marital and dependents status, etc.
    b. <u>Crew Skills</u>: Tent camp/cantonment construction, quonset hut construction, butler building construction, pontoon assembly, bolted steel tank assembly, pile driving, quarry operations, fire fighting, well drilling, etc.
    c. <u>Special Training</u>: A or B Schools, SEABEE Team, etc.
    d. <u>Individual Technical Skill for each Rating</u>: EO, EA, CE skills, etc.
    e. <u>Individual General Skills</u>: Drivers license, instructor, career counseling, typing, planning and estimating, etc.
    f. <u>Military Schools and Skills</u>: Squad Leaders school, Counterinsurgency school, special weapons training, communications operations, demolitions, etc.
    g. <u>Miscellaneous Training and Skills</u>: Disaster Recovery, foreign language, etc.

All the information pertaining to specific skills is based on the proficiency level of the individual. The proficiency levels are defined as follows:

    a. <u>Proficiency 0</u>: Individual who is capable of performing manual tasks with no proficiency in a specific skill.
    b. <u>Proficiency 1</u>: An individual capable of performing a basic technical task or skill under direct supervision.
        (1) Graduation from Class "A" School plus demonstration of proficiency in the shop/field.
        (2) Non-Class "A" School personnel who have attended Special SEABEE Training Courses and demonstrate proficiency in the shop/field.
    c. <u>Proficiency 2</u>: (Journeyman) - An individual capable of performing a specific technical task with only indirect supervision. Attainment of proficiency 1 criteria plus a minimum of one (1) year related experience in the specific skill or completion of proficiency level 2. Special SEABEE Training Courses

FIGURE 114
KEYSORT CARD

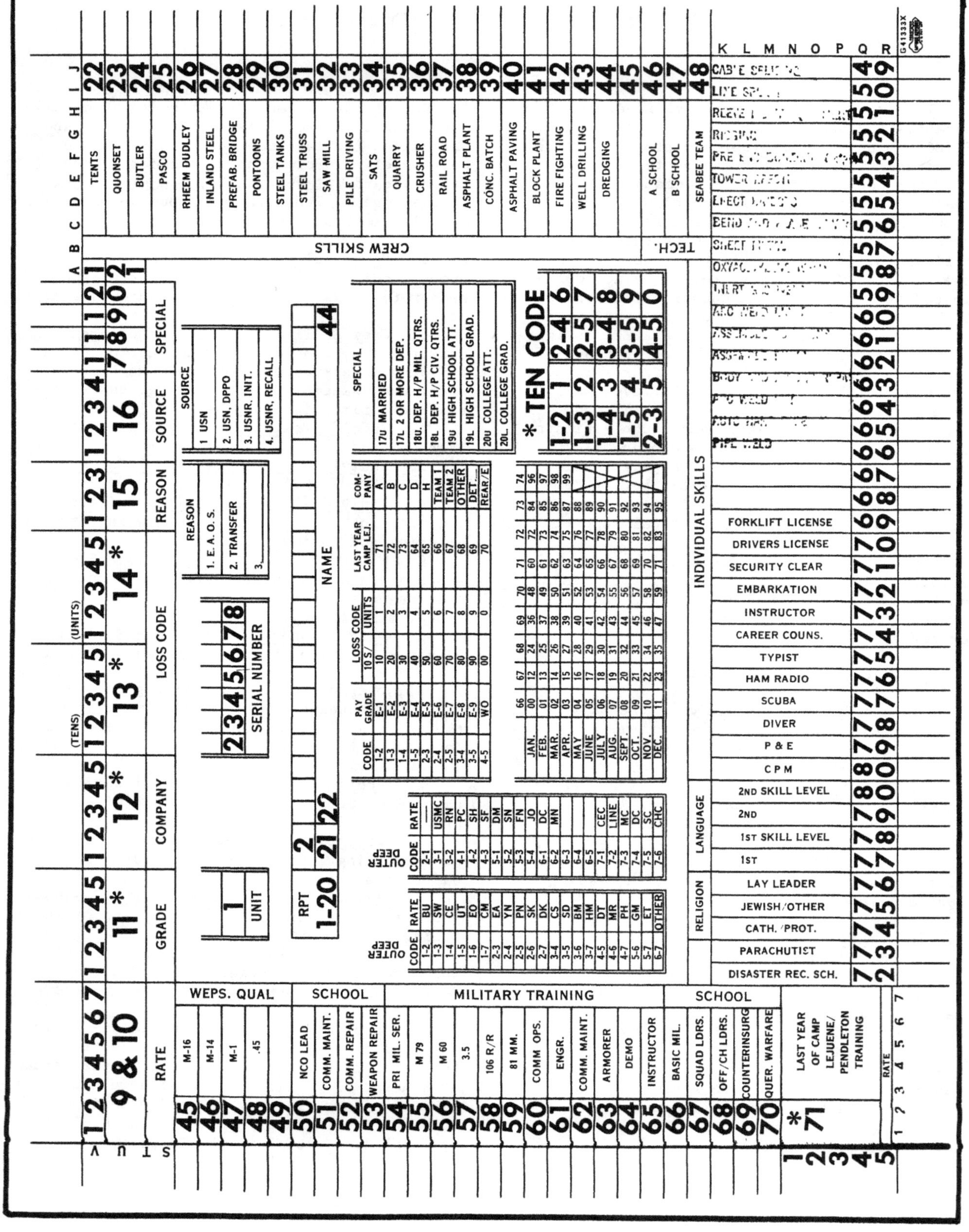

CREW SKILLS

22	TENTS
23	QUONSET
24	BUTLER
25	PASCO
26	RHEEM DUDLEY
27	INLAND STEEL
28	PREFAB. BRIDGE
29	PONTOONS
30	STEEL TANKS
31	STEEL TRUSS
32	SAW MILL
33	PILE DRIVING
34	SATS
35	QUARRY
36	CRUSHER
37	RAIL ROAD
38	ASPHALT PLANT
39	CONC. BATCH
40	ASPHALT PAVING
41	BLOCK PLANT
42	FIRE FIGHTING
43	WELL DRILLING
44	DREDGING
45	A SCHOOL
46	B SCHOOL
47	
48	SEABEE TEAM

K L M N O P Q R   G41133X

TECH.

RATE — GRADE — COMPANY — (TENS) — (UNITS)

9 & 10	
11	GRADE
12	COMPANY
13	LOSS CODE
14	
15	REASON
16	SOURCE

SPECIAL

SERIAL NUMBER: 2 3 4 5 6 7 8

UNIT: 1

RPT 1-20 21 22

SOURCE
1 USN
2. USN, DPPO
3. USNR. INIT.
4. USNR. RECALL

REASON
1. E.A.O.S.
2. TRANSFER
3. ___

NAME

SPECIAL
17u MARRIED
17L 2 OR MORE DEP.
18u DEP. H/P MIL. QTRS.
18L DEP. H/P CIV. QTRS.
19u HIGH SCHOOL ATT.
19L HIGH SCHOOL GRAD.
20u COLLEGE ATT.
20L COLLEGE GRAD.

* TEN CODE

1-2	1	2-4	6
1-3	2	2-5	7
1-4	3	3-4	8
1-5	4	3-5	9
2-3	5	4-5	0

COM-PANY	
A	96
B	97
C	98
D	99
H	
TEAM 1	
TEAM 2	
OTHER	
DET.	
REAR/E	

INDIVIDUAL SKILLS

K L M N O P Q R
CAB'E SPLIC'NG
LINE SPL...
RE...E ....R..
RIG'NG
PRE ... D...... ...
TOWER .....
ERECT .....
BEND ... .. .E ....
SHEET .....
OXYAC... ....
INERT ....
ARC ..LD ....
ASS ....
ASS.......
B... ... .
P.. WELD
AUTO... ...
PIPE .....

FORKLIFT LICENSE	
DRIVERS LICENSE	
SECURITY CLEAR	
EMBARKATION	
INSTRUCTOR	
CAREER COUNS.	
TYPIST	
HAM RADIO	
SCUBA	
DIVER	
P & E	
C P M	
2ND SKILL LEVEL	
2ND	
1ST SKILL LEVEL	
1ST	
LAY LEADER	
JEWISH/OTHER	
CATH./PROT.	
PARACHUTIST	
DISASTER REC. SCH.	

LANGUAGE — RELIGION

PAY GRADE
E-1, E-2, E-3, E-4, E-5, E-6, E-7, E-8, E-9, WO

LOSS CODE 10S / UNITS

Months: JAN. FEB. MAR. APR. MAY JUNE JULY AUG. SEPT. OCT. NOV. DEC.

CODE	RATE		CODE	RATE
1-2	BU		2-1	USMC
1-3	SW		3-1	RN
1-4	CE		4-1	PC
1-5	UT		4-2	SH
1-6	EO		4-3	DM
1-7	CM		5-1	SF
2-3	EA		5-2	SN
2-4	YN		5-3	FN
2-5	PN		5-4	JO
2-6	SK		6-1	DC
2-7	DK		6-2	MN
3-4	CS		6-3	
3-5	SD		6-4	
3-6	BM		6-5	
4-5	DT		7-1	CEC
4-6	HM		7-2	LINE
4-7	MR		7-3	MC
5-6	PH		7-4	DC
5-7	GM		7-5	SC
6-7	ET		7-6	CHC
	OTHER			

OUTER DEEP / DEEP OUTER

RATE	WEPS. QUAL / SCHOOL / MILITARY TRAINING / SCHOOL
45	M-16
46	M-14
47	M-1
48	.45
49	
50	NCO LEAD
51	COMM. MAINT.
52	COMM. REPAIR
53	WEAPON REPAIR
54	PRI MIL. SER.
55	M 79
56	M 60
57	3.5
58	106 R/R
59	81 MM.
60	COMM. OPS.
61	ENGR.
62	COMM. MAINT.
63	ARMORER
64	DEMO
65	INSTRUCTOR
66	BASIC MIL.
67	SQUAD LDRS.
68	OFF./CH LDRS.
69	COUNTERINSURG.
70	QUER. WARFARE
*71	LAST YEAR OF CAMP LEJEUNE/PENDLETON TRAINING

RATE 1 2 3 4 5 6 7

plus a minimum of six (6) months skill experience (or graduation from "B" or "C" School) plus demonstration of proficiency.

d. <u>Proficiency 3</u>: (Supervisor) - An individual capable of supervising crews who will perform tasks of Journeyman quality. Attainment of proficiency 2 plus demonstration of ability to supervise in the specific skill.

At the unit level, information contained on the PRCP Keysort cards is rapidly retrieved using a manual sorting machine called a Keysort selector. For example, personnel with the following criteria could be rapidly selected:

a. All personnel in "A" Company.

b. Steelworkers highly proficient in inert gas welding and experienced in SATS matting installation.

c. CM1s that will be transferred in June 1967 whose families live in Navy housing.

5. <u>Concept of Operation</u>. Once a sort is made of the "Keysort" cards for any desired item of data or any combination of two or more items of data, any type of desired list (single, carbon, or stencil) may be prepared at the unit level using the addressograph 205 printer and addressograph plates for each individual. This printer, provided to all field units, is also manually operated and easily portable.

Keysort cards and addressograph plates are normally kept and maintained by the S-2. Each time a man is transferred, his Keysort card is transferred with his service record.

In addition to providing easily accessible information to the unit commander's, the Keysort card is designed so that all the information can be readily transferred to two standard 80 column Electronic Accounting Machine

(EAM) cards for use at the Headquarters level. Individual units initially submit the information to headquarters (CBPAC/CBLANT) on an input data sheet and periodically update the information by marking up a computer printout run by Headquarters.

With detailed information available, the Headquarters can, utilizing an electronic computer, obtain data required for effective and efficient management of personnel throughout the SEABEE community. Using PRCP, Headquarters can accurately predict any specific date in the future:

a. Construction and military capabilities.

b. Personnel, training and logistics requirements.

c. Berthing, messing, and housing requirements.

d. Contingency planning and capability.

When accurate predictions of future capabilities are compared with the desired future capabilities, deficiencies and requirements are quantified. From this data, planning and plans are formulated by Headquarters and forwarded to the commands involved. With continued verifying and updating to the date of execution of plans, redundancies will be eliminated and maximum usage will be made of the facilities and time available.

The Personnel Readiness Capability Program (PRCP) has tremendous potential of application throughout the SEABEE community and STINGER simulation. The future use and effectiveness of the system, however, is dependent on the individual units and the continual, accurate updating of the computer print-outs from Headquarters.

UNIT ABBREVIATIONS

ACB or PHIBCB	Amphibious Construction Battalion (P-1A)
CBBU	Construction Battalion Base Unit (Obsolete)
CBC	Construction Battalion Center
CBMU	Construction Battalion Maintenance Unit (P-5)
CHB	Cargo Handling Battalion (F-1)
COMCBLANT	Commander, Construction Battalions, U.S. Atlantic Fleet
COMCBPAC	Commander, Construction Battalions, U.S. Pacific Fleet
COMPHIBLANT	Commander, Amphibious Force, U.S. Atlantic Fleet
COMPHIBPAC	Commander, Amphibious Force, U.S. Pacific Fleet (Type Commander for ACB)
COMSERVLANT	Commander, Service Force, U.S. Atlantic Fleet
COMSERVPAC	Commander, Service Force, U.S. Pacific Fleet (Type Commander for MCB)
CTU	Construction Training Unit
MCB	Mobile Construction Battalion (P-25)
NAVFAC	Naval Facilities Engineering Command
NAVSCOLCONST	Naval Schools Construction
NCB	Naval Construction Brigade
NCB	Naval Construction Battalion (P-1) (Obsolete)
NCF	Naval Construction Forces
NCR	Naval Construction Regiment
RMCB	Reserve Mobile Construction Battalion
STAT	SEABEE Technical Assistance Team (Obsolete, now SEABEE Teams)

GLOSSARY OF SEABEE ABBREVIATIONS

AABFS	–	Amphibious Assault Bulk Fuel System
A/B	–	Advanced Base
ABFC	–	Advanced Base Functional Component
ABIOL	–	Advanced Base Initial Outfitting List
A/C	–	Aircraft
ADCON	–	Administrative Control
A&E	–	Architect and Engineering Firm
AKA	–	(Ship) Attack Cargo Transport
ALUSNA	–	U.S. Naval Attache
AMMI	–	New SEABEE Bridge and Pontoon Designs conceived by Dr. A. Amirikian, NAVFAC.
Amtrac	–	(Landing Craft) Amphibious Tractor
ANGLICO	–	(Marine Unit) Air/Naval Gunfire Liaison Company
AOA	–	Amphibious Objective Area (Beachhead)
APA	–	(Ship Attack Personnel Transport
AR	–	Automatic Rifleman in a Fire Team
ASP	–	Ammunition Supply Point
BCD	–	Bad Conduct Discharge
BEEP	–	Battalion Equipment Evaluation Program
BJU	–	(Navy Unit) Beach Jumper Unit
BLT	–	(Marine Unit) Battalion Landing Team
BU	–	(SEABEE Rating) Builder
CamLej	–	(Marine Base) Camp LeJeune, North Carolina
CamPen	–	(Marine Base) Camp Pendleton, California
CDO	–	Command Duty Officer
CE	–	(SEABEE Rating) Construction Electrician
CEC	–	Civil Engineer Corps
CED	–	Construction Equipment Department, the Department of the CBC that performs 5th Echelon Maintenane of MCB Equipment
CG	–	Commanding General (of a Marine Unit)

ChiCom	–	Chinese Communists Forces
CHIEF	–	An E-7, E-8, E-9 or the Chief of Civil Engineers
CHOP	–	Change of Operational Control
CI	–	Counterinsurgency
CINCPACFLT	–	Commander in Chief, U.S. Pacific Fleet
CINCLANTFLT	–	Commander in Chief, U.S. Atlantic Fleet
C&LA	–	(Embarkation Form) Cargo and Loading Analysis
CM	–	(SEABEE Rating) Construction Mechanic
CMN	–	Chief of Naval Material
CNO	–	Chief of Naval Operations
CO	–	Commanding Officer
co	–	Company
COCOA	–	An aid in remembering the important items to note when setting in a defense position: (1) Critical Terrain; (2) Obstacles; (3) Cover and Concealment; (4) Observation and Fields of Fire; (5) Avenues of approach.
CommO	–	Communications Officer
COMRATS	–	Commuted Rations. When a man elects to subsist at home the Navy gives him the amount in dollars that it would have cost to fee him (about $1.14/day).
CONUS	–	Continental United States
COSAL	–	Coordinated Ships Allowance List
COT	–	Commander of Troops (for a Parade or Review)
CP	–	Command Post
CPX	–	Command Post Exercise
C/S	–	Chief of Staff to a Marine Commanding General
CTF	–	Commandter Task Force
CTR	–	Central Tool Room, the place where battalion tools are issued and repaired.
CUCM	–	(SEABEE Rate) Master Chief Constructionman (E-9)
DAG	–	Meaningless letters used by CBC to prefix a control number assigned to assemblies of an ABIOL.

DD	– Can mean Destroyer or Dishonorable Discharge
DEFCON	– Defense Condition
DMR	– (Supply Term) Date Material Required
DOD	– Department of Defense
DOS	– (Embarkation Term) Days of Supply e.g., Rations – 5 DOZ
DPPO	– Direct Procurement Petty Officer
DZ	– Drop Zone
EA	– (SEABEE Rating) Engineering Aide
EAM	– Electronic Accounting Machine
EAOS	– (Personnel Administration Term) Expiration of Active Obligated Service
EEI	– (Intelligence Term) Essential Elements of Information
ELINT	– Electronic Intelligence
EO	– (SEABEE Rating) Equipment Operator
EPDO	– (BUPERS Field Activity) Enlisted Personnel Distribution Office
EQCM	– (SEABEE Rate) Master Chief Equipmentman (E-9)
ERG	– Emergency Recovery Group
ERS	– Emergency Recovery Section
ERU	– Emergency Recovery Unit
ETA	– Estimated Time of Arrival
ETD	– Estimated Time of Departure
FEBA	– Forward edge of the battle area. An imaginary line joining the forward edges of the most advanced defensive positions of the battle area.
FEG	– (Marine Unit) Force Engineer Group (part of the Marine Expeditory Force)
FEX	– Field Exercise (entire unit moves into the field for training, whereas in CPX only the command element and staff are involved)
FM	– Field Manual
FMF	– Fleet Marine Force (e.g. FMFPAC)
ForTrps	– Force Troops (Troops not organic to the Marine Division or Wings but which provide support to them)
FSN	– (Supply Term) Federal Stock Number
FSR	– (Marine Unit) Force Service Regiment (Part of Force Troops)
FY	– Fiscal Year
GED	– General Educational Development Tests, See Page 3 in Enlisted Service Record.

GITMO/GTMO	– (Place) Guantanamo Bay, Cuba (There are many such abbreviations)
GROUP VIII	– (SEABEES) The Construction Ratings.
HQ	– Headquarters
ICP	– Inventory Control Point
ID	– Identification (e.g. ID Card, ID Tags, etc.)
ITR	– (Marine Unit) Infantry Training Regiment (These units conduct advanced infantry training)
JAG	– Judge Advocate General
JANAP	– Joint Army, Navy, Air Force Publication
JCS	– Joint Chiefs of Staff
JOOD	– Junior Officer of the Day
KIA	– Killed in Action
LCM	– Landing Craft Medium (Carries about 40 men ship to shore; commonly referred to as a "Mike Boat")
LCU	– Landing Craft Utility (carries about 4 dozers during ship to shore movements)
LCVP	– Landing Craft Vehicles and Personnel (Smaller than LCM)
LDO	– Limited Duty Officer
LOC	– Line of Communication – roads, etc.
LPM	– Landing Party Manual
LSD/LPD	– Landing Ship Dock
LST	– Landing Ship, Task
LVT	– (Landing Craft) Landing Vehicle Tracked (Amtrac)
LZ	– Landing Zone
MAA	– Master-At-Arms
MAAG	– Military Assistance Advisory Group
MABS	– (Marine Unit) Marine Air Base Squadron (The unit that operates and maintains the air base)
MAC	– Military Airlift Command or Military Advisory Command
MAF	– Marine Amphibious Force
MAG	– Military Advisory Group or Marine Air Group
MAL	– Mobilization Allowance List (most commonly used when referring to the list of repair parts for automotive and construction equipment)
MarCor	– The U.S. Marine Corps
MarDiv	– A Marine Division

MATS	– Military Air Transportation Service	
MCB	– Mobile Construction Battalion	
MCON	– Military Construction Funds	
M/D	– Man Day	
MEB	– Marine Expeditionary Brigade (Usually a reinforced infantry regiment and an air group)	
MEF	– Marine Expeditionary Force (An Infantry Division and an Air Wing)	
MIA	– Missing in Action	
MLR	– Main Line of Resistance (the forward edge of a series of interlocking defensive positions – now called the FEBA)	
MOB	– A term NEVER used to refer to an MCB since it connotes an unruly crowd without a leader.	
MOCC	– Mount Out Control Center (A system used to assist the MCB to secure construction operations and embark aboard ship or aircraft in a rapid and orderly manner)	
MOS	– Military Occupational Specialty (A Marine Term which indicates skills not defined by pay grade – comparable to the Navy NEC.	
MP,N	– Military Personnel, Navy	
MRO	– Movement Report Office. All Fleet Units make Movement Reports to this Office.	
MSG	– Message	
MSR	– Main Supply Route (used in the invasion area to indicate the main roads leading to the front line positions)	
MSTS	– Military Sea Transportation Service	
MT	– Motor Transport	
MTO	– Material Take-Off	
M/T	– (Embarkation Term) Measurement Ton (40 cubic feet)	
M-14	– This rifle is the basic SEABEE armament. (It is one of the NATO family of weapons which fire 7.62 MM ammunition. It replaces the M1 and the BAR)	
M-16	– A 5.56 MM lightweight rifle, currently replacing the M-14 in Construction Battalions.	
M-37	– Designation for a 3/4 Ton Cargo Truck with a 4 Wheel Drive built to Military Specifications (There is a series of these designators	

	– for trucks and trailers in the ACB and MCB Allowance.
M-60	– A Machine Gun (Also of the NATO Family of Weapons)
M-79	– A 40 MM, Single-Shot, Shoulder-held, grenade launcher presently used by the SEABEES.
NAVFAC	– Naval Facilities Engineering Command
NAVSTA	– U.S. Naval Station
NBC	– Nuclear, Biological, and Chemical Warfare
NEC	– (Personnel Administration) Navy Enlisted Classification (Identification by skills not defined by rate)
NEGDEF	– Navy Emergency Ground Defense Force (A Force of Naval personnel taken from the regular complement of a Shore Station and organized as infantry)
NJP	– Non-Judicial Punishment (Awarded under Article 15 of the UCMJ. Called Captain's Mast in the Navy Office Hours in the Marine Corps)
NOK	– Next of Kin
NSD	– Naval Supply Depot
NWP	– Naval Welfare Publication
ODCR	– (Personnel Report) Officer Distribution Control Report (monthly summary of information about attached officers sent to each command to assist in personnel administration)
OinC	– Officer-in-Charge
OJT	– On-the-Job Training
O&MN	– Operation and Maintenance, Navy (Appropriations)
OOD	– Officer of the Day
OP	– Observation Post (A vantage point from which to observe enemy activity in front of the FEBA)
OPCON	– Operational Control
OPLAN	– Operation Plan
OPN	– Other Procurement, Navy (Appropriations)
OPNAV	– The Office of the Chief of Naval Operations
OPORD	– Operations Order (an OPLAN becomes an OPORD when ordered executed)
OPS	– Operations Department/Officer
OPSTAT	– Operational Status

OPTAR	– Operating Target, referring to SEABEE Operating Funds		SERE	– Survival, Evasion, Resistance and Escape

OPTAR — Operating Target, referring to SEABEE Operating Funds

PAMI — (BUPERS Field Activity) Personnel Accounting Machine Installation

PCS — (Type of Orders) Permanent Change of Station

PHIBEX — Amphibious Exercise

PM — Preventive Maintenance

POD — Plan of the Day

POE — Port of Entry

POL — (Fuel Supply) Petroleum, Oil and Lubricants

POW — Prisoner of War

PRC-25 — Back-Carried Radio which is part of the Allowance. (There are many other abbreviations for communications gear)

PRCP — Personnel Readiness Capability Program

PWRR — Prepositioned War Reserve Requirement

PWRS — Prepositioned War Reserve Stock

P-1 — The Functional Component which was the NCB (Obsolete)

P-1A — The Functional Component which is the ACB

P-25 — The Functional Component which is the MCB

P-25A — The Obsolete Table of Allowance for the Active Duty MCBs (it is a modification of the P-25)

RCT — (Marine Unit) Regimental Combat Team (A reinforced infantry regiment)

RDC — (Personnel Form) Rotation Data Card (The Navy Enlisted Man requests his choice of duty station on this card)

RDT&E — Research, Development, Test and Evaluation

RF — Radio Frequency

RR — Recoilless Rifle

SATS — Short Airfield for Tactical Support

SCORE — (Personnel Program) Selected Conversion and Retention Program (Navy Program to increase career input into critical ratings by cross rating qualified enlisted men)

SCUBA — Self-Contained Underwater Breathing Apparatus

SEAVEY/ SHOREVEY — (Personnel Program) Navy System for Rotating Enlisted Personnel from Sea to Shore & Shore to Sea.

SERE — Survival, Evasion, Resistance and Escape

SITREP — Ships Loading Characteristics Pamphlet

S/M — Serviceman

SMEAC — The first letters of the titles of the paragraphs in an OPORD, Situation, Mission, Execution, Administration and Logistics, Command and Communications.

SOAP — Supply Overhaul Assistance Program

SOG — Sergeant of the Guard

SOP — Senior Officer Present or Standard Operating Procedures

S/R — (Personnel Form) Navy Enlisted Service Record

SRO — Shore Repair Order (An Authorization to do work on a Navy Vehicle)

SSB — (Radio) Single side band (There is a family of radios that operate on single side band)

SSEO — SEABEE Systems Engineering Office

STAR — (Personnel Program) Selected Training and Retnetion Program (A Navy Program to increase Career Retention)

STINGER — SEABEE Tactically Installed Navy-Generated Engineer Resources

SW — (SEABEE Rating) Steelworker

SYSCOM — Systems Command

S-1 — The Personnel and Administration Officer or Section of a Staff of an Organization smaller than a Division.

S-2 — The Intelligence Officer/Section

S-3 — The Operations Officer/Section

S-4 — The Logistics Officer/Section

T/A — Table of Allowance

TAD — Temporary Additional Duty

TAFDS — Tactical Airfield Fuel Dispensing Systems

TAOR — Tactical Area of Responsibility

T/E — Table of Equipment (Obsolete)

TECHCON — Technical Control

TM — Technical Manual (Army)

T/O — Table of Organization (Obsolete)

TOA — Table of Allowance

TRK — Truck

TRLR — Trailer

TSFC — Tactical Support Functional Components

UCMJ	– Uniform Code of Military Justice	XO	– Executive Officer of a Commissioned Unit
UDT	– (Navy Unit) Underwater Demolition Team	2x6	– A man who enlisted for six years in the Naval Reserve under a Program whereby any two of those six years must be spent on active duty. This fulfills his military obligation.
UNCLAS	– Unclassified		
UP&TT	– (Embarkation Form) Unit Personnel and Tonnage Table		
URG	– Underway Replenishment Group	3.5	– A 3.5" Rocket Launcher (Many times incorrectly called a Bazooka)
UT	– (SEABEE Rating) Utilitiesman		
UTCM	– (SEABEE Rate) Master Chief Utilitiesman (E-9)	782 Gear	– Infantry Gear signed for on a Form #782
USAF	– United States Air Force	81 MM	– An 81 MM Mortar
USAFI	– United States Armed Forces Institute. Personnel may enroll and take Correspondence courses for credit.	1080	– (Personnel Form) PAMI sends an EAM Listing monthly to each command that contains information on each Enlisted Member. Similar to ODCR for Officers.
VFR	– Visual Flight Rule		
VS&PT	– (Embarkation Form) Vehicle Summary & Priority Table	Modifier 97	Kit-A Packup of Repair Parts Common (nuts, bolts, gasket material)
W/; W/o	– With; Without		
WIA	– Wounded in Action	Modifier 98	Kit-A Packup of Repair Parts Peculiar (specific parts to repair specific items of equipment)
WW-II	– World War Two		
WW-III	– The World War we are trying to avoid.		

SEABEE MUSIC

THE SONG OF THE SEABEES

We're the SEABEES of the Navy

We can build and we can fight

We'll pave a way to victory

And guard it day and night

And we promise

That we'll remember

The "Seventh of December"

We're the SEABEES of the Navy

BEEs of the Seven Seas

SEABEES VERSE TO THE NAVY HYMN

Lord, stand beside the men who build

With courage, strength and speed and skill

Who labor dauntless every day,

Their aim to please along the way;

Lord, hear our prayer for all SEABEES

Where ere they be on land or seas

256

www.ingramcontent.com/pod-product-compliance
Lightning Source LLC
Chambersburg PA
CBHW080521220326
41599CB00032B/6162